METHODS IN MOLECULAR BIOLOGY

Series Editor
John M. Walker
School of Life and Medical Sciences
University of Hertfordshire
Hatfield, Hertfordshire, UK

For further volumes:
http://www.springer.com/series/7651

For over 35 years, biological scientists have come to rely on the research protocols and methodologies in the critically acclaimed *Methods in Molecular Biology* series. The series was the first to introduce the step-by-step protocols approach that has become the standard in all biomedical protocol publishing. Each protocol is provided in readily-reproducible step-by-step fashion, opening with an introductory overview, a list of the materials and reagents needed to complete the experiment, and followed by a detailed procedure that is supported with a helpful notes section offering tips and tricks of the trade as well as troubleshooting advice. These hallmark features were introduced by series editor Dr. John Walker and constitute the key ingredient in each and every volume of the *Methods in Molecular Biology* series. Tested and trusted, comprehensive and reliable, all protocols from the series are indexed in PubMed.

Proteus mirabilis

Methods and Protocols

Edited by

Melanie M. Pearson

Department of Microbiology and Immunology, University of Michigan Medical School, Ann Arbor, MI, USA

Humana Press

Editor
Melanie M. Pearson
Department of Microbiology
and Immunology
University of Michigan Medical School
Ann Arbor, MI, USA

ISSN 1064-3745 ISSN 1940-6029 (electronic)
Methods in Molecular Biology
ISBN 978-1-4939-9600-1 ISBN 978-1-4939-9601-8 (eBook)
https://doi.org/10.1007/978-1-4939-9601-8

Cover caption: Scanning electron micrograph of *P. mirabilis* swarming over an agar surface. Swarm fronts were captured during active migration using a in situ vapor fixation technique. This revealed the organization of flagellar filaments during swarming, showing these are interwoven in phase to form helical connections between adjacent swarmer cells. Image provided by Dr. Brian V. Jones.

This Humana imprint is published by the registered company Springer Science+Business Media, LLC, part of Springer Nature.
The registered company address is: 233 Spring Street, New York, NY 10013, U.S.A.

Preface

Proteus mirabilis was named by Gustav Hauser in 1885, who observed bacteria cultured on gelatin that morphed from short, nearly spherical rods to lively, moving long bacilli. He named the bacterium after Proteus from Homer's Odyssey, who was able to change shape at will to avoid capture and questioning. Over 60 years later, in 1949, Johannes Kvittingen summed up continued efforts to study *P. mirabilis* with the conclusion: "In the course of the time this work on the life-cycle of *Proteus* has been going on, and as new details have been clarified, the feeling has increased that the full understanding of the biology and sociology of the microbe becomes increasingly evasive." Williams and Schwarzhoff in 1978, while noting advances in the field, wrote "the swarming phenomenon continues to escape our complete understanding."

P. mirabilis is an opportunistic bacterial pathogen, most often known as a causative agent of complicated urinary tract infection. It is also a model organism for swarming motility, wherein bacteria move as a coordinated population across a surface. Despite its genetic relatedness to widely recognized species such as *Escherichia coli*, specialized handling techniques are often necessary. This volume contains essential protocols for researchers who study, or hope to study, *P. mirabilis*. With renewed appreciation for the medical impact and environmental adaptability of this organism, coupled with continued fascination for its dynamic behavior, we present this volume to aid further pursuit of answers.

Ann Arbor, MI, USA *Melanie M. Pearson*

Contents

Contributors

CHELSIE E. ARMBRUSTER • *Department of Microbiology and Immunology, Jacobs School of Medicine and Biomedical Sciences, State University of New York at Buffalo, Buffalo, NY, USA*
BRUCE P. AYATI • *Department of Mathematics, University of Iowa, Iowa City, IA, USA*
GRZEGORZ CZERWONKA • *Department of Microbiology, Institute of Biology, Jan Kochanowski University in Kielce, Kielce, Poland*
KATARZYNA DURLIK • *Department of Microbiology, Institute of Biology, Jan Kochanowski University in Kielce, Kielce, Poland*
VALERIE S. FORSYTH • *Department of Microbiology and Immunology, University of Michigan Medical School, Ann Arbor, MI, USA*
JÓZEF GAWĘDA • *Świętokrzyskie Rheumatology Centre, St. Lukas Hospital, Końskie, Poland*
KARINE A. GIBBS • *Department of Molecular and Cellular Biology, Harvard University, Cambridge, MA, USA*
JOANNA GLEŃSKA-OLENDER • *Department of Microbiology, Institute of Biology, Jan Kochanowski University in Kielce, Kielce, Poland; Regional Science and Technology Center, Świętokrzyski Biobank, Chęciny, Poland*
MARÍA JOSÉ GONZÁLEZ • *Departamento de Microbiología, Instituto de Investigaciones Biológicas Clemente Estable, Montevideo, Uruguay*
STEPHANIE D. HIMPSL • *Department of Microbiology and Immunology, University of Michigan Medical School, Ann Arbor, MI, USA*
KRISTEN E. HOWERY • *Department of Microbiology and Immunology, Emory University, Atlanta, GA, USA*
VICTORIA IRIBARNEGARAY • *Departamento de Microbiología, Instituto de Investigaciones Biológicas Clemente Estable, Montevideo, Uruguay*
BRIAN V. JONES • *Department of Biology and Biochemistry, University of Bath, Bath, UK*
WIESŁAW KACA • *Department of Microbiology, Institute of Biology, Jan Kochanowski University in Kielce, Kielce, Poland*
IWONA KONIECZNA • *Department of Microbiology, Institute of Biology, Jan Kochanowski University in Kielce, Kielce, Poland*
KEEHWAN KWON • *The J. Craig Venter Institute, Rockville, MD, USA*
KRISTIN LITTLE • *Department of Molecular and Cellular Biology, Harvard University, Cambridge, MA, USA*
AGNIESZKA MASZEWSKA • *Department of Biology of Bacteria, Faculty of Biology and Environmental Protection, University of Lodz, Lodz, Poland*
SCARLET MILO • *Department of Chemistry, University of Bath, Bath, UK*
HARRY L. T. MOBLEY • *Department of Microbiology and Immunology, University of Michigan Medical School, Ann Arbor, MI, USA*
JONATHAN NZAKIZWANAYO • *Department of Biology and Biochemistry, University of Bath, Bath, UK*
MELANIE M. PEARSON • *Department of Microbiology and Immunology, University of Michigan Medical School, Ann Arbor, MI, USA*
HARRIET PELLING • *Department of Biology and Biochemistry, University of Bath, Bath, UK*
REMBERT PIEPER • *The J. Craig Venter Institute, Rockville, MD, USA*

PHILIP N. RATHER • *Department of Microbiology and Immunology, Emory University, Atlanta, GA, USA; Research Service, Atlanta VA Medical Center, Decatur, GA, USA*
SHAWN RICHMOND • *Department of Biological Sciences, California Polytechnic State University, San Luis Obispo, CA, USA*
ANTONI RÓŻALSKI • *Department of Biology of Bacteria, Faculty of Biology and Environmental Protection, University of Lodz, Lodz, Poland*
PAOLA SCAVONE • *Departamento de Microbiología, Instituto de Investigaciones Biológicas Clemente Estable, Montevideo, Uruguay*
SŁAWOMIR SĘK • *Faculty of Chemistry, Biological and Chemical Research Centre, University of Warsaw, Warsaw, Poland*
SARA N. SMITH • *Department of Microbiology and Immunology, University of Michigan Medical School, Ann Arbor, MI, USA*
ALEJANDRA YEP • *Biological Sciences Department, California Polytechnic State University, San Luis Obispo, CA, USA*
YANBAO YU • *The J. Craig Venter Institute, Rockville, MD, USA*
PAULINA ŻARNOWIEC • *Department of Microbiology, Institute of Biology, Jan Kochanowski University in Kielce, Kielce, Poland*
PABLO ZUNINO • *Departamento de Microbiología, Instituto de Investigaciones Biológicas Clemente Estable, Montevideo, Uruguay*

Chapter 1

Proteus mirabilis Overview

Harry L. T. Mobley

Abstract

Proteus mirabilis, a Gram-negative bacterium, commonly causes catheter-associated urinary tract infections, wound infections, gastroenteritis and, in some cases, bacteremia. The phenotypic hallmarks of this bacterium include swarming motility, urease and hemolysin production, and synthesis of numerous adherence fimbriae. While routine bacteriological methodology may often be used to study this pathogen, frequently one requires specialized techniques to investigate the pathogenesis of this species. Here, a brief overview of the discoveries associated with this fascinating bacterium illuminates the need for the development of specialized techniques to further probe the biology of *P. mirabilis*.

Key words *Proteus mirabilis*, History, Swarming, Adherence, Iron acquisition, Dienes phenomenon

Proteus mirabilis, a fascinating Gram-negative polymorphic swarming bacterium, causes infection of the urinary tract and has been known to cause infections of wound, the eye, the bloodstream, and the gastrointestinal tract. The vast majority of the literature, however, has focused on infection of the catheterized urinary tract.

When I first opened my own research lab at the University of Maryland School of Medicine, my colleague John Warren and colleagues had just published a landmark study that followed urinary tract infection in 100 long-term catheterized nursing home patients at two separate nursing homes, for a 1-year period [1]. My long-term lab manager at the time, Gwynn Chippendale, isolated all culturable bacteria from all specimens over a wide range of CFU/mL of urine. After analyzing the data, the research group found an average of more than four bacterial species per sample. Thus, these infections were usually polymicrobial. The most common organism, by far, was *Proteus mirabilis.* Epidemiologically, there was a strong correlation with the development of catheter encrustration and blockage, and development of urinary stones in these patients [2]. *P. mirabilis* and *Providencia stuartii* were found to be persistent colonizers; that is, once the organisms gained a foothold in these patients, these two organisms persisted at high

Melanie M. Pearson (ed.), *Proteus mirabilis: Methods and Protocols*, Methods in Molecular Biology, vol. 2021,
https://doi.org/10.1007/978-1-4939-9601-8_1,

titers week after week. Undoubtedly, these infections took a terrible toll on these patients, requiring frequent catheter replacement and antibiotic treatment for systemic symptoms. Indeed, an immediate autopsy study demonstrated that many of these patients died with acute pyelonephritis, that is, infection of the kidney (upper urinary tract) [3]. Despite these observations, only a few pathogenesis studies had previously been conducted.

From 1940 until 1983, PubMed lists 1840 articles on *Proteus mirabilis*. Most of these articles were clinical in nature with only a modest number devoted to pathogenesis. From 1984 (my personal landmark) to the present, 4394 articles have been published on this bacterium, and a significant number of these address the molecular mechanisms of pathogenesis.

Studies that had been completed included a thorough survey of hemagglutination patterns elicited by in vitro-cultured *P. mirabilis*. Investigators used erythrocytes from numerous vertebrate sources including chickens, guineas pigs, horses, ox, and humans to measure adherence properties of the bacterium [4]. These studies implied that *P. mirabilis* produced numerous adhesins. Decades later, sequencing of the genome revealed the capacity to synthesize 17 fimbriae in what has now become the type isolate of the species, HI4320, isolated in the original nursing home study [5].

Other researchers, including Abraham Braude [6], implicated urease as a key factor in urolithiasis (stone formation) and its importance in the development of acute pyelonephritis. It was also noted in the literature that *P. mirabilis* was hemolytic for human and other vertebrate species' erythrocytes.

The phenomenon of swarming across an agar plate fascinated researchers, who noted that the bacterium morphs from a short bacillus to a highly elongated, highly flagellated form that could move across the agar in groups of swarmers. When two different isolates were placed at the opposite ends of the agar plate, they would form a clear line as the isolates swarmed across the plate and met at the middle. This was noted by Louis Dienes in 1946 [7] with no molecular basis discovered for 62 years [8, 9].

All of these observations were ultimately relevant to pathogenesis and the molecular physiology of *P. mirabilis,* but the molecular mechanisms of action remained undefined.

With that as background, fortunately our work began at the advent of the molecular pathogenesis era. With a modest variety of restriction enzymes, tedious DNA sequencing methodology, a limited number of cloning vectors (but no PCR yet), we and others were able to attack the molecular basis of pathogenesis. Inspired by the work of Stanley Falkow, then at the University of Washington, and one of his many trainees, Jim Kaper, who had just established his lab at Maryland, we began to investigate the molecular pathogenesis of *Proteus mirabilis* urinary tract infection [10]. Since that time we have worked consistently to investigate urease, hemolysin, iron and zinc acquisition, adherence mediated by fimbriae and other adhesins, autotransported proteases, swarming and the basis for Dienes line formation (type 6 secretion system), integrative and conjugative elements, the transcriptome during bladder infection, transporters, metabolism as it relates to colonization, and of course, the complete genome sequence of strain HI4320 [11].

In addition, a relatively small number of other talented laboratories joined the quest to understand what makes *Proteus mirabilis* tick. We all used established techniques outlined in the literature for our investigations. However, it quickly became clear that we needed to develop techniques that were specific to studying *P. mirabilis*. Most notably, the ability to construct mutations by allelic exchange was not straightforward and we could not use established techniques. This limited progress. Iron acquisition systems that included siderophores were undetectable by standard methods. The phenotype of swarming motility had been studied for decades; however, the advent of molecular biology techniques began to shed some light on the mechanism of transformation from the short adherent form to the swarming morphotype [12, 13]. This riddle is still not solved.

Herein is assembled a wealth of methodology, in a volume expertly edited by Melanie Pearson, that has either been developed de novo or has been significantly modified to study *P. mirabilis*. These include culture methods for the bacterium in general or to assess motility and to prevent and induce swarming motility (Chapters 2, 3, and 4). Because motility is so central to the lifestyle of the bacterium, techniques for flagellin purification are included (Chapter 5). When different strains of *P. mirabilis* swarm together, they often form "Dienes" lines, now known to result from one or both strains killing each other. We now know that the formation of Dienes lines is the result of the action of the type 6 secretion system. Techniques for assessment of Dienes line formation are provided in Chapter 6.

Mutagenesis of *P. mirabilis* is more complex than *E. coli*, for example. These techniques, including the use of group II introns (targetrons) (Chapter 7) and conjugation and allelic exchange (Chapter 8), are described.

Besides swarming motility, other virulence factors of *P. mirabilis* have been studied in some detail. For example, methods for measuring urease activity [the driver of urolithiasis (stone formation)] (Chapter 9) and iron chelation (Chapter 10) are outlined. Adherence is a critical first step for infection; working with fimbriae (Chapter 11) is detailed here as well. At least in one case, phase variation of the MR/P fimbria is controlled by a promoter found on an invertible element. Methods for determining whether the switch is on or off are provided in detail (Chapter 12).

A number of models are presented for the study of the pathogenic potential of *P. mirabilis*. These include methods for the use of artificial bladder models (Chapter 14), adherence to cultured or naturally shed epithelial cells (Chapter 13), and the well-used UTI mouse model (both catheterized and uncatheterized) of ascending urinary tract infection (both independent challenge and cochallenge) (Chapters 15, 16 and 17). The mouse model has also been used for vaccine development and testing, described in (Chapter 18).

Additionally, critical methods for characterization of LPS (Chapter 19) and isolation of bacteriophage (Chapter 20) are provided, as well as those for detecting neutrophil activity in urine (Chapter 21), proteomic analysis for inflammation in urine (Chapter 22), antigen assessment (Chapter 23), mathematical modeling of swarming (Chapter 24), and Tn-seq (Chapter 25). This volume represents a treasure trove of relevant methodology adapted or developed specifically for the study of this fascinating bacterium. I hope that this volume is useful in your studies.

References

1. Warren JW, Tenney JH, Hoopes JM, Muncie HL, Anthony WC (1982) A prospective microbiologic study of bacteriuria in patients with chronic indwelling urethral catheters. J Infect Dis 146(6):719–723
2. Mobley HL, Warren JW (1987) Urease-positive bacteriuria and obstruction of long-term urinary catheters. J Clin Microbiol 25 (11):2216–2217
3. Warren JW, Muncie HL Jr, Hall-Craggs M (1988) Acute pyelonephritis associated with bacteriuria during long-term catheterization: a prospective clinicopathological study. J Infect Dis 158(6):1341–1346
4. Old DC, Adegbola RA (1982) Haemagglutinins and fimbriae of *Morganella*, *Proteus* and *Providencia*. J Med Microbiol 15(4):551–564. https://doi.org/10.1099/00222615-15-4-551
5. Pearson MM, Sebaihia M, Churcher C, Quail MA, Seshasayee AS, Luscombe NM, Abdellah Z, Arrosmith C, Atkin B, Chillingworth T, Hauser H, Jagels K, Moule S, Mungall K, Norbertczak H, Rabbinowitsch E, Walker D, Whithead S, Thomson NR, Rather PN, Parkhill J, Mobley HLT (2008) Complete genome sequence of uropathogenic *Proteus mirabilis*, a master of both adherence and motility. J Bacteriol 190 (11):4027–4037
6. Braude AI, Siemienski J (1960) Role of bacterial urease in experimental pyelonephritis. J Bacteriol 80:171–179
7. Dienes L (1946) Reproductive processes in *Proteus* cultures. J Bacteriol 51:585
8. Alteri CJ, Himpsl SD, Pickens SR, Lindner JR, Zora JS, Miller JE, Arno PD, Straight SW, Mobley HL (2013) Multicellular bacteria deploy the type VI secretion system to preemptively strike neighboring cells. PLoS Pathog 9 (9):e1003608. https://doi.org/10.1371/journal.ppat.1003608
9. Gibbs KA, Urbanowski ML, Greenberg EP (2008) Genetic determinants of self identity and social recognition in bacteria. Science 321 (5886):256–259. https://doi.org/10.1126/science.1160033
10. Jones BD, Lockatell CV, Johnson DE, Warren JW, Mobley HL (1990) Construction of a urease-negative mutant of *Proteus mirabilis*: analysis of virulence in a mouse model of ascending urinary tract infection. Infect Immun 58(4):1120–1123
11. Armbruster CE, Mobley HLT, Pearson MM (2018) Pathogenesis of *Proteus mirabilis* infection. EcoSal Plus 8(1). https://doi.org/10.1128/ecosalplus.ESP-0009-2017
12. Allison C, Lai HC, Hughes C (1992) Co-ordinate expression of virulence genes during swarm-cell differentiation and population migration of *Proteus mirabilis*. Mol Microbiol 6(12):1583–1591
13. Belas R, Erskine D, Flaherty D (1991) *Proteus mirabilis* mutants defective in swarmer cell differentiation and multicellular behavior. J Bacteriol 173(19):6279–6288

Chapter 2

Culture Methods for *Proteus mirabilis*

Melanie M. Pearson

Abstract

Proteus mirabilis is generally easy to culture, but its tendency to swarm on a wide variety of media can interfere with isolation of single colonies or identification of other species in a sample. Therefore, specialized media may be needed to control swarming or to study the bacteria under chemically defined conditions. Here, methods are described for routine culture of *P. mirabilis*, isolation of *P. mirabilis* from mixed cultures, and culture of *P. mirabilis* on physiologically relevant media.

Key words *Proteus mirabilis*, Culture, Swarming, Antiswarming, Swarming prevention, Artificial urine, Urine agar

1 Introduction

Proteus mirabilis can be isolated from a wide variety of environments, including humans, other mammals, reptiles, fish, insects, and sewage [1]. Therefore, it is perhaps unsurprising that *P. mirabilis* is relatively easy to culture. Like its cousin *Escherichia coli*, it can be cultured on a variety of common complex and chemically defined minimal media. Key differential features of *P. mirabilis* include Gram-negative, motile, lactose-negative, indole-negative, H_2S-producing on triple sugar iron (TSI) agar, urease-positive, ornithine decarboxylase-positive, and maltose-negative [2]. *P. mirabilis* is a facultative anaerobe, and, indeed, expresses at least one virulence factor, MR/P fimbriae, optimally in a microaerobic environment [3]. But, this species is both celebrated and cursed for its ability to rapidly swarm across many types of culture media. Other bacteria in the mixture will be obscured beneath the swarm; likewise, isolation of individual colonies is nearly impossible when swarming occurs. Unless one wishes to study swarming motility (*see* Chapter 3 of this volume), steps must be taken to prevent *P. mirabilis* from taking over the entire agar surface. This can be accomplished in a variety of ways. The purpose of this chapter is to review some of the options to obtain

Melanie M. Pearson (ed.), *Proteus mirabilis: Methods and Protocols*, Methods in Molecular Biology, vol. 2021,
https://doi.org/10.1007/978-1-4939-9601-8_2,

single colonies of *P. mirabilis* on agar, or to culture this species in physiologically relevant conditions.

2 Routine Culture

The media presented in this section are recommended for routine culture of *P. mirabilis* in a research setting. All of these media will allow culture of *P. mirabilis* without swarming, which is a requirement for many molecular biology and routine microbiological applications (*see* **Note 1**). For all agar types, autoclave to sterilize, cool to 55 °C (add any necessary autoclave-incompatible components at this point), and pour into agar plates. Let plates sit at room temperature overnight, then place in plastic bags and store at 4 °C.

1. Low salt LB. Per liter: 10 g of tryptone, 5 g of yeast extract, 0.5 g of NaCl, 15 g of agar (*see* **Notes 2** and **3**). LB works well as an all-purpose culture medium. Omit the agar to generate broth for liquid cultures (*see* **Note 4**) (Fig. 1a).
2. LSW^- agar [4]. Per liter: 10 g of tryptone, 5 g of yeast extract, 5 mL of glycerol, 0.4 g of NaCl, 20 g of agar. The glycerol, reduced salt, and increased agar all contribute to inhibit swarming.
3. *Proteus* minimal salts medium (PMSM), also called Minimal A [4]. Per liter: 10.5 g of K_2HPO_4, 4.5 g of KH_2PO_4, 0.47 g of sodium citrate, 1.0 g of $(NH_4)_2SO_4$; autoclave to sterilize and add 1 mL of 1 M $MgSO_4$, 10 mL of 20% glycerol, and 1 mL of 1% nicotinic acid (*see* **Notes 5–7**).

3 *P. mirabilis* in Mixed Cultures

P. mirabilis is commonly part of a polymicrobial community; examples include indwelling urinary catheters, the gastrointestinal tract, or sewage. A single colony of *P. mirabilis* can rapidly swarm over an entire agar plate, making isolation and identification of individual species or strains difficult if not impossible. This will occur on most types of rich media, including blood and chocolate agars, which are commonly used with clinical microbiology samples. The media and additives presented in this section are particularly useful for isolation of *P. mirabilis* when multiple species are likely to be present (*see* **Note 8**).

1. MacConkey agar. Per liter, 17 g of peptone, 3 g of proteose peptone, 10 g of lactose, 1.5 g of bile salts, 5 g of NaCl, 0.03 g of neutral red, 0.001 g of crystal violet, 13.5 g of agar; adjust pH to 7.1 ± 0.2; also available commercially (*see* **Note 9**). MacConkey agar differentiates lactose-fermenting bacteria,

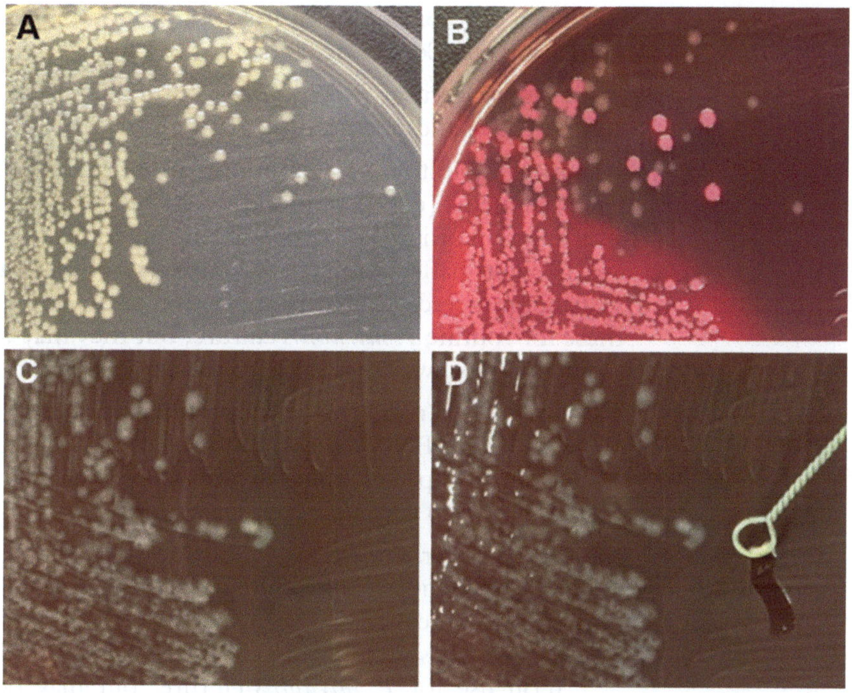

Fig. 1 *P. mirabilis* colonies on agar. (**a**) Low salt LB. (**b**) A mix of *P. mirabilis* (white) and *E. coli* (pink) on MacConkey agar. (**c**, **d**) A mix of *P. mirabilis* and *E. coli* on blood agar. What appears to be isolated colonies in panel **c** is actually covered with swarming *P. mirabilis*, which becomes evident when a loop is drawn across the background (**d**). In all panels, the *P. mirabilis* strain is HI4320. In panels **b–d**, the uropathogenic *E. coli* strain is CFT073

which will form colonies with a bright pink color; *P. mirabilis* does not ferment lactose and remains colorless (Fig. 1b). This agar also inhibits growth of most Gram-positive bacteria. Bile salts are the component of MacConkey that inhibit swarming (*see* **Note 10**). Some *P. mirabilis* strains can still swarm on this medium [5]; reducing or omitting NaCl can help.

2. Cysteine-, lactose-, and electrolyte-deficient (CLED) agar [6, 7]. Per liter: 10 g of lactose, 4 g of pancreatic digest of gelatin, 4 g of pancreatic digest of casein, 3 g of beef extract, 0.128 g of L-cystine, 0.02 g of bromothymol blue, and 15 g of agar, final pH 7.1–7.5; also available commercially. CLED agar is particularly useful for distinguishing multiple species commonly found during UTI, by colony color, size, and opacity (*P. mirabilis* will produce translucent blue colonies). Unlike MacConkey, CLED agar supports growth of Gram-positive bacteria.

4. Increased agar concentration. Typical solid media contain 1.5% agar. Raising the agar concentration to 2–3% will greatly decrease swarm migration on permissive media [8], and may

be sufficient to prevent it entirely in borderline cases (e.g., low salt LB or MacConkey agar) (*see* **Note 11**).

5. Antibiotic selection. *P. mirabilis* is naturally resistant to tetracycline [2, 9] and polymyxins [10, 11].
6. Chemical inhibitors of swarming can be added to standard bacterial media (e.g., LB, blood, or chocolate agar) to prevent swarming on otherwise permissive formulations. Most of these are likely to affect aspects of bacterial physiology beyond swarming, including growth inhibition. I do not recommend these for routine *P. mirabilis* culture, but they may be useful in settings where swarm-permissive media need to be used:
   (a) 100 μg/mL of *p*-nitrophenylglycerol (PNPG) [12, 13]. PNPG seems not to affect growth of other bacterial species, but it also represses expression of *P. mirabilis* virulence factors (*see* **Note 12**) [12].
   (b) Many other additives have been reported over the years. These include 1:500 chloral hydrate, 1:100 boric acid, sulfonamides, 1% charcoal, ethanol, triclosan, fatty acids, and 0.01% sodium azide. These are not generally in common use for *P. mirabilis* culture, but may be useful in specific settings. Azide in particular is a potent poison, and this concentration is also growth inhibitory (*see* **Note 13**). *See* [5, 14–18] for detailed lists of these additives.

4 Physiological Media

P. mirabilis can be isolated from the urinary tract, feces, blood, and wounds. It may therefore be occasionally useful to culture this organism on media that capture aspects of these physiological niches. Here are some options to consider.

1. Urine agar [19] (*see* **Notes 14** and **15**). Collect and pool urine from 3–5 volunteers (*see* **Note 16**). Sterilize urine using a 0.2 μm pore vacuum filter. To half of the urine, add 3% agar; autoclave and then cool to 55 °C (*see* **Note 17**). Warm the remaining portion of urine to 37 °C, and mix 1:1 with the agar. Pour into petri dishes (Fig. 2).
2. Artificial urine (*see* **Note 14**). Human urine can vary considerably, depending on host fluid intake, diet, time of day, and health. Therefore, artificial urine media have been developed to allow reproducible experimentation. Both of the formulations shown here support *P. mirabilis* growth.
   (a) Method 1, by Stickler et al. (*see* **Note 18**) [20, 21]. Per liter: 0.49 g of calcium chloride, 0.65 g of magnesium

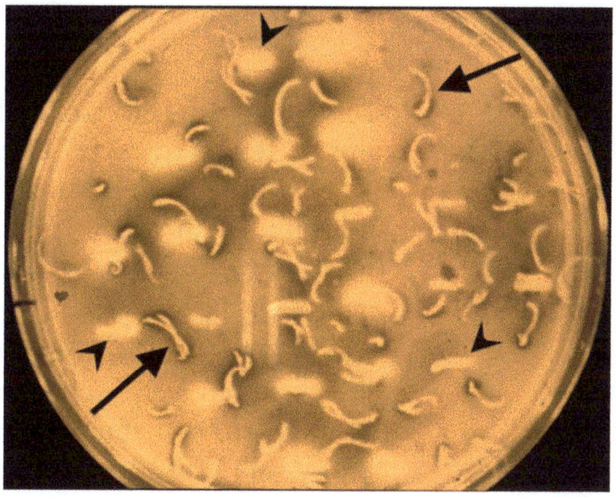

Fig. 2 Urine agar. *P. mirabilis* was streaked (arrowheads) onto agar made from human urine. Struvite crystals (arrows), produced as a result of urease activity, form in the vicinity of the bacteria. Reprinted with permission from the American Society for Microbiology [32]

chloride hexahydrate, 4.6 g of sodium chloride, 2.3 g of sodium sulfate, 0.65 g of trisodium citrate dihydrate, 0.02 g of disodium oxalate, 2.8 g of potassium dihydrogen phosphate, 1.60 g of potassium chloride, 1 g of ammonium chloride, 25 g of urea, and 5 g of gelatin. Adjust pH to 6.1, then sterilize using a 0.2 μm pore vacuum filter. Separately sterilize tryptone soya broth by autoclaving, and add to basal medium to a final concentration of 10 g/L.

(b) Method 2: artificial urine medium (AUM) [22]. Per liter: 1 g of peptone, 0.005 g of yeast extract, 0.1 g of lactic acid, 0.4 g of citric acid, 2.1 g of sodium bicarbonate, 10 g of urea, 0.07 g of uric acid, 0.8 g of creatinine, 0.37 g of calcium chloride dihydrate, 5.2 g of sodium chloride, 0.0012 g of iron II sulfate heptahydrate, 0.49 g of magnesium sulfate heptahydrate, 3.2 g of sodium sulfate decahydrate, 0.95 g of potassium dihydrogen phosphate, 1.2 g of dipotassium hydrogen phosphate, and 1.3 g of ammonium chloride. Use hydrochloric acid to adjust the pH to 6.5 and sterilize using a 0.2 μm pore vacuum filter (*see* **Note 19**).

3. Blood agar, typically 5% sheep's blood in trypticase soy agar, commercially available [23]. *P. mirabilis* is routinely detected on blood agar in clinical microbiology settings; however, it swarms robustly on this medium (Fig. 1c, d).

5 Notes

1. In general, low electrolyte levels lead to no or reduced swarming [6, 18]. However, *P. mirabilis* strains vary considerably in swarming characteristics, and some strains are more easily inhibited than others. Some strains will swarm on swarming-inhibitory media, particularly in humid environments.
2. This formulation is also known as LB-Luria, as opposed to LB-Lennox or LB-Miller, which have 5 or 10 g of NaCl per liter, respectively. The reduced salt is key to preventing *P. mirabilis* swarming. Already-grown colonies of *P. mirabilis* type strain HI4320 will start to swarm on LB plates left at room temperature, but not on plates incubated at 37 °C or stored at 4 °C.
3. *P. mirabilis* cultured on LB agar can be stored at 4 °C for up to a month. Wrap plates in paraffin sealing film or place in plastic bags to prevent agar from drying out. Most *P. mirabilis* strains will slowly turn the agar a deeper shade of orange during storage at 4 °C.
4. *P. mirabilis* will grow faster in LB formulations with 10 g or 5 g/L of NaCl, but I prefer to conduct all experiments in 0.5 g/L so that broth and agar experiments are consistent.
5. *P. mirabilis* requires the addition of nicotinic acid in a minimal medium [24].
6. For a detailed protocol to identify additives that stimulate swarming on this normally nonpermissive medium, *see* Chapter 4 of this volume.
7. The citrate in this medium binds iron. For experiments investigating iron utilization by *P. mirabilis*, a modified version Neidhardt MOPS medium is more suitable. Per liter of standard Neidhardt MOPS medium [25], add 10 mL of 20% glycerol, 1 mL of 1 M $MgSO_4{\cdot}7H_2O$, 1 mL of 1% nicotinic acid, and 1 mL of 20% casamino acids. *See* also Chapter 11 for iron utilization protocols.
8. The media listed here are suggestions for identifying and isolating *P. mirabilis* from mixed cultures. While these options support a variety of bacterial species, some non-*Proteus* species will be excluded; consider this carefully when surveying mixed cultures.
9. This is a modification of MacConkey's original formulation [26], but this modification is the version most commonly used today [27].
10. *See* Lominski and Lendrum [5] for a detailed discussion of bile salts and other antiswarming agents.

11. Hayward and Miles [28] found that 6–8% agar was sufficient to prevent swarming by all *Proteus* isolates, but this concentration of agar can be difficult to work with.
12. Similar results have been reported for 60 μg/mL of resveratrol (3,5,4-trihydroxy-*trans*-stilbene) [29].
13. Addition of 0.005% NaN_3 has also been reported to be growth inhibitory for *P. mirabilis*, and yet swarming still occurs at this concentration [30].
14. In the presence of urea, *P. mirabilis* will quickly alkalinize the medium due to urease activity and ammonia release (*see* Chapter 9 of this volume for more about urease). In urine, this causes precipitation of minerals and formation of struvite crystals (Fig. 2). Urease also causes a pH increase that can quickly become toxic in a closed environment, such as a petri dish or culture tube. Two options to avoid this pH problem are to omit urea from synthetic urine, or to conduct experiments using a urease mutant [19, 31].
15. Urine may also be used as a liquid culture medium [31]. In this case, pool urine from volunteers and filter-sterilize. Urine may be used immediately, stored at 4 °C short-term, or frozen at −20 °C for longer-term storage. Storage and freezing may lead to degradation of labile components; however, pooling and storage of aliquots may improve reproducibility of results.
16. Collection of human urine may require approval from the appropriate institutional review board. Volunteers should not have taken antibiotics recently, as many antibiotics will be excreted into the urine and will affect bacterial growth.
17. Autoclaved urine produces a pungent odor, as does urine agar when inoculated with *P. mirabilis*. Working in a well-ventilated area and/or away from others is highly recommended.
18. A modified version of this medium is described in Chapter 14 of this volume.
19. Precipitation may occur at pH 7.2 [22].

Acknowledgments

Thank you to Stephanie Himpsl for constructive discussions and Chelsie Armbruster for refining the urine agar method.

References

1. Drzewiecka D (2016) Significance and roles of *Proteus* spp. bacteria in natural environments. Microb Ecol 72(4):741–758. https://doi.org/10.1007/s00248-015-0720-6
2. O'Hara CM, Brenner FW, Miller JM (2000) Classification, identification, and clinical significance of *Proteus*, *Providencia*, and *Morganella*. Clin Microbiol Rev 13(4):534–546

3. Lane MC, Li X, Pearson MM, Simms AN, Mobley HLT (2009) Oxygen-limiting conditions enrich for fimbriate cells of uropathogenic *Proteus mirabilis* and *Escherichia coli.* J Bacteriol 191(5):1382–1392
4. Belas R, Erskine D, Flaherty D (1991) Transposon mutagenesis in *Proteus mirabilis.* J Bacteriol 173(19):6289–6293
5. Lominski I, Lendrum AC (1942) The effect of surface-active agents on *B. proteus.* J Pathol Bacteriol 54(4):421–433. https://doi.org/10.1002/path.1700540403
6. Sandys GH (1960) A new method of preventing swarming of *Proteus* sp. with a description of a new medium suitable for use in routine laboratory practice. J Med Lab Technol 17:224–233
7. Mackey JP, Sandys GH (1966) Diagnosis of urinary infections. Brit Med J 1(5496):1173. https://doi.org/10.1136/bmj.1.5496.1173
8. Rauprich O, Matsushita M, Weijer CJ, Siegert F, Esipov SE, Shapiro JA (1996) Periodic phenomena in *Proteus mirabilis* swarm colony development. J Bacteriol 178 (22):6525–6538
9. Stock I (2003) Natural antibiotic susceptibility of *Proteus* spp., with special reference to *P. mirabilis* and *P. penneri* strains. J Chemotherapy 15(1):12–26. https://doi.org/10.1179/joc.2003.15.1.12
10. Russell FE (1963) Synergism between sulphonamide drugs and antibiotics of the polymyxin group against *Proteus sp. in vitro.* J Clin Pathol 16(4):362. https://doi.org/10.1136/jcp.16.4.362
11. Sud IJ, Feingold DS (1970) Mechanism of polymyxin B resistance in *Proteus mirabilis.* J Bacteriol 104(1):289–294
12. Liaw SJ, Lai HC, Ho SW, Luh KT, Wang WB (2000) Inhibition of virulence factor expression and swarming differentiation in *Proteus mirabilis* by *p*-nitrophenylglycerol. J Med Microbiol 49(8):725–731
13. Senior BW (1978) *p*-nitrophenylglycerol--a superior antiswarming agent for isolating and identifying pathogens from clinical material. J Med Microbiol 11(1):59–61
14. Alwen J, Smith DG (1967) A medium to suppress swarming of *Proteus* species. J Appl Bacteriol 30(2):389–394. https://doi.org/10.1111/j.1365-2672.1967.tb00313.x
15. Hernandez E, Ramisse F, Cavallo JD (1999) Abolition of swarming of *Proteus.* J Clin Microbiol 37(10):3435–3435
16. Firehammer BD (1987) Inhibition of growth and swarming of *Proteus mirabilis* and *Proteus vulgaris* by triclosan. J Clin Microbiol 25 (7):1312–1313
17. Liaw SJ, Lai HC, Wang WB (2004) Modulation of swarming and virulence by fatty acids through the RsbA protein in *Proteus mirabilis.* Infect Immun 72(12):6836–6845
18. Naylor PGD (1964) The effect of electrolytes or carbohydrates in sodium chloride deficient medium on formation of discrete colonies of *Proteus* and the influence of these substances on growth in liquid culture. J Appl Bacteriol 27 (3):422–431. https://doi.org/10.1111/j.1365-2672.1964.tb05050.x
19. Armbruster CE, Hodges SA, Mobley HLT (2013) Initiation of swarming motility by *Proteus mirabilis* occurs in response to specific cues present in urine and requires excess L-glutamine. J Bacteriol 195(6):1305–1319. https://doi.org/10.1128/JB.02136-12
20. Stickler DJ, Morris NS, Winters C (1999) Simple physical model to study formation and physiology of biofilms on urethral catheters. Methods Enzymol 310:494–501
21. Griffith DP, Musher DM, Itin C (1976) Urease. The primary cause of infection-induced urinary stones. Investig Urol 13(5):346–350
22. Brooks T, Keevil CW (1997) A simple artificial urine for the growth of urinary pathogens. Lett Appl Microbiol 24(3):203–206
23. Smibert RM, Krieg NR (1994) Phenotypic characterization. In: Gerhardt P (ed) Methods for general and molecular bacteriology. American Society for Microbiology, Washington, DC, pp 642–647
24. Fildes P (1938) The growth of *Proteus* on ammonium lactate plus nicotinic acid. Br J Exp Pathol 19(4):239–244
25. Neidhardt FC, Bloch PL, Smith DF (1974) Culture medium for enterobacteria. J Bacteriol 119(3):736–747
26. MacConkey AT (1908) Bile salt media and their advantages in some bacteriological examinations. J Hyg (Lond) 8(3):322–334
27. Holt JG, Krieg NR (1994) Enrichment and isolation. In: Gerhardt P (ed) Methods for general and molecular bacteriology. American Society for Microbiology, Washington, DC, p 205
28. Hayward NJ, Miles AA (1943) Inhibition of *Proteus* in cultures from wounds. Lancet 2:116–117
29. Wang WB, Lai HC, Hsueh PR, Chiou RY, Lin SB, Liaw SJ (2006) Inhibition of swarming and virulence factor expression in *Proteus mirabilis* by resveratrol. J Med Microbiol 55 (Pt 10):1313–1321. https://doi.org/10.1099/jmm.0.46661-0

30. Alteri CJ, Himpsl SD, Engstrom MD, Mobley HLT (2012) Anaerobic respiration using a complete oxidative TCA cycle drives multicellular swarming in *Proteus mirabilis*. mBio 3(6). https://doi.org/10.1128/mBio.00365-12
31. Pearson MM, Yep A, Smith SN, Mobley HLT (2011) Transcriptome of *Proteus mirabilis* in the murine urinary tract: virulence and nitrogen assimilation gene expression. Infect Immun 79(7):2619–2631. https://doi.org/10.1128/IAI.05152-11
32. Mobley HLT (2000) Virulence of the two primary uropathogens - *Escherichia coli* and *Proteus mirabilis* exhibit distinct mechanisms of pathogenesis when causing urinary tract infections. ASM News 66(7):403–410

Chapter 3

Methods for Studying Swarming and Swimming Motility

Melanie M. Pearson

Abstract

Proteus mirabilis is well known for using its flagella to swim through liquids or swarm across solid surfaces. Both phenomena are easy to observe. Described here are two agar-based assays for studying both swimming and swarming behavior, and considerations that affect the outcome.

Key words Swimming, Swarming, Flagella, Motility, Semisolid agar

1 Introduction

Many species of bacteria use flagella to propel themselves. The ability to move toward food or away from noxious stimuli confers a survival advantage in a variety of environments, including urinary tract infection mediated by *Proteus mirabilis* [1]. The most common type of flagellum-mediated motility is swimming, where bacteria use one or several flagella to move through a liquid. *P. mirabilis*, like other members of the *Enterobacteriales*, typically has several peritrichous flagella that it uses to swim [2].

A second type of flagella-mediated motility is swarming, which occurs when a population of bacteria differentiate into long, hyperflagellated cells and move together across a surface [3, 4]. *P. mirabilis* is particularly well known for its vigorous swarming ability, and is able to swarm under conditions that are not permissive for most other bacteria. When *P. mirabilis* swarms, it usually forms a bull's-eye pattern, which is caused by repeated cycles of swarming outward, stopping and dedifferentiating into shorter rods (consolidation), and morphing back into motile swarmer cells [1]. These bacteria are also able to swarm across nonnutritive surfaces, such as urinary catheters [5], which may contribute to *Proteus*' prevalence during catheter-associated urinary tract infections [6, 7].

Here, two assays are described for measuring flagella-mediated motility. First, a method for studying swimming through semisolid

Melanie M. Pearson (ed.), *Proteus mirabilis: Methods and Protocols*, Methods in Molecular Biology, vol. 2021, https://doi.org/10.1007/978-1-4939-9601-8_3,

agar is presented. In this assay, bacteria swim outward from an initial point of inoculation. The outward migration is aided by chemotaxis, as the bacteria consume nutrients in the agar and move toward gradients of unconsumed food. Deletion of flagella or the flagellar motor results in bacteria that remain at the point of inoculation, while mutation of chemotaxis genes results in slower, undirected motion outward. In addition to motility by wild-type strains [8], this assay has been used to show the contributions of different media additives [9] and regulators [10, 11] to *P. mirabilis* motility.

Second, a protocol to generate consistent swarming across solid agar is described. *P. mirabilis* is, in general, going to swarm on complex agar. However, swarming rate, periodicity (transition between swarming and consolidation phases), and appearance (ring width, radial patterns, etc.) is highly variable, depending on assay conditions and the strain being tested [12]. Some variables that affect swarming behavior are shown in Table 1. Obtaining reproducible results can be tricky; consistency is key! The protocol presented here is designed to generate highly reproducible swarms. Variations of this assay have been used to investigate gene expression during swarming [13–15], basic nutritional requirements of swarming [9, 16, 17], and swarming across urinary catheter segments [5]. Numerous mutations that affect *P. mirabilis* swarming have also been described [1]. In addition to increasing or decreasing swarm rates, these mutations often result in unusual swarm patterns, aside from deletion of flagella, which completely prevents both swimming and swarming motility.

Table 1
Factors that affect swarming[a]

Variable	Increase	Decrease
Temperature	37 °C	23 °C
NaCl	10 g/L	0.5 g/L
Humidity	High	Low
Medium	Complex	Minimal
Agar	≤15 g/L	>20 g/L
pH	7	5 or 9
Cell density[b]	10^6 CFU	10^2 CFU
Culture type[b]	Swarm	Broth

[a]Swarming may occur in any of the conditions listed in this table, depending on the strain of *P. mirabilis*. For a discussion about entirely preventing swarming by *P. mirabilis*, *see* Chapter 2 of this volume

[b]Cell density and culture method affect the timing of swarming initiation, but not necessarily the appearance or rate of swarming [23, 25]

2 Materials

All reagents used with bacteria should be sterile.

2.1 Swimming Motility Assay

1. *P. mirabilis* strain(s) to be tested (*see* **Note 1**).
2. Lysogeny broth (LB): per liter, 10 g of tryptone, 5 g of yeast extract, and 0.5 g of NaCl. Autoclave to sterilize.
3. Motility agar: per liter, 10 g of tryptone, 5 g of NaCl, and 2.5 g of agar. Autoclave to sterilize, then use a pipet to transfer 25 mL of agar per 100 mm petri dish. Make agar the day of the experiment (*see* **Note 2**).
4. Sterile culture tubes.
5. Incubator set to desired survey temperature (typically 30 or 37 °C).
6. Inoculating needle.
7. Large beaker or tray of water.
8. Spectrophotometer and 1.5 mL semi-micro cuvettes.
9. Ruler.
10. Camera or other documentation system.

2.2 Swarming Motility Assay

1. *P. mirabilis* strain(s) to be tested (*see* **Note 1**).
2. Lysogeny broth (LB): per liter, 10 g of tryptone, 5 g of yeast extract, 0.5 g of NaCl. Autoclave to sterilize.
3. Swarm agar: per liter, 10 g of tryptone, 10 g of NaCl, 5 g of yeast extract, and 15 g of agar. Autoclave to sterilize. Pipet 25 mL of agar per 100 mm petri dish. Make swarm agar the day before it is needed and allow plates to sit at room temperature overnight (*see* **Note 3**).
4. Sterile culture tubes.
5. Incubator set to desired survey temperature (typically 30 or 37 °C).
6. Spectrophotometer and 1.5 mL semi-micro cuvettes.
7. Ruler or calipers.
8. Camera or other documentation system.
9. Optional, for viewing colony and bacterial morphology: Inverted microscope, standard microscope, Gram's crystal violet, glass microscope slide.

3 Methods

For both swimming and swarming assays, standardization is key. Motility will be affected by temperature, humidity, agar thickness, incubation time, bacterial strain, cell density, and culture conditions.

3.1 Swimming Motility Assay

1. Culture *P. mirabilis* overnight in 5 mL of LB at 37 °C, with aeration.
2. Make motility agar plates.
3. Place a beaker or tray of distilled H_2O in a 30 °C incubator, so that the interior is humidified (*see* **Note 4**).
4. Inoculate 3 mL of LB in a culture tube with 30 μL of the overnight culture (that is, a 1:100 dilution).
5. Culture bacteria at 37 °C, with aeration, until culture density reaches approximately OD_{600} 1.0.
6. Adjust the culture density to $OD_{600} = 1.0$.
7. Sterilize an inoculating needle by heating it in a flame, then allow to cool.
8. Dip the sterile needle into the standardized culture. The needle should be wet, but not dripping (*see* **Note 5**).
9. Carefully insert the needle vertically into motility agar. Avoid pushing the needle all the way through to the bottom of the petri dish (*see* **Notes 6** and **7**).
10. Withdraw the needle, taking care to keep the needle vertical. Do not wiggle the needle during insertion or withdrawal (*see* **Note 8**).
11. Incubate motility agar plates in a humid 30 °C incubator overnight (approximately 19 h) (*see* **Note 9**). Do not invert them (that is, leave the agar side down) (*see* **Note 10**).
12. After a few hours, the bacteria will appear as a diffuse colony embedded in the agar and spreading outward from the inoculation point (*see* **Note 11**). The optimal timing must be independently determined for the strain and lab space.
13. Measure the colony diameter with a ruler.
14. If desired, photograph swim plates (Fig. 1) (*see* **Note 12**).

3.2 Swarming Motility Assay

1. Make swarming agar plates the day before they are needed (*see* **Note 13**).
2. Culture *P. mirabilis* overnight in 5 mL of LB at 37 °C, with aeration.
3. Inoculate 3 mL of LB in a culture tube with 30 μL of the overnight culture (that is, a 1:100 dilution).

Fig. 1 Swimming motility. Two different strains of *P. mirabilis* were inoculated into motility agar and incubated at 30 °C for 19 h. In this example, ampicillin was added to the agar to maintain plasmids. Top, *P. mirabilis* HI4320 with an empty vector. Bottom, *P. mirabilis* HI4320 expressing motility repressor *mrpJ* from the same plasmid [11]

4. Culture bacteria at 37 °C, with aeration, until culture density reaches approximately OD_{600} 1.0.
5. Adjust the culture density to $OD_{600} = 1.0$.
6. Carefully pipet 5 μL onto the center of the agar surface. Avoid puncturing the agar with the pipet tip (*see* **Notes 14** and **15**).
7. Allow liquid to dry (~10–15 min; can be longer in humid environments) (*see* **Note 16**). The spot will expand slightly as it dries.
8. Invert plate and place in a 30 °C incubator overnight (*see* **Notes 17** and **18**).
9. Observe the swarm colony expanding from the inoculation spot. After an initial lag and outward swarm, most strains will begin repeated cycles of swarming and consolidation, which results in a bull's-eye pattern (*see* **Note 19**).
10. Measure the swarm radius from the edge of the inoculation perimeter with a ruler (*see* **Note 20**). Determine best time for this measurement empirically, since *P. mirabilis* strains vary considerably in swarming appearance and rate of spread. For *P. mirabilis* HI4320 at 30 °C, 17 h postinoculation is a reliable time point (*see* **Note 21**).
11. Photograph the swarm plates (Fig. 2) (*see* **Notes 12** and **22**).

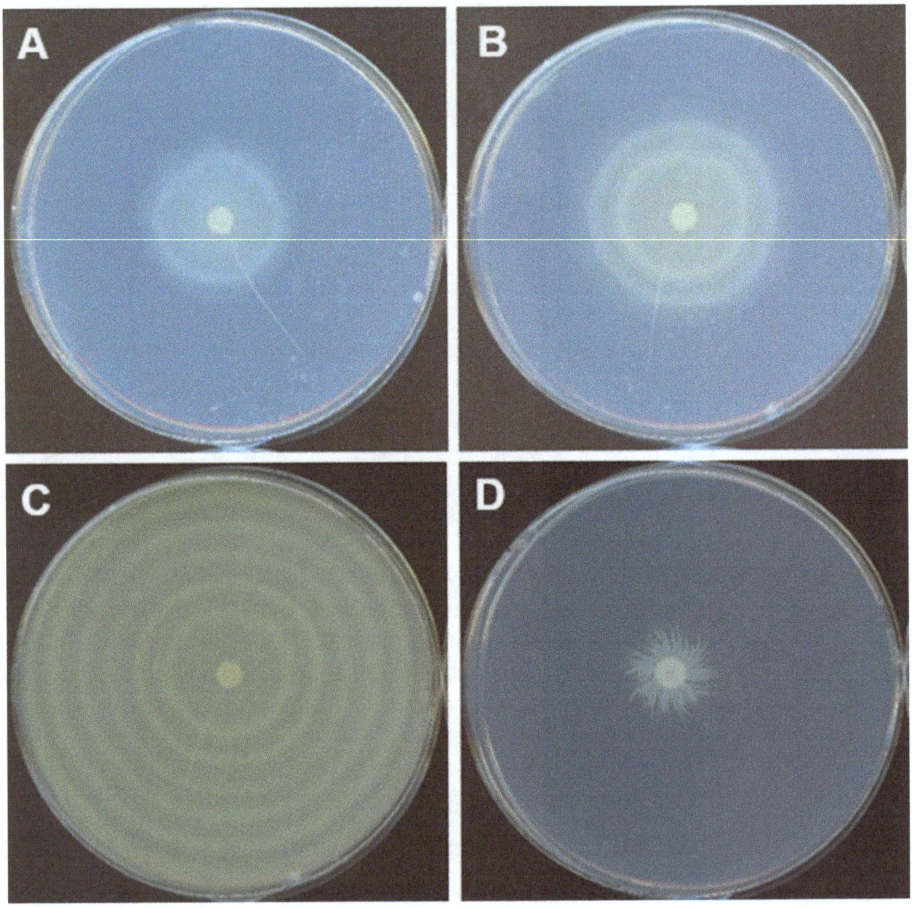

Fig. 2 Swarming motility. *P. mirabilis* was spotted onto swarming agar and incubated at 30 °C. (**a–c**) The same swarm plate of *P. mirabilis* HI4320 at 17 h postinoculation, 24 hpi, or 4 days postinoculation. (**d**) *P. mirabilis* expressing the swarming repressor *mrpJ* from a plasmid, 24 hpi

12. If you want to record the expansion of a swarming colony over time, make repeated measures of the swarm radius, and plot the values. This will result in a stepwise line, because the rate of expansion changes as the bacteria morph between swarming and consolidation phases (*see* **Note 23**) [18, 19].
13. If desired, directly observe swarming bacteria using an inverted microscope. Place the plate under 50× magnification. During active swarming, the colony edge will be smooth, and the bacteria within will be actively writhing [13]. During consolidation, the edge will have a fringed appearance, and there will be little apparent motion (Fig. 3a, b). The smooth/fringed appearance may be visible to the naked eye; hold a plate up to a light to see this.
14. If desired, observe bacterial morphology. Add a loopful of distilled water to a microscope slide. Mix a small amount of

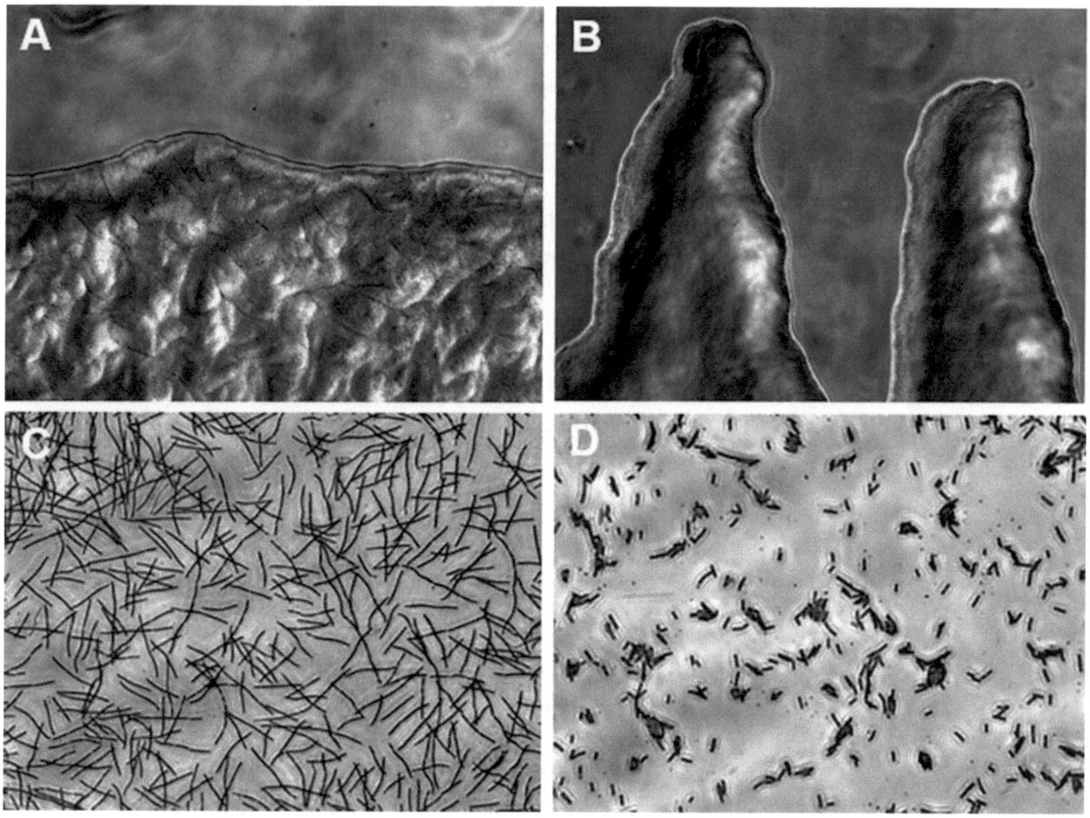

Fig. 3 Microscopy of swarming bacteria. (**a**, **b**) A swarming colony of *P. mirabilis* was observed at 50× magnification. (**a**) Swarming phase. (**b**) Consolidation phase. (**c**, **d**) Bacteria from the edge of a swarming colony were stained with crystal violet and observed under 400× magnification. (**c**) Swarming phase. (**d**) Consolidation phase [13]

bacteria from the edge of the swarm colony into the water, and let dry. Heat-fix, then flood with commercial Gram's crystal violet solution for 1 min. Rinse slide with tap water, pat-dry, and observe under 400× magnification (Fig. 3c, d) (*see* **Note 24**).

4 Notes

1. Most *P. mirabilis* isolates are capable of flagella-mediated motility, although some are not. In particular, the swarming characteristics of a given strain vary tremendously. The protocols described here are tailored for *P. mirabilis* HI4320 [20], and may need to be adjusted for other strains.
2. *P. mirabilis* will swim through many common medium formulations. The key here is the very low agar concentration, which

allows bacteria to swim through the agar, instead of swarming on top of it. The water content of the plate is essential, and using older, drier plates will reduce the observed motility. Likewise, using a pipet to transfer a precise amount of agar into each plate will increase the consistency of results.

3. This is "standard" LB medium. However, almost any rich medium (e.g., blood agar, nutrient agar, BHI, even a minimal medium supplemented with 1% casamino acids) will be permissive for *P. mirabilis* swarming. If the goal is to demonstrate swarming as a principle, any of these media will work well. The protocol here is standardized to allow for generation of reproducible data, and for this, consistency is essential. Pour and dry plates in a single layer. Note any unusual conditions in the laboratory, especially fluctuating humidity. Using a pipet to transfer a precise amount of agar into each plate will increase the consistency of results. For a discussion on individual swarm cues, *see* Chapter 4 of this volume.
4. 37 °C is also often used for swimming and swarming experiments. In this case, the bacteria will grow and move out more quickly; shorten the incubation times accordingly.
5. If the inoculation needle is too hot, there will be a hissing sound when the needle touches the culture and steam will form. Simply dip the needle into the culture again, before inoculating the motility agar, to ensure the needle is coated with live bacteria.
6. Inoculation of bacteria to the petri dish interior surface may allow twitching motility to occur, where bacteria move in between the agar and plastic surfaces without depending on flagella [21].
7. It is possible to measure three swimming colonies on one plate by spacing the inoculation points about halfway between the center and edge of the agar surface, but this should be optimized for strains other than HI4320. Do not crowd motility agar with too many inocula, because swimming motility may be impeded by an adjacent expanding swimmer colony.
8. The technique is: straight down, straight up. Deviating from a vertical inoculation, or wiggling the needle, will create uneven spread from the inoculation point. This can also happen if an asymmetrical object, such as a flat toothpick, is used to inoculate motility agar.
9. Choose the temperature and timing that work best with your experimental needs. For example, there might be more flagella at 30 °C, but the bacteria grow faster at 37 °C. Bacteria may overgrow at 37 °C overnight, so inoculation from an overnight culture and shorter incubation times may give better results. Motility will also occur at room temperature, but it will happen

more slowly. For *P. mirabilis* HI4320, 19 h at 30 °C generates reproducible swim colonies and allows up to three inoculation points per 100 mm petri dish.

10. Motility agar is semisolid, and is not likely to stay in place if the plate is inverted.
11. Sometimes one or two concentric rings may be observed on the interior of the swimming colony. These are chemotaxis rings [22]. Their appearance depends on the strain of bacteria used, the precision of the inoculation technique, and the composition of the medium.
12. A black background makes the rings more visible. One method that works well is placing the plates on a scanner, and then placing a box lined with black velvet over the plates. An automated colony counter with a camera equipped may allow adjustment of light from the top or bottom of the plate and also works well for photographing plates.
13. Swarming is very responsive to agar wetness. If the plates are too wet, the bacteria may swarm very rapidly, often lacking the bull's-eye pattern. If plates are too dry; swarming will likely still happen, but the extent will be more difficult to reproduce. Pouring the plates the day before the experiment, and allowing them to dry in a single layer at ambient temperature, will generate more consistent results.
14. Use a steady hand to leave a single deposit of culture in one spot, without bubbles or splashes. An air bubble will cause an asymmetric swarm to form. Splashes will cause smaller swarms to initiate from each splash point.
15. Multiple strains can be inoculated on one plate. However, nonisogenic strains will usually stop where they meet and form lines; *see* Chapter 6 for more information about this phenomenon.
16. If plates are taking too long to dry, they can be put in a laminar flow hood with the lids off. Remove plates as soon as the spot has dried. Do not dry too long, since overdrying of plates will reduce swarming. An overdried plate will have a wrinkly agar surface.
17. Many experimenters use 37 °C for swarm assays. I find that 30 °C allows for easier timing of swarm/consolidate stages, and also causes slightly wider terraces, which may make collection of distinct stages easier. Others have found that swarm agar made with 5 g of NaCl per liter (instead of 10 g) leads to larger diameter swarm rings when HI4320 is incubated at 37 °C [9]. *P. mirabilis* will also swarm in anaerobic conditions [16]. Choose the temperature and timing that work best with your experimental needs.

18. If a humidity-controlled environment is available for inoculation and incubation of swarm agar, this will increase the consistency of results. One suggestion is 35% humidity [23].
19. At 30 °C, once swarming has initiated, *P. mirabilis* HI4320 converts between swarming and consolidation stages about every 2 h. *P. mirabilis* initiates swarming more quickly when incubated at 37 °C; for example, strain BB2000 initiates the first round of swarming approximately 3 h postinoculation [23].
20. Some researchers prefer to measure the diameter, which is also acceptable. I prefer to measure from the edge of the point of inoculation, so a nonswarming colony has a measurement of zero.
21. Swarm initiation and swarm–consolidation cycles both occur more rapidly at 37 °C. For example, *P. mirabilis* BB2000 incubated at 37 °C completes two swarm cycles approximately 8.5 h postinoculation [23].
22. Leaving plates in the incubator longer will eventually lead to the swarm colony covering the entire plate; although this is not useful for comparing swarm radii, it may intensify the appearance of swarm rings (Fig. 2c).
23. Swarming distance divided by time will provide swarming velocity. Examples showing the change in velocity over time for *P. mirabilis* strains BB2000 and U6450 at 37 °C may be found in references [23, 24].
24. Crystal violet can be substituted with a variety of other bacterial dyes.

Acknowledgments

Thank you to Stephanie Himpsl and Chelsie Armbruster for adding suggestions to these methods.

References

1. Schaffer JN, Pearson MM (2015) *Proteus mirabilis* and urinary tract infections. Microbiol Spectr 3(5). https://doi.org/10.1128/microbiolspec.UTI-0017-2013
2. Macnab RM (1996) Flagella and motility. In: Neidhardt FC (ed) *Escherichia coli* and *Salmonella*: Cellular and molecular biology, vol 1, 2nd edn. ASM Press, Washington, DC, pp 123–145
3. Harshey RM (2003) Bacterial motility on a surface: many ways to a common goal. Annu Rev Microbiol 57:249–273. https://doi.org/10.1146/annurev.micro.57.030502.091014
4. Kearns DB (2010) A field guide to bacterial swarming motility. Nat Rev Microbiol 8 (9):634–644. https://doi.org/10.1038/nrmicro2405
5. Jones BV, Young R, Mahenthiralingam E, Stickler DJ (2004) Ultrastructure of *Proteus mirabilis* swarmer cell rafts and role of swarming in catheter-associated urinary tract infection. Infect Immun 72(7):3941–3950

6. Armbruster CE, Prenovost K, Mobley HL, Mody L (2017) How often do clinically diagnosed catheter-associated urinary tract infections in nursing homes meet standardized criteria? J Am Geriatr Soc 65(2):395–401. https://doi.org/10.1111/jgs.14533
7. Warren JW, Tenney JH, Hoopes JM, Muncie HL, Anthony WC (1982) A prospective microbiologic study of bacteriuria in patients with chronic indwelling urethral catheters. J Infect Dis 146(6):719–723
8. Tittsler RP, Sandholzer LA (1936) The use of semi-solid agar for the detection of bacterial motility. J Bacteriol 31(6):575–580
9. Armbruster CE, Hodges SA, Mobley HLT (2013) Initiation of swarming motility by *Proteus mirabilis* occurs in response to specific cues present in urine and requires excess L-glutamine. J Bacteriol 195(6):1305–1319. https://doi.org/10.1128/JB.02136-12
10. Li X, Rasko DA, Lockatell CV, Johnson DE, Mobley HLT (2001) Repression of bacterial motility by a novel fimbrial gene product. EMBO J 20(17):4854–4862
11. Pearson MM, Mobley HLT (2008) Repression of motility during fimbrial expression: identification of 14 *mrpJ* gene paralogues in *Proteus mirabilis*. Mol Microbiol 69(2):548–558
12. Coetzee JN, Sacks TG (1960) Morphological variants of *Proteus hauseri*. J Gen Microbiol 23:209–216. https://doi.org/10.1099/00221287-23-2-209
13. Pearson MM, Rasko DA, Smith SN, Mobley HLT (2010) Transcriptome of swarming *Proteus mirabilis*. Infect Immun 78(6):2834–2845. https://doi.org/10.1128/IAI.01222-09
14. Allison C, Lai HC, Hughes C (1992) Co-ordinate expression of virulence genes during swarm-cell differentiation and population migration of *Proteus mirabilis*. Mol Microbiol 6(12):1583–1591
15. Walker KE, Moghaddame-Jafari S, Lockatell CV, Johnson D, Belas R (1999) ZapA, the IgA-degrading metalloprotease of *Proteus mirabilis*, is a virulence factor expressed specifically in swarmer cells. Mol Microbiol 32(4):825–836
16. Alteri CJ, Himpsl SD, Engstrom MD, Mobley HLT (2012) Anaerobic respiration using a complete oxidative TCA cycle drives multicellular swarming in *Proteus mirabilis*. mBio 3(6). doi:https://doi.org/10.1128/mBio.00365-12
17. Allison C, Lai HC, Gygi D, Hughes C (1993) Cell differentiation of *Proteus mirabilis* is initiated by glutamine, a specific chemoattractant for swarming cells. Mol Microbiol 8(1):53–60
18. Rauprich O, Matsushita M, Weijer CJ, Siegert F, Esipov SE, Shapiro JA (1996) Periodic phenomena in *Proteus mirabilis* swarm colony development. J Bacteriol 178(22):6525–6538
19. Kvittingen J (1949) Studies of the life-cycle of Proteus Hauser. Acta Pathol Mic Sc 26(1):24–50
20. Pearson MM, Sebaihia M, Churcher C, Quail MA, Seshasayee AS, Luscombe NM, Abdellah Z, Arrosmith C, Atkin B, Chillingworth T, Hauser H, Jagels K, Moule S, Mungall K, Norbertczak H, Rabbinowitsch E, Walker D, Whithead S, Thomson NR, Rather PN, Parkhill J, Mobley HLT (2008) Complete genome sequence of uropathogenic *Proteus mirabilis*, a master of both adherence and motility. J Bacteriol 190(11):4027–4037
21. Daum B, Gold V (2018) Twitch or swim: towards the understanding of prokaryotic motion based on the type IV pilus blueprint. Biol Chem 399(7):799–808. https://doi.org/10.1515/hsz-2018-0157
22. Hazelbauer GL, Mesibov RE, Adler J (1969) *Escherichia coli* mutants defective in chemotaxis toward specific chemicals. Proc Natl Acad Sci U S A 64(4):1300–1307
23. Belas R, Schneider R, Melch M (1998) Characterization of *Proteus mirabilis* precocious swarming mutants: identification of *rsbA*, encoding a regulator of swarming behavior. J Bacteriol 180(23):6126–6139
24. Gygi D, Rahman MM, Lai HC, Carlson R, Guard-Petter J, Hughes C (1995) A cell-surface polysaccharide that facilitates rapid population migration by differentiated swarm cells of *Proteus mirabilis*. Mol Microbiol 17(6):1167–1175
25. Matsuyama T, Takagi Y, Nakagawa Y, Itoh H, Wakita J, Matsushita M (2000) Dynamic aspects of the structured cell population in a swarming colony of *Proteus mirabilis*. J Bacteriol 182(2):385–393

Chapter 4

Testing the Ability of Compounds to Induce Swarming

Chelsie E. Armbruster

Abstract

One of the most distinctive features of *Proteus mirabilis* is its ability to undergo differentiation from short, rod-shaped vegetative cells with peritrichous flagella to massively elongated swarm cells that express hundreds to thousands of flagella. The unique bull's-eye pattern that forms from cycles of active swarming and consolidation back to the vegetative state has long been a distinguishing characteristic of this species. Many factors involved in regulation of flagellar synthesis and swarm cell differentiation have been characterized, but the exact conditions sensed by *P. mirabilis* that send a signal to initiate differentiation and motility have yet to be fully elucidated. Here we describe a method for using several types of media to investigate compounds that induce swarming motility under conditions that would not normally be permissive.

Key words Swarming, Motility, Humidity, Amino acids, pH, Surface tension

1 Introduction

On a solid surface and under permissive conditions, *Proteus mirabilis* undergoes a unique differentiation to form swarm cells that are 20- to 50-fold elongated (20–80 μm in length), multinucleated, and express hundreds to thousands of flagella [1]. In contrast to the swarming motility of other bacteria, *P. mirabilis* is capable of swarming on higher agar concentrations (1.5%) across a wide range of temperatures and results in a characteristic bull's eye pattern on an agar plate. The formation of the bull's-eye pattern is the product of sequential rounds of the differentiation process, which can be divided into three events: differentiation into swarm cells, migration across a surface, and consolidation back into a vegetative or swimmer cell morphology [2].

It was originally hypothesized that *P. mirabilis* swarming was driven by cycles of nutrient depletion at the site of inoculation and chemotaxis up the nutrient gradient to reestablish growth once a permissively rich nutrient environment was achieved [3]. An alternative hypothesis was proposed in which negative chemotaxis drives

Melanie M. Pearson (ed.), *Proteus mirabilis: Methods and Protocols*, Methods in Molecular Biology, vol. 2021,
https://doi.org/10.1007/978-1-4939-9601-8_4,

swarming down a gradient of toxic metabolites and secreted waste products, to be repeated once there was sufficient bacterial growth for the waste products to again accumulate to toxic levels [4]. However, both hypotheses were challenged by the finding that swarm cells harvested off an agar plate swarm immediately upon inoculation onto a new agar plate, rather than consolidating and resuming normal growth until encountering nutrient depletion or accumulation of a waste product [5, 6].

The signaling cascade that ultimately results in swarming is now thought to involve a combination of numerous factors, including perturbation of the cell surface, inhibition of flagellar rotation, nutrient composition (particularly availability of certain metabolites and amino acids), factors that influence proton motive force and membrane potential, cell-cell communication, temperature, and regulatory processes that ultimately influence regulation of the *flhDC* master operon and cell division [7–10]. However, many questions remain regarding the exact cues to which *P. mirabilis* responds by inducing the differentiation and swarming process. Herein we describe a standardized method for determining the ability of test compounds to induce swarming on rich medium (*see* **Note 1**) [11].

2 Materials

2.1 Bacterial Culture and Propagation

1. Low-salt lysogeny broth (LB): 10 g/L tryptone, 5 g/L yeast extract, 0.5 g/L NaCl. Autoclave to sterilize.
2. Low-salt LB agar: LB with 15 g/L agar to solidify the medium.
3. Sterile inoculating loops.
4. Sterile 100 mm petri dishes.
5. Sterile culture tubes.
6. 37 °C stationary incubator and 37 °C shaking incubator.

2.2 Determination of Optimal Salt Concentration in LB Agar

1. Water bath, set to 56 °C.
2. LB agar, made with decreasing concentrations of NaCl (*see* **Note 2**). Dispense exactly 20 mL of agar per 100 mm petri dish (*see* **Note 3**).
3. Stationary-phase broth cultures of *Proteus mirabilis* isolates to be tested.

2.3 Testing Ability of Compounds to Induce Swarming

1. Stationary-phase broth cultures of *P. mirabilis* isolates to be tested.
2. LB agar, made with the NaCl concentration identified in Subheading 2.2 and dispensed 20 mL per 100 mm petri dish.
3. Compounds to be tested, dissolved in an appropriate solvent and filter sterilized (*see* **Note 4**).

4. Plate spreader.
5. Petri dish turntable.
6. Beaker with 70% ethanol.

3 Methods

3.1 Determination of Optimal Salt Concentration

1. Prepare agar plates with a range of NaCl concentrations (three plates per concentration per bacterial strain to be tested). To minimize variability, autoclave the LB agar, cool for ~1 h in a water bath set to 56 °C, cool at room temperature for an additional 15–30 min, and dispense 20 mL each into 100 mm petri dishes. Allow agar to set at room temperature overnight (*see* **Note 5**).
2. Inoculate 5 mL of LB medium with a *P. mirabilis* isolate and incubate at 37 °C with aeration (225 rpm) for 16–20 h.
3. Inoculate LB agar plates from the prior day by placing a 5 μL droplet of the *P. mirabilis* culture onto the center of the plate (*see* **Note 6**).
4. Allow the plate to set face-up at room temperature for 10–15 min until the droplet has fully soaked into the plate and there is no liquid remaining on the surface.
5. Invert the plate and incubate at 37 °C for 16–20 h (*see* **Note 5**).
6. Measure the diameter of bacterial growth on the plate. If no swarming has occurred, there will typically be a single, large colony approximately 7–10 mm in diameter (Fig. 1).
7. For each *P. mirabilis* isolate of interest, determine the NaCl concentration at which no swarming occurs and there is no visible ruffling around the edge of the colony in the center of plate, as well as the NaCl concentration at which distinct swarm rings become visible (*see* **Note 7**). An example is shown in Fig. 1.

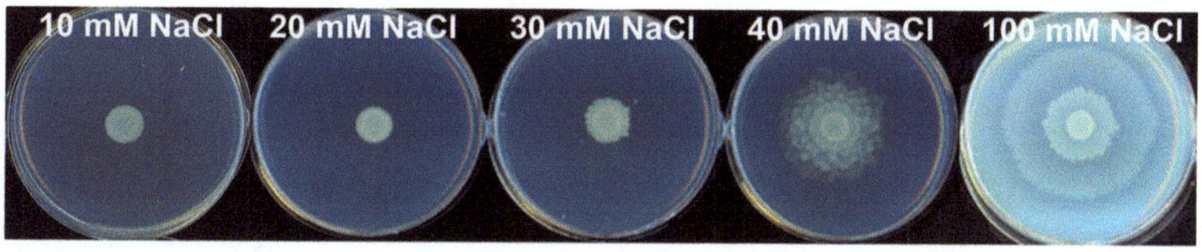

Fig. 1 Determination of LB NaCl concentration to use for testing inducers of swarming motility and maximum concentration of test compounds. Depicted are images of *P. mirabilis* strain HI4320 growth and swarming on LB agar with increasing NaCl concentrations. *P. mirabilis* does not swarm on LB agar with NaCl concentrations below 40 mM, although a ruffled edge is visible on 30 mM NaCl. The ideal concentration for testing induction of swarming by test compounds is therefore 10 mM NaCl, and the maximum concentration of test compounds is 20 mM to keep the combined total of NaCl and test compounds below 30 mM. Reproduced, with permission, from Armbruster et al., Journal of Bacteriology [11]

3.2 Testing Ability of Compounds to Induce Swarming

1. Prepare agar plates (with low salt concentration) as above.
2. Inoculate 5 mL of LB medium with *P. mirabilis* and incubate at 37 °C with aeration (225 rpm) for 16–20 h.
3. After allowing agar to set overnight at room temperature, spread-plate 100 μL of a range of concentrations of the compound to be tested, as well as the solvent (three plates per condition, plus one extra for testing surface tension if this is the first time the compound has been tested—*see* **Note 8**). Crack the lids of the plates and allow to incubate for 30–60 min to ensure the 100 μL have fully soaked in. Alternatively, *see* **Note 9**.
4. Inoculate plates as in **step 3** in Subheading 3.1, Allow the plate to set face-up at room temperature for 10–15 min until the droplet has fully soaked into the plate and there is no liquid remaining on the surface, then invert the plate and incubate at 37 °C for 16–20 h.
5. Measure the diameter of bacterial growth on each plate and compare to the diameter on a plate with solvent alone. Figure 2 shows an example of *P. mirabilis* HI4320 growth on a control plate (LB with 10 mM NaCl and water) compared to growth or swarming on plates supplemented with individual amino acids. Supplementation with arginine, glutamine, or histidine clearly allowed for development of distinct swarm rings and a diameter that is greater than that of solvent alone (water) by Student's *t* test, indicating that these compound are likely to be inducers of swarming initiation (*see* **Note 10**). Supplementation with asparagine, lysine, or methionine allowed for development of visible ruffling of the colony edge and a slight (though statistically significant) increase in diameter but did not result in distinct swarm rings and therefore likely do not represent inducers of swarming initiation.

4 Notes

1. Other media can be utilized for testing the ability of compounds to induce swarming, such as minimal media or even agar plates made from filter-sterilized human urine, but a similar protocol for ruling out variables in swarming motility should be followed. Minimal media are particularly useful for testing suspected nutrient or metabolism-related swarming cues. For urine agar plates, a urease mutant may need to be used to circumvent the impact of urea and pH on swarming [11]. Agar concentration can also be adjusted if necessary.
2. For *P. mirabilis* strain HI4320, the NaCl concentrations tested were 2 g/L, 1.5 g/L, 1 g/L, and 0.5 g/L. Reducing the salt

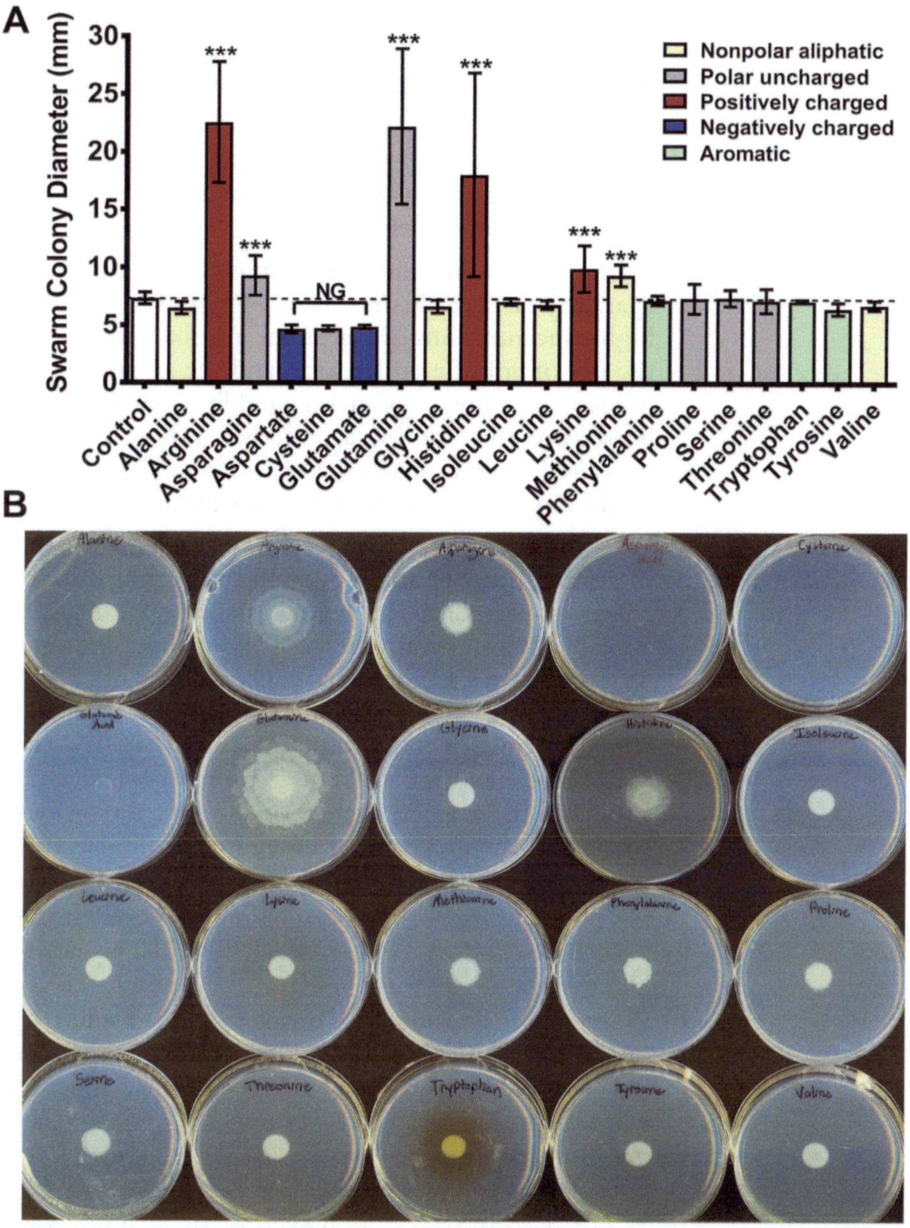

Fig. 2 Measurement of swarm colony diameter to identify inducers of swarming motility. *P. mirabilis* HI4320 was incubated on LB agar plates prepared with 10 mM NaCl vs supplemented with individual amino acids to a final concentration of 20 mM. (**a**) Data represent mean ± standard deviation for three independent experiments, three replicates each. (**b**) Representative images of growth and swarming on 60 mm agar plates containing individual amino acids. Images were reproduced, with permission, from Armbruster et al., Journal of Bacteriology [11]

concentration to 0.5 g/L prevented swarming during a 20 h incubation at 37 °C and was therefore chosen for testing induction of swarming.

3. Dispensing 20 mL of agar per petri dish reduces potential variability from differences in agar volume and drying. If 60 mm dishes are used as an alternative, 8 mL of agar should be dispensed into each dish.
4. Swarming motility and induction of swarming on low salt LB agar are both greatly influenced by pH [11, 12]. For instance, high concentrations of L-arginine are alkaline and will increase the pH of unbuffered LB agar, and the ability of L-arginine to induce swarming is optimal at a mildly acidic pH. It is therefore imperative that the impact of each test solution on pH be measured by diluting in the test medium (low salt LB broth) to the final concentration that will be used for induction of swarming.
5. Swarming motility and induction of swarming are both influenced by humidity. If a humidity-controlled incubator is not available, it is imperative to control for this as much as possible by developing a standardized procedure for the volume of agar being autoclaved, the amount of time the agar is allowed to cool, the amount of the time the agar is allowed to set in the petri dishes, and the amount of time for which the plates set with the lids propped slightly open to let the test compounds soak in. If the incubator is in an area in the lab that tends to experience changes in humidity, the humidity can be adjusted by setting a beaker full of water in the incubator to increase humidity, or by adding a tray with a drying agent such as Dry-Rite to decrease humidity.
6. When dispensing the 5 μL droplet of *P. mirabilis,* it is best to slowly and gently expel the volume of the pipette directly above the center of the agar plate, then touch the droplet to the plate without pushing past the volume stop of the pipette. This prevents creation of splatter droplets, which would form mini-swarm colonies on the plate.
7. *P. mirabilis* isolates vary widely in their ability to swarm at 37 °C for a given concentration of NaCl. It is therefore imperative to determine the optimal NaCl concentration and maximum test compound concentration for each new isolate to be tested. Strains that exhibit hyper-swarming phenotypes may need to be tested on plates solidified with a higher concentration of agar, or on a minimal medium as per **Note 1**.
8. If a test compound changes surface tension, this alone can impact the likelihood of swarming and may give a false-positive result. A simple way to test this is by placing a droplet of water on the surface of a plate and observing the contact angle of

droplet of water to the agar plate. If the droplet is spread-out and has a more acute contact angle, the solvent or compound may act like a surfactant and allow for motility due to this property alone. If the droplet is more compact and rounded and has a contact angle closer to 90°, the solvent or compound may actually perturb motility. This should be tested for a plain low-salt plate, one with solvent alone, and plates with increasing concentrations of the test compound to ensure that neither the solvent nor the compound influence surface tension.

9. Depending on the solvent, test compounds can also be added directly to the agar when cool enough to pour. If this method is used, be sure to withhold an appropriate volume of water from the LB agar recipe to account for the volume of the test compound to be added.
10. To confirm each potential inducer of swarming, repeat the experiment with the desired range of compound concentrations vs. solvent for a total of at least three independent experiments with at least three replicates each, to account for inherent variability in swarming motility. If the compound holds up in all three experiments, it is important to determine if it is a general inducer of motility or a swarming-specific inducer. This can be accomplished by adding the compound vs. solvent alone to motility agar plates (10 g/L of tryptone and 5 g/L of NaCl solidified with 0.3% agar, made the day of use), stab-inoculating *P. mirabilis* into the center of the plate, and measuring the diameter of the swimming bacteria after a ~16 h incubation at 30 °C [11].

Acknowledgments

The author would like to thank Harry Mobley and members of the Mobley laboratory for their helpful comments and critiques during the initial development of this protocol. This work was supported by the National Institute of Diabetes Digestive and Kidney Disorders (R00 DK105205 to C.E.A.).

References

1. Hoeniger JFM (1965) Development of flagella by *Proteus mirabilis*. J Gen Microbiol 40 (1):29–42. https://doi.org/10.1099/00221287-40-1-29
2. Williams FD, Schwarzhoff RH (1978) Nature of the swarming phenomenon in *Proteus*. Annu Rev Microbiol 32:101–122
3. Moltke O (1928) Contributions to the characterization and systematic classification of bac. *Proteus vulgaris* (Hauser). J Am Med Assoc 91 (17):1311–1311. https://doi.org/10.1001/jama.1928.02700170075042
4. Lominski I, Lendrum AC (1947) The mechanism of swarming of *Proteus*. J Pathol Bacteriol 59(4):688–691
5. Kvittingen J (1949) Studies of the life-cycle of *Proteus hauser*. Acta Pathol Microbiol Scand 26 (1):24–50

6. Williams FD, Anderson DM, Hoffman PS, Schwarzhoff RH, Leonard S (1976) Evidence against the involvement of chemotaxis in swarming of *Proteus mirabilis*. J Bacteriol 127 (1):237–248
7. Rauprich O, Matsushita M, Weijer CJ, Siegert F, Esipov SE, Shapiro JA (1996) Periodic phenomena in *Proteus mirabilis* swarm colony development. J Bacteriol 178 (22):6525–6538
8. Armbruster CE, Mobley HLT (2012) Merging mythology and morphology: the multifaceted lifestyle of *Proteus mirabilis*. Nat Rev Micro 10 (11):743–754
9. Morgenstein RM, Szostek B, Rather PN (2010) Regulation of gene expression during swarmer cell differentiation in *Proteus mirabilis*. FEMS Microbiol Rev 34(5):753–763. https://doi.org/10.1111/j.1574-6976.2010.00229.x
10. Lee YY, Belas R (2015) Loss of FliL alters *Proteus mirabilis* surface sensing and temperature-dependent swarming. J Bacteriol 197(1):159–173. https://doi.org/10.1128/JB.02235-14
11. Armbruster CE, Hodges SA, Mobley HLT (2013) Initiation of swarming motility by *Proteus mirabilis* occurs in response to specific cues present in urine and requires excess L-glutamine. J Bacteriol 195(6):1305–1319. https://doi.org/10.1128/JB.02136-12
12. Armbruster CE, Hodges SA, Smith SN, Alteri CJ, Mobley HL (2014) Arginine promotes *Proteus mirabilis* motility and fitness by contributing to conservation of the proton gradient and proton motive force. Microbiology Open 3 (5):630–641. https://doi.org/10.1002/mbo3.194

Chapter 5

Purification of Native Flagellin

María José González, Victoria Iribarnegaray, Pablo Zunino, and Paola Scavone

Abstract

Flagella are effective organelles of locomotion and one of several virulence factors in *Proteus mirabilis*. To study their properties and role in virulence, we describe a protocol to extract and purify the native flagellin of *P. mirabilis*. Purified flagellin can be visualized by SDS-PAGE or immunoblot and is suitable for downstream applications such as immunization.

Key words *Proteus mirabilis*, Native flagellin, Flagellum, SDS-PAGE, Western blot

1 Introduction

The prokaryotic flagellum is best known as a motility organelle responsible for bacterial movement and necessary for chemotaxis [1]. Flagella are extremely effective organelles of locomotion that permit bacteria to achieve speeds exceeding many cell body lengths per second [2]. Ongoing research is still revealing surprises in various aspects, from assembly to function [3, 4].

The flagellum can be subdivided into three substructures that are assembled in a temporal sequence [5]. The first component to be assembled is the basal body, which anchors the flagellum to the cell membrane, provides the power for rotation, and secretes the more distal components. The next component is the hook, which is connected to the basal body and serves as a flexible universal joint changing the angle of flagellar rotation. The third structure is the helical filament, which is composed primarily of the protein flagellin, one of the most abundant proteins made by the cell [5]. Flagellin is an activator of the immune system and has been recognized as a pathogen-associated molecular pattern (PAMP). The receptors TLR5 on extracellular media and NLRC4/Naip5 in the cytoplasm recognize flagellin. Flagellin has been widely studied as an adjuvant and has been proved to be effective in animal models [6].

Melanie M. Pearson (ed.), *Proteus mirabilis: Methods and Protocols*, Methods in Molecular Biology, vol. 2021, https://doi.org/10.1007/978-1-4939-9601-8_5,

Like many bacteria, *Proteus mirabilis* uses flagella to swim through liquids and toward chemical gradients [7]. In *P. mirabilis*, the flagellum has an important role in virulence. Flagella participate in the swarming motility of *Proteus*, which involves a multiflagellated and elongated cellular morphotype [8]. Several studies found that flagella contribute to ascending urinary tract infection (UTI) [9]. The importance of flagella to host cell invasion and ascending UTI was directly assessed by disrupting the gene encoding the flagellar filament (*flaD*). Production of the flagellar filament was found to facilitate the invasion of human cells in vitro and may also promote internalization through other mechanisms [9]. Expression of flagellar genes is regulated during experimental UTI. In the first 24 h postinfection, flagellar gene expression is repressed, but after 3 days the repression starts to relieve [10]. However, the role of flagella is still under debate. *P. mirabilis* native flagella are highly antigenic but not protective when mice were challenged intravenously [11]. This result can be explained because native flagellin did not enhance humoral or cellular immune response and could possibly exert a depressive instead of stimulatory effect [12].

Here, a protocol is described for purification of native flagellin from one *P. mirabilis* strain (Pr2921) by mechanical shearing and centrifugation; however, this procedure could be applied to any strain [13]. To corroborate the efficiency of extraction and purification, SDS-PAGE and subsequent western blot analysis are described. For western blotting, a rabbit/mouse polyclonal flagella antiserum, previously generated in our laboratory, was used [11]. This purified native flagellin can be stored until ready to use.

2 Materials

Prepare all solutions using ultrapure water and analytical grade reagents. Prepare and store all reagents at room temperature, unless indicated otherwise. Before preparing any material, carefully read the chemical material safety data sheets and wear appropriate lab safety equipment. Diligently follow all waste disposal regulations when disposing of waste materials.

2.1 Native Flagellin Extraction and Purification

1. *Proteus mirabilis* strain (*see* **Note 1**).
2. 1.5% BHI agar. Dissolve the agar mix according to the manufacturer's instructions and heat to 100 °C for 5–10 min to dissolve the solids. Sterilize by autoclaving at 121 °C for 15 min. Mix well and pour into sterile petri dishes.
3. Phosphate-buffered saline (PBS), per liter: 0.41 g of potassium phosphate monobasic (KH_2PO_4), 1.7 g of sodium phosphate dibasic (Na_2HPO_4), and 8.2 g of sodium chloride (NaCl),

pH 7.4. Sterilize by autoclaving at 121 °C for 15 min. Store at 4 °C indefinitely, and warm to room temperature before use.

4. Spectrophotometer.
5. Centrifuge and centrifuge tubes/bottles suitable for 100 mL volumes.
6. Ultracentrifuge and rotor capable of 40,000 × *g*.
7. High-speed blender (e.g., Waring Blender with stainless steel container).
8. Protein quantification method, such as Bradford assay.

2.2 SDS–Polyacrylamide Gel

1. Separating buffer (1.5 M Tris–HCl, pH 8.8). Dissolve 91 g of Tris in 300 mL of water. Mix and adjust pH to 8.8 with HCl. Add water to a final volume of 500 mL. Filter the solution through a 0.45 μm filter. Store at 4 °C.
2. Stacking buffer (1.5 M Tris–HCl, pH 6.8): weigh 18 g of Tris and prepare a 100 mL solution as in the previous step. Store at 4 °C.
3. Acrylamide solution (30% acrylamide–0.8% bisacrylamide): weigh 30 g of acrylamide monomer and 0.8 g *N,N'*-methylenebisacrylamide and transfer to a 100 mL graduated cylinder containing about 40 mL of water. Complete to 100 mL with water and filter through a 0.45 μm filter. Store at 4 °C, in a bottle wrapped with aluminum foil (*see* **Note 2**).
4. APS (ammonium persulfate). Add 25 mg of ammonium persulfate to 100 mL of H_2O. Store at 4 °C (*see* **Note 3**).
5. *N,N,N',N'*-tetramethylethylenediamine (TEMED). Store at 4 °C (*see* **Note 4**).
6. 10× Electrophoresis buffer: per liter, 30 g of Tris base, 144 g of glycine, 10 g of SDS. Dilute to 1× for working solution. Store up to 1 month at 4 °C. Do not adjust the pH of the stock solution, as the solution is pH 8.3 when diluted.
7. Bromophenol blue (BPB) solution. Dissolve 0.1 g of BPB in 100 mL of water.
8. 5× Sample buffer: 1.5 M Tris–HCl (pH 6.8), 10% SDS, 25% β-mercaptoethanol, 0.1% bromophenol blue, 30% glycerol. Leave one aliquot at 4 °C for current use and store the remaining aliquots at −20 °C (*see* **Note 5**).
9. 0.2% Coomassie blue staining solution: 0.5 g of Coomassie blue G-250 in 200 mL of destain solution. Mix for 1 h and filter through Whatman no. 1 paper. Store at room temperature indefinitely.
10. Destain solution: 450 mL of 95% ethanol, 450 mL of water and 100 mL of acetic acid.
11. SDS-PAGE electrophoresis unit and power supply.

2.3 Western Blot

1. 0.45 μm Nitrocellulose membrane, Whatman paper, fiber pads and cooling unit.
2. Western blot transfer buffer: 25 mM Tris, 192 mM glycine, 20% methanol, pH 8.3, keep at 4 °C. Do not adjust the pH; it will range from pH 8.1–8.4 depending on the quality of chemicals. Methanol should be analytical reagent grade.
3. TBS-Tween: 20 mM Tris–HCl, 200 mM NaCl, 0.5% Tween 20.
4. Blocking solution: 3% skim milk in TBS-Tween.
5. Diluent solution: 1% skim milk in TBS-Tween.
6. Nitro blue tetrazolium (NBT): 30 mg/mL of NBT in 70% dimethylformamide (*see* **Notes 6** and 7).
7. 5-bromo-4-chloro-3-indolyl phosphate (BCIP): 15 mg/mL of BCIP in 100% dimethylformamide (*see* **Note** 7).
8. Enzyme substrate buffer: 100 mM Tris pH 9.5, 100 mM NaCl, 50 mM $MgCl_2$, 1% NBT, 1% BCIP. Prepare this solution just before use.
9. Stop solution: 0.05% EDTA in water.
10. Antibodies against *P. mirabilis* flagellin raised in rabbit or mouse [11] (*see* Subheading 3.4).
11. Anti-rabbit or anti-mouse antibody coupled to alkaline phosphatase (e.g., Sigma).
12. Electrophoretic transfer cell for western blotting.

2.4 Generation of Polyclonal Flagella Antiserum in Mice

1. Freund's complete adjuvant (Sigma).
2. Freund's incomplete adjuvant (Sigma).
3. Purified and quantified flagellin.
4. 0.5 mL centrifuge tubes.
5. 25 G needles and syringes.

3 Methods

Carry out all procedures at room temperature unless otherwise specified.

3.1 Flagellin Extraction and Purification

1. Culture *P. mirabilis* overnight on BHI agar.
2. Use a sterile loop to scrape up swarm bacteria from the plate. Prepare a bacterial suspension in PBS with an $OD_{600} = 0.8$.
3. Pipet 10 μL of the *P. mirabilis* suspension onto the middle of a BHI agar plate.
4. Repeat **step 3** for 19 additional BHI plates (*see* **Note 8**).

5. Incubate plates for 24 h at 37 °C. The bacteria will swarm from the center spot.
6. Weigh an empty centrifuge tube/bottle.
7. Suspend the cells in 100 mL of cold PBS by scraping the bacteria off the plates with a loop.
8. Centrifuge bacteria at 5000 × *g* for 15 min at 4 °C. Discard the supernatant.
9. Weigh the centrifuge tube/bottle with the pellet, and calculate the pellet weight.
10. Resuspend the pellet in PBS, using 100 mL of PBS per 6 g of the pellet (wet weight) (*see* **Note 9**).
11. Detach the flagella mechanically in a blender at high speed (19,000 rpm) for 30 s (*see* **Note 10**).
12. Centrifuge the suspension at 16,000 × *g* for 15 min at 4 °C.
13. Discard the sediment. Centrifuge the supernatant at 40,000 × *g* for 1 h at 4 °C.
14. Resuspend the pellet with 8 mL of distilled H_2O.
15. Add the sample to 36 kDa cutoff cellulose membrane dialysis tubing.
16. Dialyze against distilled H_2O for 24 h at 4 °C.
17. Repeat **step 16** with fresh H_2O.
18. Heat flagellin samples at 80 °C for 5 min to obtain monomer suspensions (*see* **Note 11**).
19. Quantify protein (e.g., Bradford assay) (*see* **Note 12**).
20. Store flagellin stock at −20 °C until use (*see* **Note 13**).

3.2 Sodium Dodecyl Sulfate–Polyacrylamide Gel Electrophoresis

1. Set up the gel casting frames: clamp two glass plates in a casting frame, with 0.75 mm separators, on a casting stand (*see* **Note 14**).
2. Make the resolving gel (10 mL) by mixing in this order: 4 mL of water, 3.3 mL of acrylamide solution, 2.5 mL of separating buffer (pH 8.8), 0.1 mL of 10% SDS, 0.1 mL of APS, and 4 μL of TEMED. Mix carefully to avoid the formation of bubbles.
3. Pour the solution between the glass plates with a pipette, leaving about 1/4 of the space free for the stacking gel.
4. Carefully cover the top of the resolving gel with 50% isopropanol, and wait until the resolving gel polymerizes (~30 min). A clear line will appear between the gel surface and the solution on top when polymerization is complete.
5. Discard the isopropanol.
6. Make the stacking gel (4 mL) by mixing in this order: 2.7 mL of water, 0.67 mL of acrylamide solution, 0.5 mL of stacking

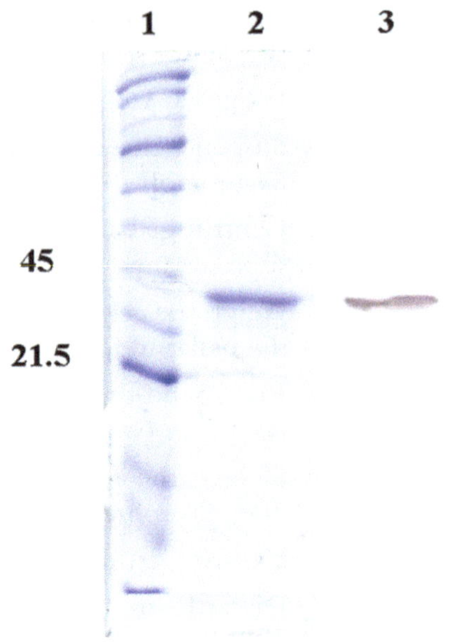

Fig. 1 10% SDS-PAGE with Coomassie stain (lanes 1 and 2) and Western blot (lane 3) of *P. mirabilis* flagellin. Lane 1, protein marker with the size of two bands noted in kilodaltons

solution, 40 μL of 10% SDS, 40 μL of APS and 4 μL of TEMED. Mix carefully with a pipette to avoid the formation of bubbles.

7. Pour stacking gel on top of resolving gel, then add a comb to form wells. Allow to polymerize for at least 30 min.
8. Prepare flagellin samples by boiling in 5× sample buffer for 5–10 min.
9. Load 5 μL of each sample and run the gel at 125 V until the bromophenol blue tracking dye reaches the bottom of the gel. Do not forget to load a protein marker into the first lane.
10. After electrophoresis is complete, shut off the power, disconnect the electrodes and remove the gel.
11. If western blotting is going to be done, proceed to Subheading 3.3 (*see* **Note 14**). Otherwise, Coomassie stain for 1 h at RT, preferably in an orbital shaker.
12. Destain gel; the major band is flagellin (Fig. 1, lanes 1 and 2).

3.3 Western Blot

1. Prepare at least 1 L of transfer buffer in advance and keep it refrigerated. It is important that the buffer temperature is 4 °C at the start of the transfer.
2. Rinse the gel in transfer buffer prior to blotting to facilitate the removal of electrophoresis buffer salts and detergents.

3. Cut a nitrocellulose membrane to the dimensions of the gel; wear gloves to avoid contamination of the membrane.
4. Use a soft pencil to identify the gel and the orientation of the membrane.
5. Wet the membrane in transfer buffer for about 15–30 min. Meanwhile, soak the precut filter paper and fiber pads in transfer buffer.
6. Assemble the cassette for protein transfer. Place a presoaked fiber pad on the back (negative) panel of the cassette and then filter paper, gel and membrane, filter paper, and fiber pad. Be sure to center all components and to align the gel in the center of the cassette (*see* **Note 15**).
7. Roll a glass pipette or test tube over the top of the membrane to eliminate all air bubbles. Keep all of the materials wet with buffer. Close the cassette, holding it firmly so that the sandwich will not move.
8. Place the gel holder in the buffer tank so that the negative panel of the holder is facing the cathode electrode panel.
9. Place an ice cooling unit in the remaining space, to maintain appropriate buffer temperature for approximately 1 h.
10. Add transfer buffer (approximately 400 mL) to fill the tank to just above the level of the top of the gel holder cassette.
11. Run the transfer for 1 h at 100 V.
12. After the transfer, disassemble the cassette. Pay attention to the side of the membrane that was in contact with the gel, which will have the proteins. Place this side up.
13. Block the membrane with TBS-T-3% skim milk for 1 h at room temperature.
14. Wash the membrane once with TBS-T.
15. Prepare a 1:100 dilution of a specific antibody against flagellin in TBS-T-1% skim milk (*see* **Note 16**). Incubate with the membrane at 37 °C for 1 h.
16. Wash once with TBS-T.
17. Prepare a 1:30,000 dilution with the secondary antibody coupled to alkaline phosphatase in TBS-T-1% skim milk. Incubate at 37 °C for 45 min.
18. Wash with TBS-T.
19. Add enzyme substrate buffer and incubate until the bands start to appear (Fig. 1, lane 3).
20. Stop the reaction with the stop solution.

3.4 Generation of Polyclonal Flagella Antiserum in Mice

Each animal procedure should be approved by the local ethical committee and must follow national legislation in animal experimentation.

1. Prepare an emulsion with flagellin and Freund's complete adjuvant in a ratio of 1:1. For each animal, calculate a maximum volume of 0.3 mL with a concentration of 30 μg per mouse of flagellin (*see* **Notes 17–19**).
2. While holding the mouse, dispense the emulsion subcutaneously with a syringe coupled to a 25 G needle.
3. Repeat **steps 1** and **2** on days 14 and 28 after the first dose, but prepare the emulsion using Freund's incomplete adjuvant.
4. Before the first dose and after day 40, take blood samples (~50 μL) (*see* **Note 20**).
5. Incubate the blood for 3 h at 4 °C.
6. Centrifuge the blood at 1600 × *g* for 10 min at room temperature.
7. Retain the serum (*see* **Note 21**). Aliquot and store at −20 °C.

4 Notes

1. Choice of *P. mirabilis* strain. In our laboratory, we have a collection of different clinical isolates and we have also used the present procedure to obtain native flagellin from them [13]. Different flagellins show some differences in the size of the band (around 41 kDa), probably associated with differences in sequence or rearrangements of flagellin sequences [14]. We also observed differences in flagellin stimulation of CacoCCL20-Luc cells supporting the observations of the flagellin size.
2. This diluted reagent may be used for about 2 weeks when kept at room temperature. Caution: Acrylamide monomer is neurotoxic and should be handled accordingly.
3. For short term (less than 12 h), store at 4 °C. For long term storage it is recommended to aliquot APS in 1 mL tubes and store at −20 °C.
4. TEMED can be stored up to 12 months.
5. For long term storage, store in 0.5 mL aliquots at −80 °C.
6. Caution: NBT dust is extremely toxic! Take safety precautions, including the use of gloves, glasses, and face mask. Avoid formation of dust and aerosols.
7. Light sensitive, so store in a brown bottle.

8. Swarming *P. mirabilis* produce a large amount of flagella. From 20 plates we obtain around 7 mg/mL of flagellin. We do not have experience on flagellin purification with other media (such as liquid media) or incubation conditions.
9. The amount obtained from 20 plates is approximately 3 g of wet weight.
10. The blender produces heat when it is working, so we recommend blending at high speed for a short period of time and also using cold buffers. This will also improve the quality of the preparation, as heat will denature the flagellin.
11. Obtaining a monomer suspension is recommended before quantification and immunization.
12. Any protein quantification procedure will be suitable for flagellin quantification.
13. Some flagellins are thermosensitive and may degrade during storage, so it is not recommended to freeze again once thawed.
14. It is recommended to run twin gels, one for Coomassie staining and the other for blotting.
15. The transfer will be incomplete for any portion that is outside of the gel holder cassette or if air bubbles are trapped between the gel and the filter paper.
16. We use polyclonal antibodies generated by our laboratory [15], so we adjusted the concentration according to this antibody. Each antibody titer or dilution for immunoblot should be independently established.
17. If native flagellin is going to be used for immunization or cell culture, first perform LPS (endotoxin) decontamination. This can be achieved with several commercial options, such as endotoxin removal spin columns (e.g., Thermo Scientific).
18. The emulsion should be mixed well, until the emulsion become white.
19. Rabbits may also be used to generate antibodies. The volume of the emulsion for rabbit immunization should be 1 mL, with a concentration of 100 μg/rabbit.
20. Consult with your local ethical committee for animal experimentation for the preferred method. We obtain blood from the tail vein. This could be done with a small incision or using a butterfly needle coupled to a 0.5 mL tube (or smaller). 50 μL of blood is enough to obtain 30 μL of serum.
21. A clot will form, so be very careful when aspirating the serum to avoid disrupting the clot and to contaminate the serum with red blood cells.

References

1. Sourjik V, Wingreen NS (2012) Responding to chemical gradients: bacterial chemotaxis. Curr Opinion Cell Biol 24(2):262–268
2. Berg HC (2003) The rotary motor of bacterial flagella. Annu Rev Biochem 72:19–54
3. Evans LD, Poulter S, Terentjev EM, Hughes C, Fraser GM (2013) A chain mechanism for flagellum growth. Nature 504 (7479):287–290
4. Rossez Y, Wolfson EB, Holmes A, Gally DL, Holden NJ (2015) Bacterial flagella: twist and stick, or dodge across the kingdoms. PLoS Path 11(1):e1004483
5. Chevance FF, Hughes KT (2008) Coordinating assembly of a bacterial macromolecular machine. Nature Rev Microbiol 6(6):455–465
6. Mizel SB, Bates JT (2010) Flagellin as an adjuvant: cellular mechanisms and potential. J Immunol 185(10):5677–5682
7. Macnab RM (2003) How bacteria assemble flagella. Annu Rev Microbiol 57:77–100
8. Rather PN (2005) Swarmer cell differentiation in *Proteus mirabilis*. Environ Microbiol 7 (8):1065–1073
9. Schaffer JN, Pearson MM (2015) *Proteus mirabilis* and urinary tract infections. Microbiology Spectrum 3(5)
10. Pearson MM, Yep A, Smith SN, Mobley HLT (2011) Transcriptome of *Proteus mirabilis* in the murine urinary tract: virulence and nitrogen assimilation gene expression. Infect Immun 79(7):2619–2631
11. Legnani-Fajardo C, Zunino P, Algorta G, Laborde HF (1991) Antigenic and immunogenic activity of flagella and fimbriae preparations from uropathogenic *Proteus mirabilis*. Canadian J Microbiol 37(4):325–328
12. Scavone P, Umpiérrez A, Rial A, Chabalgoity JA, Zunino P (2014) Native flagellin does not protect mice against an experimental *Proteus mirabilis* ascending urinary tract infection and neutralizes the protective effect of MrpA fimbrial protein. Antonie van Leeuwenhoek 105 (6):1139–1148
13. Umpiérrez A, Scavone P, Romanin D, Marqués JM, Chabalgoity JA, Rumbo M, Zunino P (2013) Innate immune responses to *Proteus mirabilis* flagellin in the urinary tract. Microb Infect 15(10–11):688–696
14. Murphy CA, Belas R (1999) Genomic rearrangements in the flagellin genes of *Proteus mirabilis*. Mol Microbiol 31:679–690
15. Legnani-Fajardo C, Zunino P, Piccini C, Allen A, Maskell D (1996) Defined mutants of *Proteus mirabilis* lacking flagella cause ascending urinary tract infection in mice. Microb Pathog 21:395–405

Chapter 6

Analysis of *Proteus mirabilis* Social Behaviors on Surfaces

Kristin Little and Karine A. Gibbs

Abstract

The opportunistic pathogen *Proteus mirabilis* engages in visually dramatic and dynamic social behaviors. Populations of *P. mirabilis* can rapidly occupy surfaces, such as high-percentage agar and latex, through a collective surface-based motility termed swarming. When in these surface-occupying swarm colonies, *P. mirabilis* can distinguish between clonal siblings (self) and foreign *P. mirabilis* strains (nonself). This ability can be assessed by at least two standard methods: boundary formation, aka a Dienes line, and territorial exclusion. Here we describe methods for quantitative analysis of swarm colony expansion, of boundary formation, and of territorial exclusion. These assays can be employed to assess several aspects of *P. mirabilis* sociality including collective swarm motility, competition, and self versus nonself recognition.

Key words Swarm analysis, Boundary assays, Dienes typing, Territorial exclusion, Self versus nonself recognition, Competition

1 Introduction

1.1 Background

Proteus mirabilis is a Gram-negative bacterial species found in several environments and in numerous animal niches including the human gut [1]. *P. mirabilis* can also cause urinary tract infections, particularly in patients with long-term catheters [2]. Numerous factors contribute to *P. mirabilis* pathogenesis [3–6], likely including the ability of cells to navigate across hard surfaces such as latex catheters [7]. In the laboratory, *P. mirabilis* populations can engage in such surface-based motility, termed swarming, on up to 2.5% LB agar medium [8]. Populations of swarming cells exhibit clear phenotypic heterogeneity. Upon contact with a hard agar surface, short (~2 μm) cells undergo a dramatic transcriptional change and within hours elongate up to 40-fold into flexible, snake-like swarmer cells that are composed of several copies of the single chromosome [9–11]. Swarmer cells can form multicellular rafts that collectively move across the surface [12]. After a period of motility, cells can divide into ~ 2-μm long cells that are not motile on the surface [9]. Colonies of swarming *P. mirabilis* populations

Melanie M. Pearson (ed.), *Proteus mirabilis: Methods and Protocols*, Methods in Molecular Biology, vol. 2021,
https://doi.org/10.1007/978-1-4939-9601-8_6,

exhibit a stereotyped bull's-eye pattern, which reflects oscillating periods of outward colony expansion on the surface punctuated by periods of nonexpansion [8].

Swarming populations of *P. mirabilis* are collectives capable of additional behaviors such as self versus nonself recognition. This self-recognition behavior is evident as follows: two approaching *P. mirabilis* swarming colonies of self strains (usually isogenic) will merge; by contrast, approaching nonself strains (usually of differing genotypes) will form a human-visible boundary between the colonies on a surface [13, 14]. *P. mirabilis* utilizes both lethal and nonlethal mechanisms to achieve this separation [14–16]. This boundary formation phenotype was first published in 1946 by Louis Dienes, resulting in the moniker of Dienes lines [13]. Dienes line formation has historically been used to type clinical isolates [17, 18]. The diversity of Dienes line forming groups is broad and does not directly correlate with the diversity of described O-antigen serotypes [14, 17, 18]. However, more accurate techniques can now be rapidly deployed for strain identification. The assay for Dienes lines formation instead provides a rapid, initial survey of whether strains recognize each other as different. As an alternative, we have recently developed a quantitative assay for in-depth study of how bacteria are able to distinguish self from nonself. In this assay, comingled nonisogenic strains engage in territorial exclusion, with one strain reaching the outer edges of the swarm colony and the other restricted to the center. This dominance in the swarm colony can be achieved through nonlethal and/or lethal mechanisms [16, 19, 20]. The outcomes of territorial exclusion are consistent, and this assay has been used to characterize the contributions of individual genes to interbacterial interactions [20].

While experimental methods for analysis of lethal mechanisms (e.g., toxin/antitoxin) can be readily found, fewer experimental methods for examination of nonlethal mechanisms are broadly applied. The territorial exclusion assay is especially applicable to nonlethal interactions in which the restricted population can persist within the collective [20]. Here we detail methods for quantitative analysis of surface-based social (cell-to-cell) interactions as applied to *P. mirabilis*. As the ability for self versus nonself recognition has been broadly found among bacterial species, these protocols can be adapted based on experimental questions and can be altered as needed for other microbes.

1.2 Overview of Methods

We first describe how to visualize and quantify swarming colonies of *P. mirabilis* (Subheading 3.1). We then detail how to conduct boundary assays (Subheading 3.2) and territorial exclusion assays (Subheading 3.3) for analysis of *P. mirabilis* social interactions on surfaces.

2 Materials

All media should be prepared with deionized, sterilized water, preferably purified to a resistivity of 18 MΩ-cm at 25 °C. All procedures should be carried out at room temperature (~22 °C) using sterile techniques, unless otherwise noted. *P. mirabilis* is a biosafety level 2 (BSL-2) human pathogen. Use appropriate BSL-2 procedures and personal protective equipment. Conduct experiments in an approved BSL-2 level facility.

2.1 General Microbiology

1. Sterilizable surface or sterile biosafety cabinet.
2. Bunsen burner.
3. Microcentrifuge.
4. Autoclave.
5. Pipettes and sterile pipette tips.
6. Sterile inoculation needle.
7. Spectrophotometer capable of reading optical density (OD) at 600 nm.
8. Spectrophotometer cuvettes.
9. Temperature-controlled incubator set at 37 °C.

2.2 Macroscale Analysis of P. mirabilis *Swarm Colonies*

1. Lennox lysogeny (LB) broth: Dissolve 10 g of tryptone, 5 g of yeast extract, and 5 g of sodium chloride (NaCl) in purified water through stirring and boiling on a hot plate. Adjust volume to 1 L with ultrapure water. For 1 L, autoclave for at least 30 min at 121 °C to sterilize.
2. Low-swarm medium (LSW^{-}) [21]: Dissolve 10 g of tryptone, 5 g of yeast extract, 5 g of glycerol, and 0.4 g of NaCl in ultrapure water through stirring and boiling on a hot plate. Adjust volume to 1 L with purified water. For 1 L, autoclave for at least 30 min at 121 °C to sterilize. Once medium has cooled to 55 °C, add 12 mL of a premixed additive solution. To prepare 50 mL of this additive, mix 40 mL of 10% sterile glycerol and 4 mL each of 1 M sterile magnesium sulfate and 1% sterile nicotinic acid.
3. Blood-agar based Oxoid CM55 medium (Oxoid, Hampshire, UK): prepare according to manufacturer's instructions. Alternatively, one may use 1.5% LB agar (*see* **Note 1**). Swarm agar plates should be prepared freshly when possible and allowed 24 h to solidify at room temperature. Sterile media can be stored in volumes of 50–250 mL for at least 1 month at room temperature and liquefied before use via microwave or hot plate. This swarm-permissive media should not be reheated more than once for the assays below as reheating leads to

evaporation and potentially other effects to the composition of the media.

4. Dye solution: A 1000× stock solution for a dye that can be used to aid visualization is 40 mg/mL of Congo red mixed with 20 mg/mL of Coomassie Blue, prepared in sterile water [22]. Dilute to 1× in autoclaved, molten agar prior to pouring plates (*see* **Note 2**).
5. Optional: 0.5% India ink for testing moisture content of plates (*see* **Note 3**).
6. Sterile 90–100 mm petri dishes.
7. Sterile wooden dowels, inoculation needle, or similar.
8. Sterile glass culture tubes.

2.3 Boundary Assay (Dienes Line Typing)

1. Swarm-permissive agar (*see* **item 3** in Subheading 2.2).
2. *P. mirabilis* culture grown in LB broth.
3. Inoculation needle.
4. Stencil or ruler for spacing of inoculation points.

2.4 Territorial Exclusion Assay

1. *P. mirabilis* strains encoding different antibiotic selection markers on the chromosome.
2. 48-pronged replica plater (i.e., "frogger").
3. Sterile microcentrifuge tubes.
4. Ethanol.
5. LB broth (*see* **item 2** in Subheading 2.1).
6. LSW$^-$ agar containing selective antibiotics (*see* **item 2** in Subheading 2.2).
7. Swarm-permissive agar (*see* **item 3** in Subheading 2.2).

3 Methods

3.1 Macroscale Analysis of P. mirabilis Swarm Colonies

Swarming is influenced by humidity, agar content, and temperature among other factors (*see* **Notes 4–7** and Chap. 3 of this volume). Similar limitations have been described for surface-based motility in *Pseudomonas aeruginosa* [23]. To generate *P. mirabilis* swarm colonies in a reproducible manner, conditions for experiments should be kept as consistent as possible across experiments. A temperature-controlled environment is recommended for these assays, even when performed at room temperature. The experiments in Subheadings 3.2 and 3.3 will rely on the method described in Subheading 3.1.

3.1.1 Generation of Swarm Colonies

1. Store *P. mirabilis* populations at −80 °C in 30% sterile glycerol in water. Streak out populations for single colonies on

LSW^- agar. Do not use 1.5% LB agar plates for maintenance as *P. mirabilis* colonies will swarm across these plates. For wild-type strains, these maintenance plates may be stored at 4 °C for up to 2 weeks.

2. Select a single, isolated colony from the maintenance plates using an inoculation needle or similar and then grow in a culture tube containing LB broth for ~16 h at 37 °C, aerated. Include antibiotics at appropriate concentrations for plasmid maintenance.
3. Prepare CM55 agar plates (*see* **Note 1**). After autoclaving, allow medium to cool to 55 °C. Add sterile dye solution (1:1000) to the medium (*see* **Note 2**). Include antibiotics at appropriate concentrations for plasmid maintenance.
4. Pour media into sterile petri dishes. We recommend adding 25 mL per 90–100 mm petri dish as a standard volume. Allow plates to cool at room temperature in a single layer for several hours until agar has solidified, or alternatively, overnight. Drying times and temperatures should remain constant across experiments to produce reproducible colony structures.
5. Shortly before use, lay plates in a sterile laminar flow hood in a single layer with lids off. Allow to dry for 30 min or until plates lose glossy sheen associated with moisture and wet spots are absent. The plates have been dried too much if the agar surface forms fissures or cracks. You may need to adjust cooling and drying times in accordance with ambient temperature and humidity (*see* **Notes 3** and **4**).
6. When you are prepared to inoculate plates, measure the optical density of each culture (OD_{600}). Normalize each culture to a standardized OD_{600} of 0.1 (*see* **Note 6**) by diluting cultures in fresh LB broth.
7. Immediately inoculate normalized cultures using a sterilized inoculation needle. Briefly dip the needle into each culture and tap on the side of the tube to remove excess liquid. Lightly touch the surface of the agar with the needle; do not stab into agar. Be careful to avoid dripping excess liquid on swarm-permissive plates.
8. Allow plates to sit upright (i.e., lid-side up) on the benchtop until any visible liquid at the inoculation site has been absorbed, usually about 30 min at room temperature (*see* **Note 7**).
9. Incubate plates upside down at 37 °C for at least 16 h (*see* **Note 5**). Swarm plates should be analyzed before the swarm colony reaches the edge of the plate. Optimize incubation times based on the specific bacterial strain and conditions.

3.1.2 Analysis of Swarm Colonies

1. Measure the swarm diameter at equidistant points around the perimeter at regular intervals by marking the plate's surface.
2. Plot swarm diameters (e.g., as a bar graph). Information about swarm cycles and motility rates can be derived from these data [16, 24, 25].
3. Photograph swarm plates, and analyze the resulting images using conventional software such as FIJI [26] and/or Matlab (MATLAB and Statistics Toolbox Release 2012b, The MathWorks, Inc., Natick, MA, USA).
4. Assess overall colony morphology in terms of presence, number, and width of swarm terraces. Make note of the colors and sheen quality of the swarm colony. Examples of swarm colony analysis methods have been previously published in [8, 16, 23–25, 27].

Swarm colony development and structure are visual outputs for several fundamental behaviors as demonstrated in Fig. 1. Most wild-type *P. mirabilis* strains form colonies with regularly spaced terraces corresponding to iterative cycles of motility and division (Fig. 1a) [8]. Populations incapable of swarm expansion should grow at the inoculation site as a single colony and fail to expand across the agar surface (Fig. 1b). Such populations should be further analyzed as warranted, as these defects can be due to perturbations in a number of swarm-associated pathways. Several reviews provide an overview of these pathways [28–32]. Swarm colonies that fail to produce concentric rings (terraces) have been associated with hyperswarming phenotypes in which there is an excess of flagella produced (Fig. 1c). Such a result is often due to disruptions in flagellar regulation [25, 33, 34]. Disruptions in social behaviors can also be reflected in the swarm colony morphology. For example, cells disrupted in self versus nonself recognition behaviors form swarm colonies composed of wild-type terraces but that are attenuated in the final swarm colony's diameter [16, 24]. Other factors, such as swarm deficiencies due to loss of cell envelope structures (Fig. 1d) [25] or overexpression of swarm-inhibitory factors such as *mrpJ* [35], could also result in similar phenotypes. In sum, gross changes in the width or number of terraces could indicate changes in swarmer cell development, population motility, and/or social interactions.

3.2 Boundary Assay

Boundary assays (aka Dienes line assays) allow for a rapid readout of recognition behaviors between strains (Fig. 2). Strains are considered to have a nonself identity when a boundary becomes visible (Fig. 2a). By contrast, strains for which the swarm colonies merge without a visible boundary are considered to have a self identity (Fig. 2b). This boundary assay requires two or more *P. mirabilis* strains to be swarmed on the same swarm-permissive agar plate.

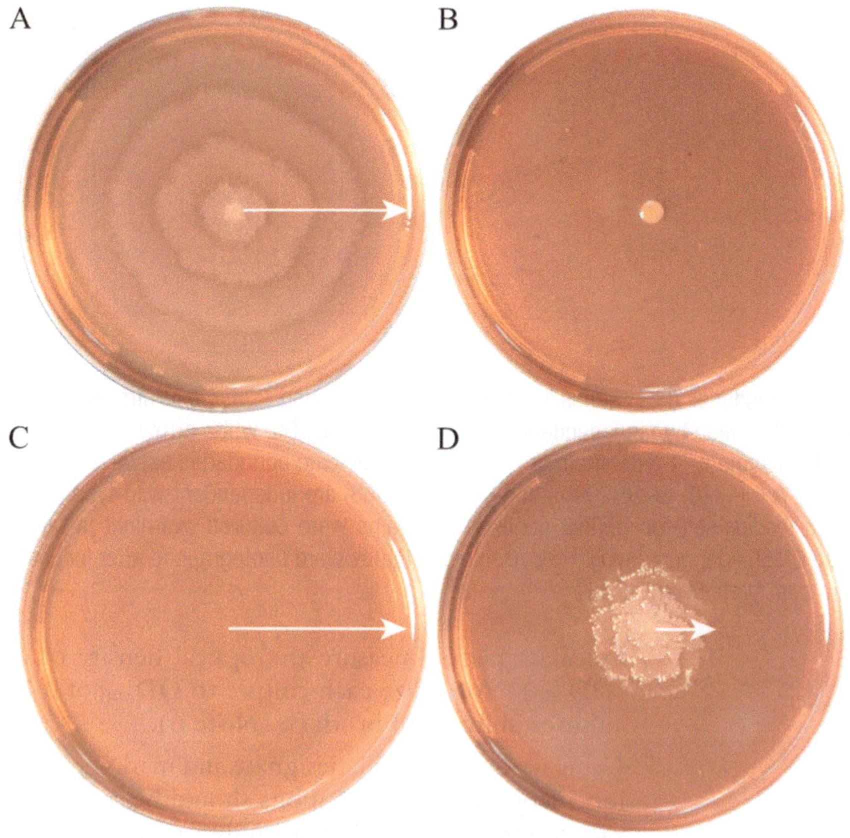

Fig. 1 *P. mirabilis* strains can form various swarm colony morphologies. (**a**) Wild-type *P. mirabilis* strain BB2000 pBBR1 [25]. (**b**) A nonswarming phenotype using BB2000Δ*rffG* pBBR1 [25]. (**c**) A hyperswarming phenotype using BB2000Δ*rcsB* pBBR1 [25]. (**d**) A restricted swarming phenotype using BB2000Δ*rffG* pP$_{lac}$-*rffG* [25]. Strains were cultured overnight in LB broth containing 35 μg mL^{-1} kanamycin and then inoculated onto CM55 agar containing 35 μg mL^{-1} kanamycin and 1× dye solution. The hyperswarming colony fails to form swarm terraces and forms a thin film on the surface; plates were photographed after incubation for 48 h at 37 °C. The other plates were photographed after incubation for 24 h at 37 °C. White arrows indicate radius of swarm colony. 90-mm petri plates were used

There are caveats to this boundary assay, which we highlight throughout the protocol. In addition, several genetic loci are now known to influence boundary formation [15, 19, 20]; as such, we do not recommend boundary assays for typing strains. Instead, there are several alternative and more accurate techniques as recently reviewed in [36].

1. Prepare swarm-permissive agar plates as described in Subheading 3.1.1, **steps 3–5**. The dye solution should be added to the plate to enhance boundary visualization.
2. Culture strains. We recommend that each boundary assay contains a positive (non-self strain) and a negative (self strain) control for boundary formation. When you are prepared to

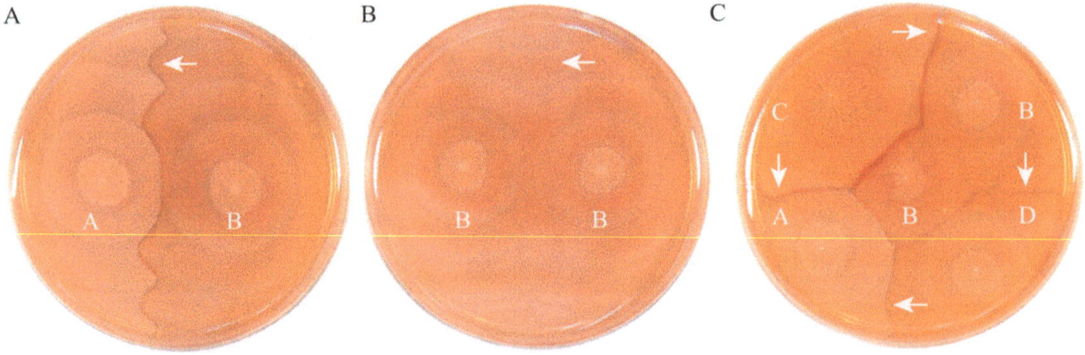

Fig. 2 Assessing boundaries (i.e., Dienes lines) versus mergers using *P. mirabilis* strains. (**a**) The white arrow marks a boundary between two independent wild-type strains. (**b**) The white arrow marks a merger between two colonies of the same wild-type strain. (**c**) White arrows indicate boundaries between swarm colonies. For all plates, strains A (HI4320), B (BB2000), and C (ATCC29906) are independent wild-type strains. Strain D is BB2000 lacking the *ids* self-recognition genes [15]. Strains were cultured overnight in LB broth and then inoculated onto CM55 agar containing 1× dye solution. Plates were photographed after incubation for 48 h at 37 °C. 90-mm petri plates were used

inoculate plates, measure the optical density of each culture (OD_{600}). Normalize each culture to OD_{600} of 0.1 by diluting cultures in fresh LB broth (*see* **Note 6**).

3. Using a ruler or stencil, designate and mark the location of each inoculation point. Populations should be spaced at even intervals across the plate. We strongly recommend that no more than five strains be grown on a single 90–100 mm plate.
4. Inoculate as described in Subheading 3.1.1, **steps 7** and **8**.
5. Incubate plates, inverted, at 37 °C until visible boundaries are apparent (*see* **Note 8**). Plates can also be incubated at room temperature until boundaries appear (*see* **Note 5**). Exact incubation times will need to be optimized depending on strain. We recommend that assays for strains be performed as consistently as possible across several days and that at least three biological replicates be performed to determine the formation of a boundary. We also recommend that the assay plates be analyzed in a double-blind manner when possible.
6. Analyze Dienes lines. Dienes lines have a binary read out: boundary (recognition of nonself) or merge (recognition of self). On undyed plates, boundaries will be visualized as macroscopic lines or gaps between colonies. On dyed plates, the dyes often stain the boundary a deeper purple color (*see* **Note 2**). The visible boundary varies broadly between conditions and samples. The boundary can sometimes appear thin and faint using one set of samples, and then separately appear broad and distinct between other samples (Fig. 2c) (*see*

Note 9). Each swarm colony should be evaluated for normal morphology. The normal morphology for each strain should be independently determined as described in Subheading 3.1.2.

3.3 Territorial Exclusion Assays

3.3.1 Coswarm Assay

We have developed a spatial and quantitative method to query social interactions, termed a territorial exclusion assay [20], that can serve as an alternative to the boundary formation assay (Fig. 3a). Briefly, this assay involves mixing two marked strains and allowing for the mixed population to swarm outward. At intervals during (or after) expansion, the spatial localization of each strain within the swarmed population can be assessed through replica plating. In general, one strain will be present at the outer edges of the swarm colony, while the other strain is constrained to the inner portion of the swarm (Fig. 3b) [20].

1. Generate two test strains encoding distinct antibiotic resistance markers to enable isolation of each strain through antibiotic selection. Alternatively, strains with different natural antibiotic resistances can be used (*see* **Note 10**).
2. Prepare swarm media compatible for both strains as described in Subheading 3.1.1, **steps 3–5**. One swarm plate per coswarm is required.
3. Prepare two LSW^- plates, each containing one of the selective antibiotics. You will require one of each selective plate per coswarm.
4. Culture each strain in LB broth overnight as described in Subheading 3.1.1, **step 2**. If strains are cultured with selective antibiotics, wash the cells with LB broth twice before proceeding with the following steps. To wash the cells, spin down 1 mL of overnight culture, resuspend in LB broth. Repeat. Spin down cells once more and resuspend in 1 mL LB broth.
5. Normalize strains to $OD_{600} = 0.1$ as described in Subheading 3.1.1, **step 6** (*see* **Note 6**).
6. In a 1.5-mL tube, combine the strains and mix well. The standard assay consists of using an initial ratio of 1:1 (*see* **Note 11**). Confirm the initial ratio by calculating colony forming units per mL of this initial ratio using standard methods; the result will be available after the assay is completed.
7. Immediately inoculate mixed strains onto a swarm agar plate as described in Subheading 3.1.1, **steps 7** and **8**.
8. Incubate plates, inverted, for at least 16 h at 37 °C (*see* **Note 5**).
9. Analyze plates when the swarm colony is ~0.5 cm from the edge of the petri plate, as follows.
10. Sterilize a 48-pronged replica plater by dipping the ends into ethanol and carefully lighting with a flame.

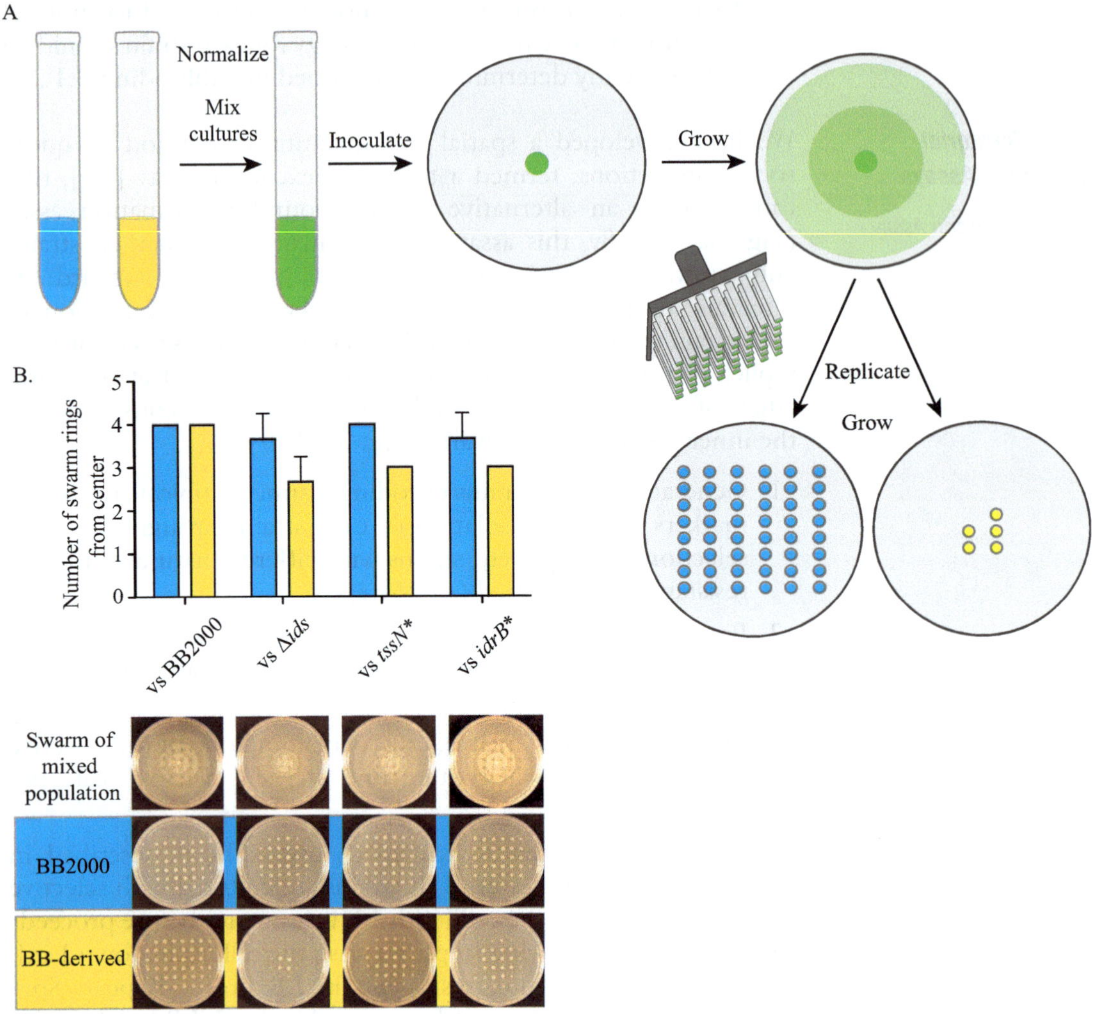

Fig. 3 Territorial exclusion assay. (**a**) Diagram of workflow for territorial exclusion assay. In brief, two strains labeled with distinct antibiotic selection markers are cultured, grown overnight, normalized, and mixed in desired ratio. The mixture is then inoculated onto a swarm permissive plate and allowed to grow at desired temperature. A 48-pronged replica plater is then used to pick up gridded regions of the swarm colony and then stamp onto two separate non-swarm-permissive plates, each selective for one strain). Plates are allowed to grow at desired temperature. Growth is then analyzed to determine the distance each strain migrated within the mixed swarm colony. (**b**) This assay can be used to distinguish between self-recognition behaviors, resulting in quantifiable metrics for analysis. This figure is modified from [20] and used with permission from *mBio*, the American Society for Microbiology, and the authors

11. Once cooled, gently rest the sterilized 48-pronged replica plater onto the plate containing the coswarmed populations. Very gently press the replica plater in plate to ensure even selection across the plate.
12. Carefully place the replica plater onto one each of the selective LSW$^-$ plates. Do not flame or rinse replica plater between

selective plates. We recommend performing several (three or more) independent biological replicates and reversing the order in which the selective plates are inoculated. Take photographs of the plates containing the coswarmed populations before and after replica plating.

13. Incubate selective plates for 16 h at 37 °C (*see* **Note 5**).
14. Take photographs of the selective plates.
15. Analyze the growth patterns on each of the selective plates. Growth on each selective plate corresponds to the presence of the associated strain within the swarm colony.
16. Using a ruler, measure the radial distance from inoculation point or diameter across the outermost dots with growth on each plate.
17. Compare the diameter on the selective plate to the diameter of the entire coswarm to establish a normalized swarm fraction.
18. Plot this normalized fraction for each strain, ultimately resulting in normalized migration distances that are applicable between strains and between experimental conditions.

We have found that disruptions in self-recognition genetic loci result in consistent, measurable differences in swarm migration (Fig. 3b) [20]. Further, we have found that distinct wild-type strains also result in predictable, differential migration pattern in which one strain is found on the outer edges of the swarm and the other is not [20].

3.3.2 Coswarm Assay with Boundary Formation

A variation on the coswarm assay allows for rapid analysis of which strain is dominant over another in swarm motility-based competitions without requiring that either strain encode antibiotic resistance [20, 37]. We have found that this combined assay often helps to resolve inconclusive interactions that are seen using the boundary assay (Subheading 3.2) alone.

1. Make the mixture of strains as described in Subheading 3.3.1, **steps 4–6**.
2. As described in Subheadings 3.1 and 3.2, inoculate a swarm-permissive plate with this mixture, and separately, with each of the individual strains (i.e., strain A and separately, strain B) contained within the mixture, forming a triangle with at least 1 cm between each inoculation point.
3. Incubate the plate inverted at 37 °C (*see* **Note 5**) until the control boundary forms.
4. A boundary between the coswarm and strain A would indicate that strain B is found at the outer edges of the swarm and that strain B outcompetes strain A. To confirm, a merge between the coswarm and strain B should also be evident.

4 Notes

1. Either CM55 or 1.5% LB agar can be used as a standard swarm-permissive condition for *P. mirabilis*. We recommend using CM55 as the resulting colonies produce crisp swarm rings. CM55 is a proprietary blend that includes 1.5% agar. To prepare LB agar: dissolve 10 g of tryptone, 5 g of yeast extract, 5 g of NaCl, and 15 g of Bacto agar in ultra-pure water through stirring and boiling on a hot plate. Adjust volume to 1 L with purified water. For 1 L, autoclave for at least 30 min at 121 °C to sterilize and allow it to cool to 55 °C. If desired, add dye solution. Pour 25 mL of medium per 90–100 mm petri dish.
2. Addition of Congo red and Coomassie blue dyes enhances contrast between the swarm colonies and the agar background to emphasize boundary formation. Some faint boundaries may not be clearly visible in the absence of dyes. Experiments should be conducted in the presence and absence of dye to ensure that dyes do not influence the results.
3. To test whether plates are too wet for use, employ an India ink spread assay [23]. As adapted from a previous report, select one plate per batch of media to test and pipette 10 μL of 0.5% India ink onto this plate. The ink will spread across the surface if the plate is too wet.
4. An alternative approach is to dry the plates with the lids ajar surrounding a flame (e.g., from a Bunsen burner); this method will take longer for the swarm-permissive plates to dry sufficiently. Plates should be visually inspected for moisture before use. Pooled moisture will allow cells to swim across the surface rather than engage in swarm motility. Allow agar plates to dry for longer if visible moisture remains pooled on the agar surface.
5. Swarm assays may be conducted at 37 °C or at room temperature. If conducting swarm assays at room temperature, increase incubation time to 48 h.
6. The OD_{600} of the initial inoculation should be consistent across experiments; however, our recommended OD_{600} of 0.1 is arbitrary. Inoculation of fewer cells will increase the amount of time required for emergence of the first swarm ring [8].
7. There are several indicators for whether inoculation techniques were sufficient. If there are multiple swarming colonies present on the surface after incubation, then droplets were splattered during inoculation. For the next time, tap dry the inoculation needle more and make sure that the agar plate is physically separate from the culture tube during inoculation. If the

inoculation point pools out beyond a tight circle after incubation, then the agar plates were likely dried insufficiently after inoculation. For the next time, increase drying times. If the swarm colony does not have visible concentric rings after incubation, then the agar was likely insufficiently dried before inoculation; increase the duration of this drying period for the next time.

8. Boundary development can take 16–48 h depending on the strains being analyzed. Longer incubations (48 h) often allow boundary formation to become more visually pronounced. Nonboundary swarm assays should only be conducted for 16–24 h so that swarm colony analysis can occur before swarm colony reaches the edge of the plate.
9. Differences in boundary morphology are readily visible. We have found that temperature, agar volume, and humidity can alter the form that a boundary takes, even between the same two isolates. In addition, we caution that at times the interface between the swarm rings of merging colonies can resemble faint boundaries; these caveats are increased when strains form visually distinctive terraces.
10. Several *P. mirabilis* strains are resistant to tetracycline [38, 39] and polymyxin B [40, 41].
11. Coswarm analyses can be conducted at multiple ratios of the two competing strains; for example, we have successfully used ratios of 10:1, 1:1, and 1:10 (Strain A–Strain B) [25].

Acknowledgments

We thank members of the Gibbs lab for thoughtful discussions and development of the protocols. The writing of this review was funded by a Smith Family Graduate Fellowship in Science and Engineering (to K.L.), the David and Lucile Packard Foundation, the George W. Merck Fund, and Harvard University.

References

1. Drzewiecka D (2016) Significance and roles of *Proteus* spp. bacteria in natural environments. Microb Ecol 72(4):741–758. https://doi.org/10.1007/s00248-015-0720-6
2. Schaffer JN, Pearson MM (2015) *Proteus mirabilis* and urinary tract infections. Microbiol Spectr 3(5):1–39. https://doi.org/10.1128/microbiolspec.UTI-0017-2013
3. Allison C, Emody L, Coleman N, Hughes C (1994) The role of swarm cell differentiation and multicellular migration in the uropathogenicity of *Proteus mirabilis*. J Infect Dis 169 (5):1155–1158. https://doi.org/10.1093/infdis/169.5.1155
4. Armbruster CE, Forsyth-DeOrnellas V, Johnson AO, Smith SN, Zhao L, Wu W, Mobley HLT (2017) Genome-wide transposon mutagenesis of *Proteus mirabilis*: essential genes, fitness factors for catheter-associated urinary tract infection, and the impact of polymicrobial infection on fitness requirements. PLoS Pathog 13(6):e1006434. https://doi.org/10.1371/journal.ppat.1006434

5. Burall LS, Harro JM, Li X, Lockatell CV, Himpsl SD, Hebel JR, Johnson DE, Mobley HLT (2004) *Proteus mirabilis* genes that contribute to pathogenesis of urinary tract infection: identification of 25 signature-tagged mutants attenuated at least 100-fold. Infect Immun 72(5):2922–2938. https://doi.org/10.1128/Iai.72.5.2922-2938.2004
6. Pearson MM, Yep A, Smith SN, Mobley HLT (2011) Transcriptome of *Proteus mirabilis* in the murine urinary tract: virulence and nitrogen assimilation gene expression. Infect Immun 79(7):2619–2631. https://doi.org/10.1128/IAI.05152-11
7. Stickler D, Hughes G (1999) Ability of *Proteus mirabilis* to swarm over urethral catheters. Eur J Clin Microbiol Infect Dis 18 (3):206–208. https://doi.org/10.1007/s100960050260
8. Rauprich O, Matsushita M, Weijer CJ, Siegert F, Esipov SE, Shapiro JA (1996) Periodic phenomena in *Proteus mirabilis* swarm colony development. J Bacteriol 178 (22):6525–6538. https://doi.org/10.1128/jb.178.22.6525-6538.1996
9. Hoeniger JF (1966) Cellular changes accompanying the swarming of *Proteus mirabilis*. II. Observations of stained organisms. Can J Microbiol 12(1):113–123. https://doi.org/10.1139/m66-017
10. Hoeniger JF (1965) Development of flagella by *Proteus mirabilis*. J Gen Microbiol 40 (1):29. https://doi.org/10.1099/00221287-40-1-29
11. Pearson MM, Rasko DA, Smith SN, Mobley HLT (2010) Transcriptome of swarming *Proteus mirabilis*. Infect Immun 78 (6):2834–2845. https://doi.org/10.1128/IAI.01222-09
12. Jones BV, Young R, Mahenthiralingam E, Stickler DJ (2004) Ultrastructure of *Proteus mirabilis* swarmer cell rafts and role of swarming in catheter-associated urinary tract infection. Infect Immun 72(7):3941–3950. https://doi.org/10.1128/IAI.72.7.3941-3950.2004
13. Dienes L (1946) Reproductive processes in *Proteus* cultures. Proc Soc Exp Biol Med 63 (2):265–270. https://doi.org/10.3181/00379727-63-15570
14. Senior BW (1977) The Dienes phenomenon: identification of the determinants of compatibility. J Gen Microbiol 102(2):235–244. https://doi.org/10.1099/00221287-102-2-235
15. Gibbs KA, Urbanowski ML, Greenberg EP (2008) Genetic determinants of self identity and social recognition in bacteria. Science 321 (5886):256–259. https://doi.org/10.1126/science.1160033
16. Saak CC, Gibbs KA (2016) The self-identity protein IdsD is communicated between cells in swarming *Proteus mirabilis* colonies. J Bacteriol 198(24):3278–3286. https://doi.org/10.1128/JB.00402-16
17. De Louvois J (1969) Serotyping and the Dienes reaction on *Proteus mirabilis* from hospital infections. J Clin Pathol 22 (3):263–268. http://dx.doi.org/10.1136/jcp.22.3.263
18. Skirrow MB (1969) The dienes (mutual inhibition) test in the investigation of *Proteus* infections. J Med Microbiol 2(4):471–477. https://doi.org/10.1099/00222615-2-4-471
19. Alteri CJ, Himpsl SD, Pickens SR, Lindner JR, Zora JS, Miller JE, Arno PD, Straight SW, Mobley HLT (2013) Multicellular bacteria deploy the type VI secretion system to preemptively strike neighboring cells. PLoS Pathog 9 (9):e1003608. https://doi.org/10.1371/journal.ppat.1003608
20. Wenren LM, Sullivan NL, Cardarelli L, Septer AN, Gibbs KA (2013) Two independent pathways for self-recognition in *Proteus mirabilis* are linked by type VI-dependent export. mBio 4(4). https://doi.org/10.1128/mBio.00374-13
21. Belas R, Erskine D, Flaherty D (1991) Transposon mutagenesis in *Proteus mirabilis*. J Bacteriol 173(19):6289–6293. https://doi.org/10.1128/jb.173.19.6289-6293.1991
22. Dietrich LE, Teal TK, Price-Whelan A, Newman DK (2008) Redox-active antibiotics control gene expression and community behavior in divergent bacteria. Science 321 (5893):1203–1206. https://doi.org/10.1126/science.1160619
23. Morales-Soto N, Anyan ME, Mattingly AE, Madukoma CS, Harvey CW, Alber M, Deziel E, Kearns DB, Shrout JD (2015) Preparation, imaging, and quantification of bacterial surface motility assays. J Vis Exp (98). https://doi.org/10.3791/52338
24. Zepeda-Rivera MA, Saak CC, Gibbs KA (2018) A proposed chaperone of the bacterial type VI secretion system functions to constrain a self-identity protein. J Bacteriol 200(14). https://doi.org/10.1128/JB.00688-17
25. Little K, Tipping MJ, Gibbs KA (2018) Swarmer cell development of the bacterium *Proteus mirabilis* requires the conserved ECA biosynthesis gene, *rffG*. J Bacteriol 200(18). https://doi.org/10.1128/JB.00230-18
26. Schindelin J, Arganda-Carreras I, Frise E, Kaynig V, Longair M, Pietzsch T, Preibisch S, Rueden C, Saalfeld S, Schmid B, Tinevez JY, White DJ, Hartenstein V, Eliceiri K, Tomancak P, Cardona A (2012) Fiji: an open-source platform for biological-image analysis.

Nat Methods 9(7):676–682. https://doi.org/10.1038/nmeth.2019
27. Morris JD, Hewitt JL, Wolfe LG, Kamatkar NG, Chapman SM, Diener JM, Courtney AJ, Leevy WM, Shrout JD (2011) Imaging and analysis of *Pseudomonas aeruginosa* swarming and rhamnolipid production. Appl Environ Microbiol 77(23):8310–8317. https://doi.org/10.1128/AEM.06644-11
28. Alavi M, Belas R (2001) Surface sensing, swarmer cell differentiation, and biofilm development. Methods Enzymol 336:29–40. https://doi.org/10.1016/S0076-6879(01)36575-8
29. Armbruster CE, Mobley HLT (2012) Merging mythology and morphology: the multifaceted lifestyle of *Proteus mirabilis*. Nat Rev Microbiol 10(11):743–754. https://doi.org/10.1038/nrmicro2890
30. Belas R (2014) Biofilms, flagella, and mechanosensing of surfaces by bacteria. Trends Microbiol 22(9):517–527. https://doi.org/10.1016/j.tim.2014.05.002
31. Morgenstein RM, Szostek B, Rather PN (2010) Regulation of gene expression during swarmer cell differentiation in *Proteus mirabilis*. FEMS Microbiol Rev 34(5):753–763. https://doi.org/10.1111/j.1574-6976.2010.00229.x
32. Rather PN (2005) Swarmer cell differentiation in *Proteus mirabilis*. Environ Microbiol 7(8):1065–1073. https://doi.org/10.1111/j.1462-2920.2005.00806.x
33. Clemmer KM, Rather PN (2007) Regulation of *flhDC* expression in *Proteus mirabilis*. Res Microbiol 158(3):295–302. https://doi.org/10.1016/j.resmic.2006.11.010
34. Morgenstein RM, Rather PN (2012) Role of the Umo proteins and the Rcs phosphorelay in the swarming motility of the wild type and an O-antigen (*waaL*) mutant of *Proteus mirabilis*. J Bacteriol 194(3):669–676. https://doi.org/10.1128/Jb.06047-11
35. Pearson MM, Mobley HLT (2008) Repression of motility during fimbrial expression: identification of 14 *mrpJ* gene paralogues in *Proteus mirabilis*. Mol Microbiol 69(2):548–558. https://doi.org/10.1111/j.1365-2958.2008.06307.x
36. Adzitey F, Huda N, Ali GR (2013) Molecular techniques for detecting and typing of bacteria, advantages and application to foodborne pathogens isolated from ducks. 3 Biotech 3(2):97–107. https://doi.org/10.1007/s13205-012-0074-4
37. Saak CC, Zepeda-Rivera MA, Gibbs KA (2017) A single point mutation in a TssB/VipA homolog disrupts sheath formation in the type VI secretion system of *Proteus mirabilis*. PLoS One 12(9):e0184797. https://doi.org/10.1371/journal.pone.0184797
38. Adamus-Bialek W, Zajac E, Parniewski P, Kaca W (2013) Comparison of antibiotic resistance patterns in collections of *Escherichia coli* and *Proteus mirabilis* uropathogenic strains. Mol Biol Rep 40(4):3429–3435. https://doi.org/10.1007/s11033-012-2420-3
39. Eisenberg GM, Weiss W, Flippin HF (1958) In vitro susceptibility of *Proteus* species to streptomycin, chloramphenicol, tetracycline, and novobiocin. Am J Clin Pathol 30(1):20–24. https://doi.org/10.1093/ajcp/20.4.325
40. McCoy AJ, Liu H, Falla TJ, Gunn JS (2001) Identification of *Proteus mirabilis* mutants with increased sensitivity to antimicrobial peptides. Antimicrob Agents Chemother 45(7):2030–2037. https://doi.org/10.1128/AAC.45.7.2030-2037.2001
41. Sud IJ, Feingold DS (1970) Mechanism of polymyxin B resistance in *Proteus mirabilis*. J Bacteriol 104(1):289–294.

Chapter 7

Insertional Mutagenesis Protocol for Constructing Single or Sequential Mutations

Melanie M. Pearson, Stephanie D. Himpsl, and Harry L. T. Mobley

Abstract

Genetic mutation enables the study of the function of specific genes, particularly when a mutant is compared against its isogenic parent. In *Proteus mirabilis* bacteria, traditional allelic exchange mutation is labor-intensive and has a high failure rate in some strains. Likewise, there is no working protocol for lambda red recombinase-based mutation in *P. mirabilis*. Here we describe an alternative method of insertional mutagenesis based on retargeting of group II introns. The protocol includes steps to generate single or multiple mutations, with the possibility to delete intervening sequences of DNA.

Key words Mutagenesis, Group II intron, Targetron, Cre-lox recombination

1 Introduction

Genetic manipulation is an essential tool for studying how individual genes contribute to bacterial phenotypes. In *P. mirabilis*, targeted gene mutations have most often been assessed in mouse urinary tract infection models or swarm assays [1, 2]. Although mutagenesis is widely employed in *P. mirabilis* research, some strains are difficult to target; this was highlighted when the first *P. mirabilis* mutation, in the urease operon, was constructed [3]. Therefore, another method, employing a programmable group II intron, has been modified to work in *P. mirabilis* [4].

Group II introns use a ribonucleoprotein complex containing an intron-encoded reverse transcriptase to specifically insert into a target sequence, a process called retrohoming. They were adapted to be retargetable into a broad range of DNA targets [5]. The engineered introns, called targetrons, use base pairing for target recognition. Identification of potential sites in target genes has been optimized using computer algorithms [6]. Successful insertions may be easily identified using an antibiotic resistance marker that only becomes active after the targetron has integrated into its target (retrotransposition-activated selectable marker, or RAM)

Melanie M. Pearson (ed.), *Proteus mirabilis: Methods and Protocols*, Methods in Molecular Biology, vol. 2021,
https://doi.org/10.1007/978-1-4939-9601-8_7,

[7]. Targetrons have been combined with Cre/lox recombination to remove the antibiotic resistance marker, thereby allowing sequential mutations and deletions of intervening sequence [8, 9]. The protocols for both single insertions and sequential mutations in *P. mirabilis* follow. The Methods are subdivided into six sections: (1) preparation of competent cells, (2) reprogramming the targetron, (3) screening for the reprogrammed targetron, (4) mutating *P. mirabilis*, (5) generating a markerless mutation, and (6) constructing double mutants and deletions.

2 Materials

This protocol was modified for *P. mirabilis* from the TargeTron Gene Knockout System (Millipore Sigma), which includes most of the materials required to construct targeted insertional mutations. Commercial sources cited here are known to work in *P. mirabilis*; alternatives may be available but will need to be empirically tested to confirm they function in the targeted strain of bacteria. Prepare all solutions in sterile, purified water, unless otherwise specified.

2.1 Bacterial Culture and Propagation

1. Lysogeny broth (LB) (per liter): 10 g of tryptone, 5 g of yeast extract, 0.5 g of NaCl. Autoclave to sterilize (*see* **Note 1**).
2. LB agar: LB plus 15 g of agar/L; autoclave to sterilize; cool to 55 °C; pour into petri dishes, approx. 20–25 mL/plate.
3. Antibiotics: 50 μg/mL of ampicillin, 25 μg/mL of kanamycin, 20 μg/mL of chloramphenicol. To make LB agar with antibiotics: after cooling autoclaved agar to 55 °C, add antibiotics and mix gently but thoroughly before pouring (*see* **Note 2**).
4. Sterile snap-cap culture tubes, 17 × 100 mm.
5. Flat toothpicks; sterilize by autoclaving.
6. Patching grids for organization of colonies (24–48 patches per plate works well).
7. 37 °C incubator, plus 37 °C and 30 °C shaker incubators.
8. The target *P. mirabilis* strain. This protocol assumes a *P. mirabilis* host strain that is sensitive to ampicillin, kanamycin, and chloramphenicol (*see* **Note 3**).

2.2 DNA Manipulation

1. PCR reagents (*Taq* DNA polymerase, dNTPs, *Taq* buffer). Store at −20 °C.
2. PCR thermocycler.
3. DNA cleanup method (removal of primers, dNTPs).
4. PCR cloning kit. Store at −20 °C.

5. Sequencing primers: Base these off of the plasmid in the previous step, and design them to read into the cloning site.
6. Restriction enzymes: HindIII-HF and BsrGI-HF (*see* **Note 4**). Store at −20 °C.
7. Ligation: T4 DNA ligase and ligase buffer. Store at −20 °C.
8. Spectrophotometer capable of reading 1 μL volumes.
9. Plasmid miniprep kit.

2.3 Transformation

1. Agarose gel electrophoresis supplies and equipment.
2. Electrocompetent *Escherichia coli*, such as DH5α, either commercial or self-prepared (*see* **Note 5**).
3. Electroporator.
4. Electroporation cuvettes, 1 mm gap.
5. X-gal (5-bromo-4-chloro-3-indolyl-β-D-galactopyranoside): dissolve 40 mg/mL in *N*,*N*-dimethylformamide. Store at −20 °C (*see* **Note 6**).

2.4 Intron-Specific Materials

1. Plasmid pACD4K-CloxP. The TargeTron Gene Knockout System (Millipore Sigma) comes with plasmid pACD4K-C. This plasmid works well in *P. mirabilis* but may only be used to construct single mutations. A second plasmid, pACD4K-CloxP, also available from Millipore Sigma, includes loxP sites that flank the kanamycin resistance gene. The latter plasmid may therefore be used to construct markerless mutants, thus allowing subsequent rounds of mutation.
2. Plasmid pAR1219, or other source of inducible T7 polymerase. This is necessary to drive induction of intron jumping in *P. mirabilis* (*see* **Note 7**).
3. Intron template. The template is supplied with the TargeTron Gene Knockout System (Millipore Sigma) (*see* **Note 8**).
4. Plasmid pQL123 (encodes *cre*-recombinase; other *cre*-encoding plasmids may also work) [10] (*see* **Note 9**).
5. 100 mM IPTG. Store at −20 °C.
6. 40% glucose, filter-sterilized.
7. EBS universal primer: 5′-CGA AAT TAG AAA CTT GCG TTC AGT AAA C-3′. Working stock is 20 μM. Store at −20 °C. This primer is supplied with the TargeTron Gene Knockout System (Millipore Sigma).
8. Primers for pACD4K-CloxP:

pACD4K-C5′	5′-CCG CGA AAT TAA TAC GAC TCA CTA-3′
pACD4K-C3′	5′-GGT ATC CCC AGT TAG TGT TA-3′

3 Methods

Carry out all procedures at room temperature, unless otherwise specified. All procedures using live bacteria should be conducted using sterile technique.

3.1 Prepare Ahead of Time: Competent Cells with Helper Plasmid

1. Transform the target strain of *P. mirabilis* with plasmid pARl219. Maintain this strain in LB supplemented with 50 μg/mL of ampicillin (*see* **Notes 7** and **10**).
2. Make electrocompetent cells of *P. mirabilis* containing pARl219 (*see* **Note 10**). Store in 50 μL aliquots at −80 °C. Thaw on ice 15–30 min before use.

3.2 Mutate the Retargeting Plasmid to Target the Gene of Interest

1. Evaluate potential mutation sites in the gene of interest using a site locator tool (for example, http://clostron.com/clostron2.php) [11] (*see* **Note 11**). The program will identify insertion sites, along with three primers (IBS, EBS2, and EBS1d), which are used with the EBS universal primer to retarget the intron.
2. From the options given by the program, choose a primer set based on the preferred insertion site. The primer sets are listed from highest to lowest probability of success. An insertion site closer to the 5′ end of the target gene is preferable to minimize chances of obtaining a truncated protein that retains functionality.
3. Synthesize the IBS, EBS2, and EBS1d primers for the target site. Use HPLC or PAGE purification (*see* **Notes 12** and **13**).
4. Design PCR primers for genomic *P. mirabilis* DNA that flank and read into the intron insertion site; these will be used to confirm the mutation at the end of the protocol. A wild-type amplicon of approximately 125–200 bp is ideal.
5. Dilute the primers in water: IBS and EBS1d to 100 μM and EBS2 to 20 μM.
6. Make a four-primer master mix:

2 μL	100 μM IBS primer
2 μL	100 μM EBS1d primer
2 μL	20 μM EBS2 primer
2 μL	20 μM EBS universal primer
12 μL	Water
20 μL	Total

7. Set up this PCR reaction (*see* **Note 14**)

23 μL	Water
1 μL	Four-primer mix
1 μL	Intron PCR template
25 μL	2× PCR master mix
50 μL	Total

8. Cycle the PCR reaction:

Initial denaturation	94 °C, 30 s
30 cycles: Denaturation Annealing Extension	 94 °C, 15 s 55 °C, 30 s 72 °C, 30 s
Final extension	72 °C, 2 min
Hold	4 °C

9. Check 5 μL of the PCR reaction on a 2% agarose gel. A major band should be present at 350 bp. Smaller bands are expected and are not a problem.
10. Clean up the PCR reaction (remove primers and dNTPs).
11. Clone the PCR product using a vector designed for this purpose (i.e., TA cloning vector) (*see* **Note 15**).
12. Transform competent *E. coli* with the ligated vector from the previous step (follow the cloning kit protocol or use self-prepared competent cells).
13. Plate transformed bacteria on LB agar with appropriate antibiotic selection and incubate overnight at 37 °C (*see* **Note 16**).

3.3 Screen for Desired Mutation in the Intron Fragment (See Note 15)

1. Patch colonies to a new LB agar plate; continue to use antibiotic selection for the plasmid. If blue–white screening was used (X-gal), choose white or light blue colonies (*see* **Note 16**). Incubate at 37 °C for at least 4 h (overnight is fine).
2. Set up colony PCR reactions to check for insert size (*see* **Note 17**):

31.5 μL	dH_2O
4.0 μL	10× Taq buffer
3.2 μL	2.5 mM dNTPs
0.4 μL	10 mM Forward primer*

(continued)

0.4 μL	10 mM Reverse primer*
0.5 μL	*Taq* DNA polymerase
40 μL	Total

Touch a sterile toothpick to a colony patch, then add bacteria to the PCR reaction.
***Item 5** from Subheading 2.2.

3. Cycle the PCR reaction:

Initial denaturation	94 °C, 30 s
30 cycles of:	
Denaturation	94 °C, 15 s
Annealing	55 °C, 30 s
Extension	72 °C, 30 s
Final extension	72 °C, 2 min
Hold	4 °C

4. Check 5–10 μL of the PCR reaction on an agarose gel. The correct product is approximately 450–550 bp, depending on how far the PCR primers are located from the insertion site.
5. Choose 4–6 plasmids with the correct-sized insert. Inoculate 5 mL of LB plus appropriate antibiotic from the patch plate and incubate at 37 °C overnight.
6. Miniprep the plasmids.
7. Quantify the plasmid yields (e.g., using a NanoDrop).
8. Sequence plasmids to confirm retargeting of the intron. Use plasmid-based primers that will read into the cloned PCR fragment (i.e., **item 5** from Subheading 2.2).
9. Analyze the sequences, and choose a plasmid that has the correct, retargeted sequence (i.e., find the IBS, EBS1d, and EBS2 primer sequences in the plasmid trace file) (*see* **Note 18**).
10. Set up a restriction digestion: (*see* **Note 4**).

25 μL	Plasmid
4 μL	10× CutSmart buffer
0.5 μL	HindIII-HF
0.5 μL	BsrGI-HF
9 μL	Water
39 μL	Total

11. Incubate reaction for 30 min at 37 °C (*see* **Note 4**).
12. Add 0.5 μL of HindIII-HF plus 0.5 μL of BsrGI-HF and incubate again at 37 °C for 30 min.

13. Incubate at 80 °C for 10 min to kill the enzymes.
14. Tape two wells together on a gel comb to make one large well. Cast a 1% agarose gel. Load the entire digest in the large well, and separate by electrophoresis. Gel purify the intron (350 bp fragment).
15. Ligate the intron with the kit vector:

2 μL	pACD4K-CloxP linear vector (40 ng)
15 μL	Gel-purified intron fragment
2 μL	10× ligase buffer
1 μL	T4 DNA ligase
20 μL	Total

16. Incubate the ligation mixture at room temperature for 5–30 min.
17. Heat-kill the ligase at 65 °C for 10 min.
18. Desalt the ligation (e.g., with a DNA-binding spin column; *see* **Note 19**).
19. Electroporate *E. coli* (e.g., DH5α) with 5 μL of the ligation reaction.
20. Recover the bacteria in 400 μL of LB at 37 °C for 1 h, with aeration.
21. Plate 50 μL, 150 μL, or 200 μL of the recovered bacteria on LB agar with chloramphenicol plus 40 μL of X-gal (spread-plate X-gal on the agar surface and allow to dry before adding bacteria; *see* **Note 20**). Incubate plates at 37 °C overnight.
22. Patch white or light blue colonies to a fresh LB plus chloramphenicol plate. Incubate at 37 °C for at least 4 h (overnight is fine).
23. Perform colony PCR as in **steps 2** and **3** of Subheading 3.3, using primers pACD4K-C 5′ and 3′ to look for plasmids with the correct size of insert. The correct amplicon is approximately 450 nt.
24. Choose one patch with the correct-sized insert. Inoculate 5 mL of LB plus chloramphenicol with the patch and incubate at 37 °C overnight with aeration.
25. Streak the same patch for isolation on LB agar supplemented with chloramphenicol. Incubate at 37 °C overnight.
26. Make a glycerol freezer stock from the agar plate.
27. Miniprep the plasmid from the overnight culture.

3.4 Mutate P. mirabilis

1. Electroporate *P. mirabilis*/pAR1219 with 1 μL of the plasmid.
2. Recover in 250 μL LB at 37 °C for 1 h with aeration.
3. Prepare 3 mL of LB supplemented with 50 μg/mL ampicillin, 20 μg/mL chloramphenicol, and 1% glucose. Ampicillin selects for pAR1219, chloramphenicol selects for pACD4K-CloxP, and glucose represses T7 polymerase on pAR1219 until the culture is ready to be induced.
4. Add 100 μL of the recovered transformation to the 3 mL of LB. Grow overnight at 37 °C with aeration (*see* **Note 7**).
5. The next day, prepare 2 mL of LB supplemented with 50 μg/mL of ampicillin, 20 μg/mL of chloramphenicol, and 1% glucose.
6. Add 40 μL of the overnight culture to the 2 mL of LB.
7. Grow culture to mid-log phase (OD_{600} 0.5–0.8; about 2–3 h) (*see* **Note 21**).
8. Add 10 μL of 100 mM IPTG to the 2 mL culture.
9. Shake at 30 °C for 30 min.
10. Spin down cells; resuspend in 400 μL of LB plus 1% glucose (do not add chloramphenicol or ampicillin).
11. Culture at 30 °C for 1 h, with aeration.
12. Plate 20 μL, 50 μL, 100 μL or 200 μL on LB supplemented with 25 μg/mL kanamycin.
13. Incubate plates at 37 °C overnight.
14. Patch colonies to LB agar supplemented with 25 μg/mL kanamycin (*see* **Note 22**).
15. Colony PCR to screen for insertion of the intron in the target gene, using primers that target *P. mirabilis* genomic DNA and flank the intron insertion site (*see* Subheading 3.2, **step 4**). The intron is 2 kb in size (Fig. 1).
16. If the mutant is confirmed, streak for isolation. Incubate at 37 °C overnight (*see* **Note 23**).
17. Double patch colonies onto LB agar supplemented with 25 μg/mL of kanamycin or 50 μg/mL of ampicillin. This is to screen for loss of pAR1219. Incubate at 37 °C overnight.
18. Pick a patch that is both kanamycin resistant and ampicillin sensitive. Streak for isolation on LB agar supplemented with 25 μg/mL kanamycin (*see* **Note 24**).
19. Make a glycerol freezer stock of the mutant *P. mirabilis* strain.

3.5 Generating a Markerless Mutant

1. Make electrocompetent cells of the mutant strain (*see* **Notes 10** and **25**).
2. Electroporate the mutant strain with pQL123; select on LB agar supplemented with 100 μg/mL of ampicillin.

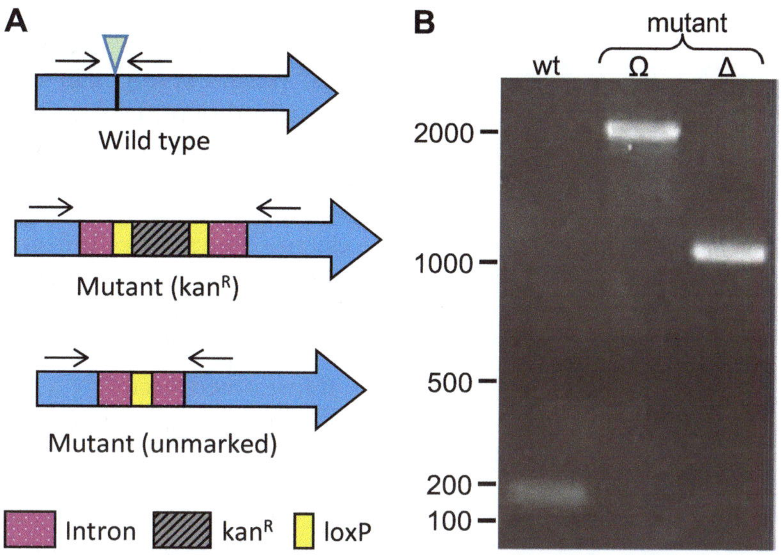

Fig. 1 PCR confirmation of targetron mutagenesis. (**a**) Depiction of primer locations (small arrows) in a target gene in wild-type, mutant, and unmarked mutant DNA. In the wild-type example, the inverted triangle indicates the location of the desired insertion. (**b**) This example shows a *dppA* mutant in *P. mirabilis* HI4320 [8] constructed using the TargeTron Gene Knockout System (Millipore Sigma). Lane 1, wild type (wt); lane 2, insertion (Ω) of the targetron adds approximately 2 kb to the product size; lane 3, deletion (Δ) of the kanamycin resistance cassette by Cre/lox recombination leaves an approximately 1 kb intron insertion in the target gene. Size markers, in nucleotides, are indicated on the left

3. Inoculate 5 mL of LB supplemented with 100 μg/mL of ampicillin and 1 mM IPTG with an ampicillin-resistant colony; grow at 37 °C overnight. IPTG induces *cre* recombinase, which causes the loxP sites to recombine and the kanamycin gene to be excised while leaving the rest of the intron in the target gene (Fig. 1).
4. Dilution plate the overnight culture onto plain LB agar. A 10^{-6} dilution typically yields well-isolated colonies.
5. Triple patch colonies onto plain LB agar, LB with 25 μg/mL of kanamycin (to screen for loss of the kanamycin cassette), and LB with 100 μg/mL of ampicillin (to screen for loss of pQL123).
6. Use colony PCR as in **step 15** of Subheading 3.4 to confirm loss of the kanamycin cassette from a patch that is sensitive to both kanamycin and ampicillin with the same PCR primers used to confirm the original targetron mutant. The product should be ~1 kb smaller now (Fig. 1). This unmarked strain is now ready to be mutated with other targetron plasmids.

7. If kanamycin-sensitive, ampicillin-resistant (kan^S amp^R) patches but no double-sensitive patches are obtained, passage a kan^S amp^R patch overnight in LB broth with no antibiotic selection to get rid of pQL123. The plasmid must be removed before pAR1219 can be reintroduced in order to make further mutations.

3.6 Double Mutants and Deletions

1. Prepare electrocompetent cells from the markerless mutant. It is essential that the kanamycin resistance cassette and pQL123 have been removed before this step.
2. Transform with pAR1219. Prepare electrocompetent cells from the transformed strain.
3. Proceed with the mutagenesis protocol in Subheading 3.2, this time targeting the second gene of interest.

In a mutant with two or more loxP sites (e.g., a double or triple mutant), there are two possible outcomes for Cre/lox recombination. The first outcome is the kanamycin resistance cassette will be removed from the most recent mutation site (Fig. 2e). The second possibility is recombination will occur between the first and second mutated genes, resulting in a deletion of the entire intervening sequence (Fig. 2f). The likelihood of the second result increases if the two mutated genes are close to each other. However, outgrowth of a deletion mutant will only occur if intervening genes are not essential.

4 Notes

1. Higher concentrations of NaCl, common in many LB preparations, will cause most *P. mirabilis* isolates to swarm and therefore make isolation of individual colonies impossible.
2. Prepare antibiotic stock solutions at 1000× concentration, filter-sterilize, portion into 1 mL aliquots, and store at −20 °C. Chloramphenicol is prepared in ethanol.
3. A *P. mirabilis* strain with intermediate levels of antibiotic resistance may still work; for example, although type strain HI4320 has low-level chloramphenicol resistance, pACD4K-C can still be selected in this strain using 20 μg/mL of chloramphenicol. Alternative plasmid sources of T7 polymerase and Cre recombinase that do not use ampicillin selection may be available; alternatively, antibiotic markers may need to be switched for compatible markers for a given *P. mirabilis* strain. It is also possible that a given plasmid is incompatible with the target *P. mirabilis* strain. *See* **Note 7** for an alternative method for *P. mirabilis* that is incompatible with T7 helper plasmid pAR1219.

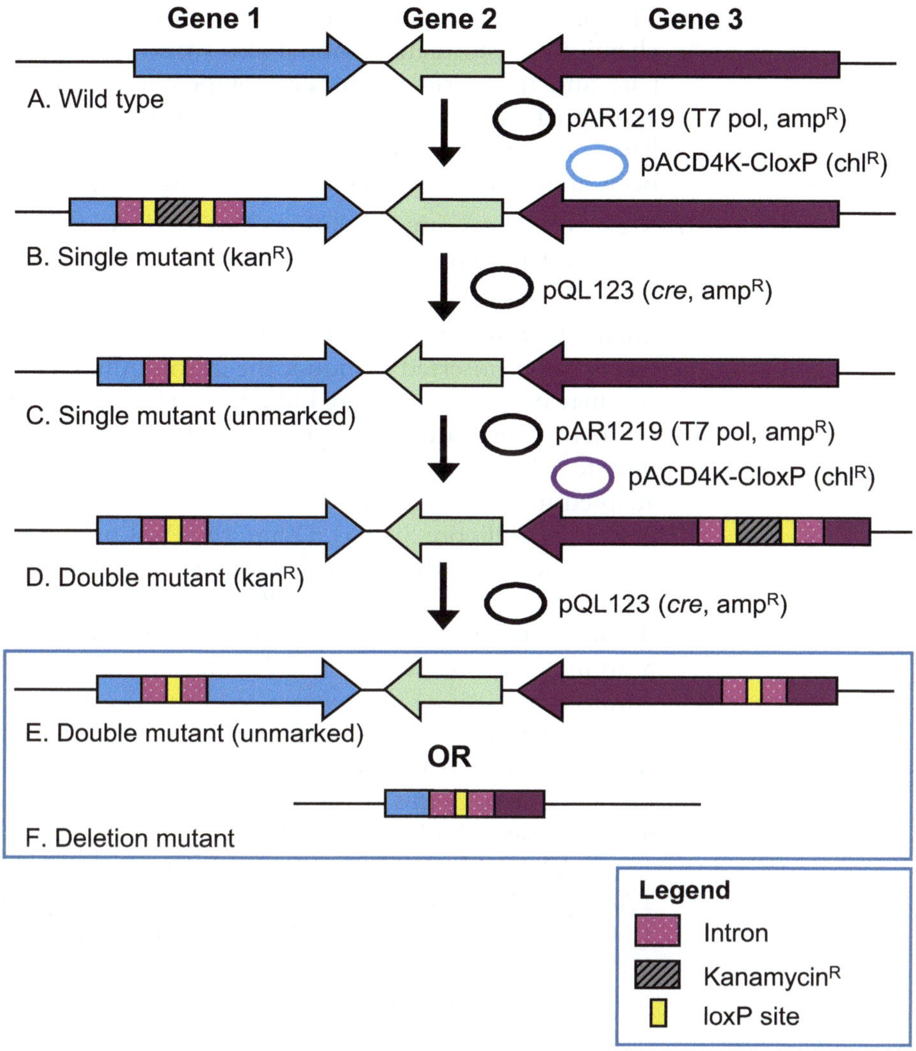

Fig. 2 Construction of double mutants or deletions using Cre/lox. A reprogrammed targetron is used to transform wild-type *P. mirabilis* (**a**) and induced to insert into gene 1, resulting in a single, kanamycin-resistant mutant (**b**). Cre recombinase is used to remove the antibiotic resistance marker, leaving behind the remainder of the targetron and a single loxP site (**c**). The unmarked mutant is then transformed with a second targetron that has been reprogrammed to insert into gene 3; induction of the intron results in a double mutant (**d**). A subsequent round of Cre/lox recombination has two possible outcomes: either a markerless double mutant (**e**) or a deletion of the intervening sequence (**f**)

4. Non high-fidelity (HF) versions of BsrGI and HindIII can be used. However, the original (non-HF) version of BsrGI requires incubation at 60 °C. If using this thermophilic version of the enzyme, first digest the plasmid with HindIII for 30 min at 37 °C, then add BsrGI and incubate at 60 °C for an additional 30 min.

5. This protocol was developed using electrocompetent cells, but chemical transformation should also work. In that event, prepare and transform chemically competent bacteria using an appropriate protocol. Chemical transformation and mating have also been used to transform *P. mirabilis*, but have not been directly tested with targetron mutagenesis.
6. X-gal is light sensitive. Solutions that have degraded will take on a brownish color.
7. For *P. mirabilis* strains (or non-DE3 *E. coli* strains) that are unable to take up the pAR1219 plasmid, an alternative promoter can be inserted to induce the intron. Alternative promoters may or may not require induction. This method using the synthetic constitutive Em7 promoter (Invitrogen) was used to mutate a *P. mirabilis* isolate from the mouse gut [12] and was adapted from a TargeTron Gene Knockout System alternative protocol (Millipore Sigma). Dilute the PCR product from the initial reaction (Subheading 3.2, **steps** 7 and **8**) 100-fold for use as a template in a second PCR reaction listed below. Design the alternative promoter (e.g., Em7) with the following modification, and order this oligonucleotide with PAGE purification:

   5′-AAA AAA GCT T-(~60 bp promoter sequence)-AAA AGA GCT TAT AAT TAT CCT TA-3′.

   For example, this is the modified constitutive Em7 promoter primer (Em7 underlined):

   5′-AAA AAA GCT T<u>CC CAT GGA CGT GTT GAC AAT TAA TCA TCG GCA TAG TAT ATC GGC ATA GTA TAA TAC GAC TCA GGG CC</u>A AAA GAG CTT ATA ATT ATC CTT A-3′.

   Set up this PCR reaction:

22 μL	dH_2O
1 μL	Alternative promoter primer (10 μM)
1 μL	EBS1d primer (10 μM)
1 μL	PCR product from initial PCR (100-fold dilution)
25 μL	2× PCR master mix
50 μL	Total

   The cycle conditions are the same as in Subheading 3.2, **step 8**. Then, continue with steps listed for Subheading 3.2, **step 9** to Subheading 3.4, **step 4**. In Subheading 3.4, **step 3**, omit addition of 50 μg/mL of ampicillin. After Subheading 3.4, **step 4**, spin down the overnight culture and resuspend in 150 μL of LB. Plate the entire 150 μL on LB supplemented with 25 μg/mL of kanamycin. Complete the protocol beginning at Subheading 3.4, **step 13**.

8. Further information about the intron, including additional suggestions and troubleshooting, may be found in both the Sigma TargeTron User Guide and at http://clostron.com/clostron2.php.
9. pQL123 may have low yields from minipreps. Therefore, a larger culture volume (e.g., 10 mL of overnight culture) may produce better results.
10. Electrocompetent *P. mirabilis* can be made by following standard *E. coli* protocols, such as Protocol 26 in *Molecular Cloning* [13]. Briefly, chill a logarithmic-phase culture on ice, collect the bacteria by centrifugation, and resuspend in 0.5× volume of ice-cold sterile purified water. Centrifuge, resuspend in 0.5× volume (i.e., half of the volume of the water used in the previous step) of ice-cold 10% glycerol, and repeat centrifugation/resuspension in half volumes until the culture is a concentrated slurry of bacteria. Aliquot in single-use 50 μL portions and freeze at −80 °C until ready to use.
11. Millipore Sigma has a proprietary algorithm for identifying potential insertion sites. Use of the algorithm is dependent on purchase of the TargeTron Gene Knockout System.
12. Primers that have only been purified by desalting will sometimes work. However, longer primers without extra purification are much more likely to contain errors that will reduce the likelihood of success.
13. Alternatively, have the entire ~350 bp reprogrammed intron fragment synthesized. This is an increasingly attractive option as the price of DNA synthesis falls. If using this approach, skip to Subheading 3.3, **step 10**. The digested fragment should be cleaned up (e.g., DNA-binding column or n-butanol precipitation), but does not need to be gel-purified before ligation with pACD4K-CloxP. The primers from Subheading 3.2, **step 4** will still need to be designed to confirm targetron insertion.
14. The intron template and 2× PCR master mix are included with the TargeTron Gene Knockout System. Consult reference [5] for more information about the intron template. The 2× PCR master mix includes DNA polymerase, dNTPs, buffer, and dye.
15. Subheading 3.2, **steps 11–13**, and Subheading 3.3 are a departure from the original Sigma protocol, which uses a direct digestion of the PCR amplicon before cloning into pACD4K-C. Omission of this modified protocol may lead to poor results in *P. mirabilis*. PCR reprogramming and restriction digestion of the linear PCR product are not always successful. The extra steps in this modified protocol both increase the success rate

and provide additional data for troubleshooting if no mutants are obtained. In addition, the purified plasmid that results from the modified protocol can be stored and used to generate mutants in other strains of *P. mirabilis*, provided the target sequence is sufficiently similar.

16. Several commercial PCR cloning vectors allow for blue–white screening via *lacZ* α-complementation. In this case, include X-gal in the agar to facilitate identification of colonies that contain the intron fragment.
17. If colony PCR is not working well, a crude lysate may provide a more reliable DNA template. Use a pipet tip to transfer a small portion of bacteria (approximately the amount from a colony) to 50 μL of water in a 0.5 mL centrifuge tube. Pipet up and down to thoroughly suspend the bacteria. Incubate bacteria at 95 °C for 10 min. Spin the tube in a benchtop microcentrifuge at full speed for 1 min. Remove 25 μL of the supernatant to a new tube and discard the rest. Use 1 μL of lysate as a template in the PCR reaction.
18. The PCR reaction frequently yields imperfectly retargeted products. It is best to obtain a plasmid that has been perfectly retargeted, but imperfect plasmids may still be able to direct the intron into the target gene. Specifically, errors in EBS1d and EBS2 may interfere with intron targeting, but the IBS region may be more tolerant of errors. Errors within a few bases (3–4) of the HindIII and BsrGI restriction sites, in IBS and EBS1d, respectively, may also be tolerated (*see* **Note 13**).
19. Butanol precipitation may be used in place of a spin column. Add 30 μL of water and 500 μL of n-butanol to DNA. Centrifuge in a benchtop microfuge at full speed for 10 min. Discard the supernatant and tap the tube on a paper towel to remove excess liquid. Add 500 μL of ice-cold 70% ethanol and centrifuge at full speed for 5 min. Remove the supernatant and air-dry the pellet. Add competent cells to the dried pellet and proceed with transformation.
20. X-gal is included here for blue–white screening because the linearized pACD4K-C vector provided in the TargeTron kit also includes a *lacZ* fragment that can be inadvertently recloned during ligation. Any blue colonies resulting from this step contain *lacZ*, while colonies that contain the retargeted intron fragment will be white or light blue.
21. Some retargeting constructs grow very slowly during this step. If this happens, continue culturing until the OD_{600} is at least 0.4. Mutants have been obtained from cultures that took 5–6 h to grow up during this step. This is especially likely to occur if the overnight culture is not as dense as typically expected after

overnight incubation, but can occur even if the overnight culture appears normal.

22. Large, healthy colonies after overnight incubation usually contain the desired mutation. If no colonies or very small colonies are visible after a 15–16 h incubation, keep the plate in the incubator for up to 24 h to see if any more colonies grow up. These little colonies will frequently not grow after patching onto agar containing kanamycin, and typically yield a wild-type PCR amplicon during screening. Even though they are usually not correct (provided that a growth defect is not anticipated), they are still worth screening, just in case.
23. If no mutants are obtained, there are several options. First, if a targetron plasmid was used that was not perfectly retargeted (*see* **Note 17**), sequence more plasmids to find one that is completely correct, or try the alternative suggested in **Note 13**. Second, order another set of retargeting primers (IBS, EBS1d, and EBS2) to target a different site in the target gene. Some retargeting plasmids are more efficient than others. Third, it is worth repeating the protocol in Subheading 3.4 with the original retargeting plasmid; if the retargeting is inefficient, the mutant may be obtained on the second or third attempt. Fourth, for a *P. mirabilis* strain that has not been used with this method before, ensure the strain is sensitive to all antibiotics used in this protocol, at the specified concentrations. Last, consider whether the target gene might be essential in *P. mirabilis*.
24. Usually a single passage in LB without selection is sufficient to lose pAR1219. However, if no patches are ampicillin-sensitive, restreak one patch from the LB kanamycin plate and repeat **steps 17** and **18**. If the problem persists, the patch may be cultured in broth overnight, dilution plated to obtain individual colonies, and then double-patched onto LB with either ampicillin or kanamycin.
25. The mutant must be cured of pAR1219 (i.e., ampicillin-sensitive) before proceeding (previous section, **steps 17** and **18**).

Acknowledgments

Thank you to Madison Fitzgerald for field-testing this protocol. This work was supported by the National Institutes of Health [R01 AI059722 to H.L.T.M.].

References

1. Armbruster CE, Mobley HLT, Pearson MM (2018) Pathogenesis of *Proteus mirabilis* infection. EcoSal Plus 8(1). https://doi.org/10.1128/ecosalplus.ESP-0009-2017
2. Schaffer JN, Pearson MM (2015) *Proteus mirabilis* and urinary tract infections. Microbiol Spectr 3(5). https://doi.org/10.1128/microbiolspec.UTI-0017-2013
3. Jones BD, Lockatell CV, Johnson DE, Warren JW, Mobley HLT (1990) Construction of a urease-negative mutant of *Proteus mirabilis*: analysis of virulence in a mouse model of ascending urinary tract infection. Infect Immun 58(4):1120–1123
4. Pearson MM, Mobley HLT (2007) The type III secretion system of *Proteus mirabilis* HI4320 does not contribute to virulence in the mouse model of ascending urinary tract infection. J Med Microbiol 56 (Pt 10):1277–1283
5. Karberg M, Guo H, Zhong J, Coon R, Perutka J, Lambowitz AM (2001) Group II introns as controllable gene targeting vectors for genetic manipulation of bacteria. Nat Biotechnol 19(12):1162–1167. https://doi.org/10.1038/nbt1201-1162
6. Perutka J, Wang W, Goerlitz D, Lambowitz AM (2004) Use of computer-designed group II introns to disrupt *Escherichia coli* DExH/D-box protein and DNA helicase genes. J Mol Biol 336(2):421–439
7. Zhong J, Karberg M, Lambowitz AM (2003) Targeted and random bacterial gene disruption using a group II intron (targetron) vector containing a retrotransposition-activated selectable marker. Nucleic Acids Res 31(6):1656–1664
8. Pearson MM, Rasko DA, Smith SN, Mobley HLT (2010) Transcriptome of swarming *Proteus mirabilis*. Infect Immun 78 (6):2834–2845. https://doi.org/10.1128/IAI.01222-09
9. Enyeart PJ, Chirieleison SM, Dao MN, Perutka J, Quandt EM, Yao J, Whitt JT, Keatinge-Clay AT, Lambowitz AM, Ellington AD (2013) Generalized bacterial genome editing using mobile group II introns and Cre-lox. Mol Syst Biol 9:685. https://doi.org/10.1038/msb.2013.41
10. Liu Q, Li MZ, Leibham D, Cortez D, Elledge SJ (1998) The univector plasmid-fusion system, a method for rapid construction of recombinant DNA without restriction enzymes. Curr Biol 8(24):1300–1309
11. Heap JT, Kuehne SA, Ehsaan M, Cartman ST, Cooksley CM, Scott JC, Minton NP (2010) The ClosTron: mutagenesis in *Clostridium* refined and streamlined. J Microbiol Methods 80(1):49–55. https://doi.org/10.1016/j.mimet.2009.10.018
12. Seo SU, Kamada N, Muñoz-Planillo R, Kim YG, Kim D, Koizumi Y, Hasegawa M, Himpsl SD, Browne HP, Lawley TD, Mobley HLT, Inohara N, Núñez G (2015) Distinct commensals induce interleukin-1b via NLRP3 inflammasome in inflammatory monocytes to promote intestinal inflammation in response to injury. Immunity 42(4):744–755. https://doi.org/10.1016/j.immuni.2015.03.004
13. Sambrook J, Russell DW (2001) Molecular cloning: a laboratory manual, 3rd edn. Cold Spring Harbor Laboratory Press, Cold Spring Harbor, NY

Chapter 8

Allelic Exchange Mutagenesis in *Proteus mirabilis*

Kristen E. Howery and Philip N. Rather

Abstract

This chapter outlines a method for making unmarked, in-frame deletion mutations in *Proteus mirabilis* by allelic replacement. This method utilizes an R6K-based suicide plasmid allowing for integration of the plasmid by homologous recombination via a cloned insert. The plasmid also contains the *sacB* gene to select for plasmid loss (excision) in the presence of sucrose to create a mutant allele. This method has been applied to the *P. mirabilis* strains PM7002 and BB2000 and should be generally applicable to other *P. mirabilis* strains. The same methods described in this chapter can be used to introduce marked or point mutations in a given gene.

Key words *Proteus mirabilis*, Allelic replacement, Suicide plasmid, Swarming

1 Introduction

Proteus mirabilis is a Gram-negative bacterium and a leading cause of urinary tract infections in patients undergoing extensive catheterization. This bacterium is noted for its robust swarming motility on agar surfaces, where it forms a characteristic bull's-eye pattern [1, 2]. This pattern is formed by the cyclic differentiation of short vegetative cells into elongated swarmer cells that migrate on surfaces, followed by dedifferentiation (consolidation) back to vegetative cells. Each cycle of differentiation/dedifferentiation results in a swarming ring on agar surfaces. Swarming is a remarkably complex process, and a large number of genes have been identified that control both swarmer cell differentiation and the cell–cell interactions required for swarming [1, 3–5].

The ability of *P. mirabilis* to cause disease can be attributed to a large number of virulence genes [3, 4]. These genes have predominantly been identified by genetic approaches such as transposon mutagenesis. Subsequent methods to address the function of a specific gene have been done by targeted mutagenesis using either a Group-II intron based strategy to program insertions (*see* TargeTron, Chap. 7 of this volume) [6] or by allelic replacement using an

Melanie M. Pearson (ed.), *Proteus mirabilis: Methods and Protocols*, Methods in Molecular Biology, vol. 2021,
https://doi.org/10.1007/978-1-4939-9601-8_8,

R6K-based suicide plasmid, pKNG101, that lacks a functional *pir* gene, encoding the key replication protein π [7, 8]. This plasmid is unable to replicate unless the *pir* gene is provided *in trans* and can be mobilized into *P. mirabilis* from *Escherichia coli* strains such as SM10 λ*pir* or S17-1λ*pir* that provide both the RP4 mobilization functions and the *pir* gene for replication [8]. The R6K-based system has been used for allelic replacement in *P. mirabilis* strains BB2000 [9], PM7002 [10], and HI4320 [11]. However, it should be noted that HI4320 contains an R6K-like plasmid that may limit the effectiveness of R6K based suicide plasmids [4, 12]. Although HI4320 often is recalcitrant to this method, Campbell-type insertions have been constructed in HI4320 with R6K-based suicide plasmids [11, 13]. Depending on the antibiotic resistance present in the strain of interest, the methods outlined below should be applicable to a wide variety of *P. mirabilis* strains.

2 Materials

2.1 Bacterial Strains and Plasmids

1. *Proteus mirabilis* American Type Culture Collection (ATCC) strain 7002 (PM7002) (*see* **Note 1**).
2. *E. coli* CC118 λ*pir*.
3. *E. coli* SM10 λ*pir*.
4. *E. coli* DH5α.
5. Plasmid pBluescript KS(−); maintain using ampicillin.
6. Plasmid pKNG101; maintain using streptomycin.

2.2 Bacterial Growth Media

1. Modified Luria–Bertani (LB) broth (per liter): 10 g of tryptone, 5 g of yeast extract, 5 g of NaCl.
2. Modified LB plates containing 1.5% agar for *E. coli* and 3.0% agar for *P. mirabilis* (to prevent swarming).
3. LSW agar (per liter): 10 g of tryptone, 5 g of yeast extract, 5 mL of glycerol, 0.4 g of NaCl, and 20 g of agar.
4. Antibiotic selection: for *E. coli*, 25 μg/mL of streptomycin, 200 μg/mL of ampicillin and 25 μg/mL of chloramphenicol. For *P. mirabilis*: 35 μg/mL of streptomycin, 15 μg/mL of tetracycline, and 20 μg/mL of kanamycin. Tetracycline and chloramphenicol are prepared in 100% ethanol. For agar plates, autoclave agar to sterilize, cool to 55 °C, then add antibiotics.
5. LSW agar plates containing 10% sucrose.
6. LB agar plates with ampicillin and X-Gal: add 60 μL of 20 mg/mL 5-bromo-4-chloro-indolyl-β-D-galactopyranoside (X-Gal) prepared in *N*,*N*-dimethylformamide to LB ampicillin agar plates before the agar has solidified.

2.3 Electroporations

1. 0.2 cm diameter cuvettes.
2. Electroporator (e.g., Bio-Rad MicroPulser) set at 2.5 mV.
3. 10% ice-cold glycerol.
4. Fresh LB medium.

2.4 Miscellaneous Reagents and Equipment

1. T4 polynucleotide kinase.
2. DNA plasmid purification kit.
3. DNA gel purification kit.
4. High-fidelity DNA polymerase, PCR reagents, and thermocycler.
5. DNA ligation kit.
6. Tris–acetate–EDTA (TAE) buffer. A 50× stock is prepared as follows (per liter): 242 g of Tris base, 57 mL of glacial acetic acid, and 100 mL of 500 mM EDTA solution.
7. 1% agarose gel in TAE buffer and electrophoresis equipment.
8. Primers to create ~1000 base pair fragments upstream and downstream of the gene of interest.
9. Primers to screen for the deletion (located upstream and downstream of the deleted gene).
10. Restriction enzymes: EcoRV, XbaI, SalI (*see* **Note 2**).
11. 37 °C incubators, static and shaking.

3 Methods

3.1 Construction of Unmarked In-Frame Deletions

1. To generate markerless, in-frame deletion mutants in *P. mirabilis*, use PCR to generate an approximately 1000 base pair fragment containing the first few codons of the gene and upstream flanking DNA (IFD1: in-frame deletion 1). In addition, use PCR to generate a second fragment of approximately the same size, which contains the last few codons of the gene and downstream flanking DNA (IFD2: in-frame deletion 2).
   (a) Use a high-fidelity DNA polymerase such as Pfu (do not use Taq polymerase) for this amplification in order to minimize mutations. In addition, this will generate blunt ends for subsequent ligations.
   (b) The reverse primer used to amplify IFD1 and the forward primer for IFD2 should be 5′-phosphorylated using T4 polynucleotide kinase using standard procedures provided by the manufacturer.
2. Following amplification of IFD1 and IFD2, confirm the size and specificity of the PCR product on a 1.0% TAE agarose gel.
3. Gel purify IFD1 and IFD2.

4. Ligate IFD1 and IFD2 at a 1:1 ratio; the concentration of each should be around 100 ng. Run the resulting ligation reaction on a 1.0% TAE agarose gel.
5. Gel purify the fragment corresponding to the ligation of IFD1 and IFD2 (approximately 2000 base pairs). Use a small volume in a PCR reaction with the IFD1 forward primer and the IFD2 reverse primer in order to amplify the ligation product in the correct orientation.
6. Gel purify the PCR-amplified ligation product of IFD1 and IFD2 (IFD: in-frame deletion) for use in cloning.

3.2 Cloning into pBluescript KS(−)

1. Clone the IFD PCR fragment into the EcoRV site of pBluescript KS(−).
2. Prepare *E. coli* DH5α host cells for electroporation by growing to mid-logarithmic phase in 30 mL of LB broth. Centrifuge cells at 5000 × *g* for 5 min at 4 °C.
3. Following centrifugation, resuspend DH5α cells in 1 mL of ice-cold 10% glycerol, transfer to a 1.5 mL microcentrifuge tube and centrifuge for 5 min at 12,000 × *g*.
4. Remove the supernatant and rewash the cells in the same manner two additional times.
5. Add 1 μL of the ligation reaction to 60 μL of cells and transfer to an ice-cold 0.2-cm diameter electroporation cuvette. To a separate cuvette, add cells without the ligation reaction as a control.
6. Electroporate cells and remove from the cuvette using 100 μL of fresh LB broth and add to 1 mL of LB broth. Incubate cells by shaking at 250 rpm for 1.5 h at 37 °C.
7. Plate cells on 1.5% LB plates containing ampicillin and X-Gal and incubate overnight at 37 °C. As this transformation produces an abundance of colonies, plate a small volume of the transformation (i.e., 50, 100, and 200 μL) and place the remaining cells at 4 °C overnight in case more cells need to be plated.
8. After overnight incubation, select white colonies and grow overnight in 2 mL of LB with ampicillin shaking at 200 rpm at 37 °C.
9. Prepare plasmid DNA and screen for the correct insertion by restriction enzyme digestion using enzymes that flank both sides of the EcoRV site.

3.3 Subcloning into pKNG101

1. Digest pBluescript KS(−) plasmids which contain the IFD insertion with XbaI and SalI (unless these restriction enzymes are also specific to sites in the IFD fragment, in which case choose another combination of enzymes).

2. Ligate the XbaI/SalI-digested IFD fragment to XbaI/SalI digested pKNG101.
3. For subsequent electroporations, grow *E. coli* CC118 λ*pir* to an $OD_{600} = 0.5$ in 30 mL of LB broth. Pellet cells by centrifugation at 5000 × *g* for 10 min at 4 °C.
4. Following centrifugation, resuspend the CC118 λ*pir* cell pellet in 1 mL of ice-cold 10% glycerol, transfer to a 1.5 mL microcentrifuge tube and centrifuge for 1 min at 11,000 × *g* at room temperature. Remove the supernatant and rewash the cells two additional times in the same manner. Resuspend the final pellet in 250 μL of 10% glycerol.
5. Add 1 μL of the ligation reaction to 60 μL of cells and transfer to an ice-cold 0.2-cm diameter electroporation cuvette. To a separate cuvette, add cells without the ligation reaction as a control.
6. Electroporate cells using 2.5 mV and remove from the cuvette using fresh LB broth. Add cells to 1 mL of LB broth and incubate by shaking at 250 rpm for 1.5 h at 37 °C.
7. Plate cells on 1.5% agar LB plates containing 25 μg/mL of streptomycin and incubate overnight at 37 °C.
8. Miniprep plasmids from individual colonies and screen for the IFD fragment by restriction analysis by a double digestion with XbaI and SalI.
9. Transform a pKNG101 derivative confirmed to have the IFD insertion into SM10 λ*pir* cells by adding 1 μL of the plasmid to 60 μL of SM10 λ*pir* cells prepared for transformation as described in Subheading 3.2, **step 5** (*see* **Note 3**).
10. Recover SM10 λ*pir* cells transformed with pKNG101-IFD for 1.5 h at 37 °C with shaking at 250 rpm. Streak-plate 200 μL and 20 μL of the transformed cells onto a 1.5% LB agar plate containing 25 μg/mL streptomycin and incubate overnight at 37 °C.
11. After overnight incubation, culture an individual SM10 λ*pir* pKNG101-IFD colony in 2 mL of LB broth overnight at 37 °C with shaking at 200 rpm.

3.4 Introducing pKNG101 Derivatives into P. mirabilis by Conjugation

1. Culture an individual SM10 λ*pir* pKNG101-IFD colony overnight in 2 mL of LB with 25 μg/mL of streptomycin at 37 °C with shaking at 200 rpm.
2. Culture the *P. mirabilis* strain to be used (e.g., PM7002) overnight in LB broth with shaking at 200 rpm.
3. Prepare overnight cultures of SM10 λ*pir* pKNG101-IFD and PM7002 for conjugation by removing 1 mL of each culture

and centrifuging the cells in a microcentrifuge at 11,000 × *g* at room temperature for 1 min.

4. Remove the supernatant and resuspend the cells in 1 mL of fresh LB. Wash the cells two additional times with LB.
5. Combine 100 μL of the washed PM7002 cells and 100 μL of the washed SM10 *λpir* pKNG101-IFD cells in a microcentrifuge tube.
6. Add the entire volume (200 μL) of the combined cells to a 1.5% LB plate that has been predried for 2 h by incubating inverted in a 37 °C incubator. Cells should be spotted in as small a surface area as possible, as conjugation is facilitated at high cell density. As controls, individually add 100 μL of the PM7002 cells and the SM10 *λpir* pKNG101-IFD cells alone to separate 1.5% LB agar plates. Incubate all plates at 37 °C for 7 h.
7. Following the conjugation, remove the cells by washing the plate with 4 mL of LB broth.
8. Plate cells on 3.0% LB agar plates containing 35 μg/mL of streptomycin (to select for pKNG101 plasmid integration) and 10 μg/mL of tetracycline (to counterselect the *E. coli* SM10 *λpir* donor) in volumes of 300, 200, and 100 μL. In addition, plate 300 μL of the individual cell donor and recipient cells as controls (*see* **Note 4**).
9. Incubate plates overnight at 37 °C.

3.5 Use of Sucrose Media to Select for Plasmid Excision

1. Restreak streptomycin-resistant PM7002 colonies arising from the above mating onto 3.0% LB agar containing 35 μg/mL of streptomycin and incubate overnight at 37 °C.
2. Restreak individual streptomycin-resistant PM7002 ex-conjugants onto LSW agar plates containing streptomycin and 10% sucrose to confirm sucrose sensitivity, which is more pronounced on LSW plates. Incubate plates overnight at room temperature for 2 days to confirm sucrose sensitivity (*see* **Note 5**).
3. Colonies that are unable to grow on streptomycin/sucrose plates should contain the integrated pKNG101 plasmid. Culture these cells to mid-logarithmic phase in LB broth without antibiotic selection to allow for the loss of pKNG101.
4. Serially dilute the LB culture in tenfold aliquots from 10^{-1} to 10^{-6} and plate on LSW agar and LSW agar containing 10% sucrose. Incubate plates for 2 days at room temperature. There should be at least a tenfold decrease in the number of colonies on the sucrose plate compared with the plate lacking sucrose.
5. Restreak sucrose-resistant ex-conjugants onto LSW/sucrose to confirm sucrose resistance.

3.6 Confirmation of In-Frame Deletions

1. Culture cells from a sucrose-resistant colony overnight in LB broth. Wash a small aliquot of cells (30–50 μL) with deionized water twice and boil for 10 min for use in colony PCR.
2. Perform PCR using primers up- and downstream of the deleted gene to confirm whether the wild-type copy or the deleted copy of the gene remained in the genome following ex-conjugation.
3. Run the colony PCR on a 1.0% agarose gel. Approximately 50% of colonies will yield wild-type results and 50% of colonies will contain the deleted gene. Freeze colonies confirmed to contain the deletion at −80 °C in 20% glycerol.

4 Notes

1. PM7002 and many other *P. mirabilis* strains are resistant to tetracycline, which works well for counterselecting the donor strain during conjugation. However, tetracycline resistance should be checked for each strain before starting.
2. The insert to be cloned should be checked for the absence of these sites.
3. The electroporation frequency of *E. coli* SM10 *λpir* is very low. Only use this strain as a host when pure plasmid DNA is used. All other cloning procedures should be done using CC118 *λpir*.
4. There can occasionally be background colonies of *P. mirabilis* on these plates. However, the number of colonies on the *P. mirabilis* only plates should be far lower than on the mating plates. For the *E. coli* only cells, background colonies are rarely observed.
5. For reasons that are unclear, we have observed that sucrose selection works better at room temperature than at 37 °C. However, on rare occasions, sucrose selection at 37 °C was required to make certain mutations. We always use room temperature first and then 37 °C as a last resort.

Acknowledgments

P.N.R. is supported by the following grants: VA Merit Award I01 BX001725, Research Career Scientist Award IK6BX004470, both from the Department of Veterans Affairs, and R01AI072219 from the National Institutes of Health.

References

1. Mobley HLT, Belas R (1995) Swarming and pathogenicity of *Proteus mirabilis* in the urinary tract. Trends Microbiol 3(7):280–284
2. Rauprich O, Matsushita M, Weijer CJ, Siegert F, Esipov SE, Shapiro JA (1996) Periodic phenomena in *Proteus mirabilis* swarm colony development. J Bacteriol 178 (22):6525–6538
3. Armbruster CE, Mobley HLT (2012) Merging mythology and morphology: the multifaceted lifestyle of *Proteus mirabilis*. Nat Rev Microbiol 10(11):743–754. https://doi.org/10.1038/nrmicro2890
4. Armbruster CE, Mobley HLT, Pearson MM (2018) Pathogenesis of *Proteus mirabilis* Infection. EcoSal Plus 8(1). https://doi.org/10.1128/ecosalplus.ESP-0009-2017
5. Morgenstein RM, Szostek B, Rather PN (2010) Regulation of gene expression during swarmer cell differentiation in *Proteus mirabilis*. FEMS Microbiol Rev 34(5):753–763. https://doi.org/10.1111/j.1574-6976.2010.00229.x
6. Pearson MM, Mobley HLT (2007) The type III secretion system of *Proteus mirabilis* HI4320 does not contribute to virulence in the mouse model of ascending urinary tract infection. J Med Microbiol 56 (Pt 10):1277–1283. https://doi.org/10.1099/jmm.0.47314-0
7. Kaniga K, Delor I, Cornelis GR (1991) A wide-host-range suicide vector for improving reverse genetics in gram-negative bacteria: inactivation of the *blaA* gene of *Yersinia enterocolitica*. Gene 109(1):137–141
8. Kolter R, Inuzuka M, Helinski DR (1978) Trans-complementation-dependent replication of a low molecular weight origin fragment from plasmid R6K. Cell 15(4):1199–1208
9. Saak CC, Gibbs KA (2016) The self-identity protein IdsD is communicated between cells in swarming *Proteus mirabilis* colonies. J Bacteriol 198(24):3278–3286. https://doi.org/10.1128/jb.00402-16
10. Morgenstein RM, Rather PN (2012) Role of the Umo proteins and the Rcs phosphorelay in the swarming motility of the wild type and an O-antigen (*waaL*) mutant of *Proteus mirabilis*. J Bacteriol 194(3):669–676. https://doi.org/10.1128/jb.06047-11
11. Jones BD, Lockatell CV, Johnson DE, Warren JW, Mobley HLT (1990) Construction of a urease-negative mutant of *Proteus mirabilis*: analysis of virulence in a mouse model of ascending urinary tract infection. Infect Immun 58(4):1120–1123
12. Pearson MM, Sebaihia M, Churcher C, Quail MA, Seshasayee AS, Luscombe NM, Abdellah Z, Arrosmith C, Atkin B, Chillingworth T, Hauser H, Jagels K, Moule S, Mungall K, Norbertczak H, Rabbinowitsch E, Walker D, Whithead S, Thomson NR, Rather PN, Parkhill J, Mobley HLT (2008) Complete genome sequence of uropathogenic *Proteus mirabilis*, a master of both adherence and motility. J Bacteriol 190(11):4027–4037. https://doi.org/10.1128/jb.01981-07
13. Howery KE, Clemmer KM, Simsek E, Kim M, Rather PN (2015) Regulation of the Min cell division inhibition complex by the Rcs phosphorelay in *Proteus mirabilis*. J Bacteriol 197 (15):2499–2507. https://doi.org/10.1128/jb.00094-15

Chapter 9

Quantification of Urease Activity

Shawn Richmond and Alejandra Yep

Abstract

Urease is one of the most distinctive virulence factors of *Proteus mirabilis* pathogenesis. Urease activity correlates with many landmark side effects of *P. mirabilis* catheter-associated urinary tract infections, such as urolithiasis and bacteremia. Here we describe two simple and inexpensive colorimetric methods for quantifying urease activity in single species cultures as well as cocultures.

Key words *Proteus mirabilis*, Urease activity, Berthelot, Phenol red

1 Introduction

Proteus mirabilis is a Gram-negative environmental and gastrointestinal commensal bacterium associated with a variety of human infections, but known mostly for its role in catheter-associated urinary tract infections (CAUTI) [1]. Among the many virulence factors of *P. mirabilis*, its potent cytoplasmic urease enzyme stands out in a central role.

Bacterial ureases are high molecular weight heteromultimeric nickel metalloenzymes with interesting active site chemistry. Urease has been thoroughly studied for decades, being both the first enzyme to be crystallized [2] and the first one shown to incorporate a nickel metallocenter [3]. Ureases catalyze the conversion of urea to ammonia and carbamic acid, which is eventually decomposed into a second molecule of ammonia and carbonic acid. Although the action of urease supplies bacteria with a readily available nitrogen source, this may not be the most important contribution of urease to *P. mirabilis* pathogenesis. The concomitant rise of pH generated by this increase in ammonia concentration is, however, central to uropathogenesis, as it has been linked to survival in the bladder [4], urinary stone formation [5], uroepithelial damage [6], access to kidneys and bloodstream [6, 7], and persistence [8], in addition to formation of crystalline biofilms [9] followed by catheter encrustation and blockage [10, 11].

Melanie M. Pearson (ed.), *Proteus mirabilis: Methods and Protocols*, Methods in Molecular Biology, vol. 2021,
https://doi.org/10.1007/978-1-4939-9601-8_9,

As we progress in elucidating the multiple roles of urease in colonization and invasion, urease activity methods have been adapted for multiple applications. These methods rely on detection of the ammonia product [12, 13] or on quantifying the increase in pH resulting from this enzymatic activity [14, 15]. In this chapter, we describe two inexpensive colorimetric methods that can be employed for measuring urease activity of *P. mirabilis* or other urease-producing bacteria such as *Providencia stuartii*, *Morganella morganii*, and others. The first method is based on a modified version of the Berthelot reaction for ammonia quantification [12] adapted for urease activity [16]. This is an endpoint assay where cells are grown in the desired medium inducing urease expression, then lysed and the activity of intracellular urease present in the cell-free extract is assayed. The second method is a continuous assay that records the urease activity of live *P. mirabilis* cells in urine as they induce a pH change in the medium that leads to changes in absorbance of the phenol red indicator [14]. As the consequences of polymicrobial interactions in *Proteus* pathogenesis are unveiled [17, 18], determination of urease activity in mixed cultures becomes necessary. We have successfully employed both methods to measure the enhancement of urease activity brought about by polymicrobial interactions between *P. mirabilis* and other urease-producing as well as non urease-producing uropathogens [17, 18].

2 Materials

2.1 Berthelot Reaction

1. Low salt LB (lsLB) broth: per liter, 10 g of tryptone, 0.5 g of NaCl, 5 g of yeast extract (*see* **Note 1**).
2. LsLB agar: per liter, 10 g of tryptone, 0.5 g of NaCl, 5 g of yeast extract, 17 g of agar (*see* **Note 1**).
3. Pooled sterilized human female urine: Collect urine from healthy nonpregnant women that are not taking antibiotics. Pool urine within 4 h of collection, and centrifuge for 15 min at 4000 × *g* to precipitate epithelial cells and mineral aggregates that could clog filters. Filter the supernatant through a 0.22 μm vacuum flask filtration apparatus to sterilize. Measure urine pH and adjust to initial pH = 6.0 with HCl or NaOH if needed (*see* **Note 2**).
4. PEM buffer (*see* **Note 3**): 20 mM sodium phosphate buffer pH 7.5, 10 mM β-mercaptoethanol, 20 mM EDTA.
5. UPEM substrate (*see* **Note 3**): 20 mM sodium phosphate buffer pH 7.5, 10 mM β-mercaptoethanol, 20 mM EDTA, 300 mM urea.
6. Liquid nitrogen, or similar method for rapid freezing.

7. Berthelot solution A: dissolve 5 g of phenol and 25 mg of sodium nitroprusside in ddH_2O for a final volume of 500 mL (*see* **Note 4**).
8. Berthelot solution B: Alkaline hypochlorite (Sigma) (0.2% sodium hypochlorite in alkali solution).
9. 100 mM NH_4Cl stock solution.
10. Purified jack bean urease (Sigma) for urease activity controls, 1 U/mL in PEM buffer (*see* **Note 5**).
11. Centrifuge and rotor suitable for 50-mL conical tubes.
12. Sonicator equipped with microtip.
13. Spectrophotometer (visible range).
14. Water bath set to 37 °C.
15. Static and shaking incubators set at 37 °C.
16. 13 × 100 mm disposable test tubes.
17. 1 or 4 mL disposable spectrophotometer cuvettes.
18. Total protein concentration assay kit (e.g., Bradford, Lowry, BCA, or other preferred method).

2.2 Phenol Red

1. Pooled sterilized human female urine as in **step 3** in Subheading 2.1 (*see* **Note 2**).
2. 0.1% (w/v) phenol red in deionized water, sterilized using a 0.22 μm syringe filter (*see* **Notes 6** and 7).
3. 5 M urea in deionized water, filter sterilized (*see* **Note** 7).
4. 0.9% saline, sterilized by autoclaving.
5. 1.5 mL sterile microfuge tubes.
6. 15- and 50-mL sterile conical tubes.
7. 96-well sterile microtiter plate with lid.
8. 25-mL sterile serological pipette.
9. Multichannel pipette (5–50 μL).
10. Water bath set at 37 °C.
11. Static and shaking incubators set at 37 °C.
12. Centrifuge and rotor suitable for 50-mL conical tubes.
13. Microcentrifuge.
14. Plate reader (visible range, capable of kinetic reads, temperature controlled).

3 Methods

3.1 Endpoint Assay of Urease Activity in Cell-Free Lysates Employing the Berthelot Reaction

3.1.1 Culture and Preparation of Cell-Free Lysate

1. Inoculate 3 mL of lsLB with an isolated *P. mirabilis* colony growing on lsLB agar (*see* **Note 1**), and incubate at 37 °C overnight with shaking at 200 rpm. Prepare cultures of other strains being tested similarly.
2. The next day, inoculate 40 mL of lsLB with 400 μL of overnight culture (1:100 dilution), and incubate at 37 °C with 200 rpm shaking until OD_{600} ~ 0.5 (typically 1.5–2.5 h). The incubation time will decrease significantly if medium is prewarmed at 37 °C prior to inoculation.
3. Remove an aliquot of the culture, determine OD_{600}, and plate appropriate dilutions for CFU/mL counts and future normalization (*see* **Note 8**).
4. Pellet cells by centrifuging for 10 min at 2500 × *g* in a 50-mL conical tube. Discard supernatant.
5. Resuspend cells in 40 mL of pooled sterilized human urine that has been prewarmed to 37 °C.
6. Incubate for 30 min at 37 °C with shaking at 200 rpm for urease induction.
7. Wash cells twice in 40 mL of PEM buffer by resuspending and pelleting again. Discard supernatant.
8. Resuspend cells by pipetting up and down in 1 mL of PEM buffer. Transfer to a 15-mL conical tube.
9. Freeze resuspended cells by dipping tube in liquid nitrogen. Thaw cells in 37 °C water bath. Keep at 37 °C, mixing frequently, only until thawed. Perform 3 cycles of freeze–thaw in this manner, vortexing vigorously after each thaw cycle (*see* **Note 9**). After this step, keep cells on ice.
10. Sonicate 3× for 10 s using a microtip at 30% amplitude or as suggested by manufacturer, allowing extract to cool on ice between sonication cycles (*see* **Note 10**).
11. Remove cell membranes by centrifuging the lysates for 5 min at 18,000 × *g* in a microcentrifuge. Promptly transfer supernatant to a clean microfuge tube. This cell-free lysate is the source of cytoplasmic urease (*see* **Note 11**).
12. Save aliquots of cell-free lysate for total protein quantification (Subheading 3.1.3). Aliquots can be kept frozen for several months (*see* **Note 11**).

3.1.2 Ammonia Calibration Curve

A calibration curve should be generated every time the assay is performed.

1. Prepare appropriate dilutions of NH_4Cl. Typically, these are 0, 0.5, 1, 2, 3, 4, and 5 mM NH_4Cl.

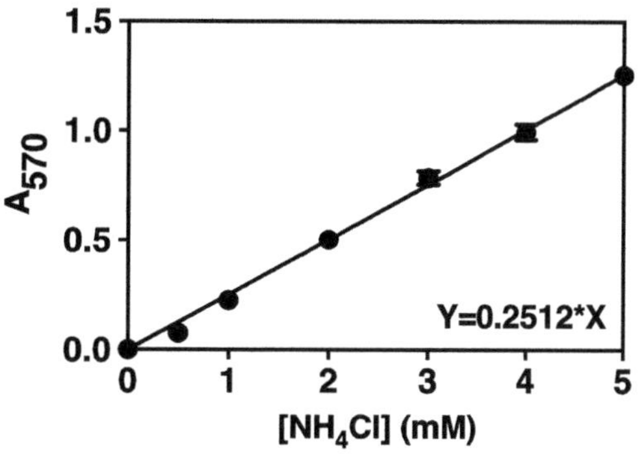

Fig. 1 Sample ammonia calibration curve for the Berthelot reaction. Assay tubes containing the indicated millimoles of NH_4Cl in 250 μL in duplicate were subjected to the Berthelot reaction as described in Subheading 3.1.2. Absorbance was read at 570 nm. Data points represent the average of two replicates, and error bars denote standard deviations. Linear regression through 0,0 was calculated using GraphPad Prism (equation shown in graph). This equation was used to calculate micromoles of ammonia produced in each reaction

2. Pipet 250 μL of each NH_4 concentration in duplicate into 13×100 mm test tubes.
3. Add 500 μL of solution A, 500 μL of solution B, and 2.5 mL of ddH_2O (*see* **Note 12**).
4. Incubate at room temperature for 5 min. The tubes will develop a blue color that correlates in intensity with the concentration of ammonia in solution (*see* **Note 13**).
5. Vortex tubes briefly to mix.
6. Transfer into 1- or 4-mL disposable cuvettes and read absorbance at 570 nm. Use the reaction with no ammonia as a blank.
7. Plot the calibration curve through 0,0 and calculate the slope. The calibration curve should be linear. Use the slope calculated in this step to convert absorbance data to ammonia concentration data (*see* Fig. 1).

3.1.3 Urease Activity Assay

1. Pipet 240 μL of UPEM substrate into a 13 × 100 mm test tube and add 10 μL of enzyme preparation (cell-free extract or purified urease control). Mix by vortexing very briefly. Incubate at room temperature for 5 min (*see* **Note 14**). If measuring urease activity of mixed cultures, *see* **Note 15**.
2. Stop the reaction by adding 500 μL of solution A, 500 μL of solution B, and 2.5 mL ddH_2O (*see* **Note 12**).
3. Incubate at room temperature for 5 min.
4. Vortex tubes briefly to mix.

5. Transfer into 1- or 4-mL disposable cuvettes and read absorbance at 570 nm (*see* **Note 13**).
6. One unit of urease activity is defined as the amount of enzyme source (purified enzyme or cell-free extract) required to release 1 μmol of ammonia after 5 min incubation at room temperature (*see* **Note 16**).

3.1.4 Normalization of Urease Activity to Total Protein Concentration of Cell-Free Extract

1. Determine total protein concentration on saved aliquots of cell-free extracts by following the manufacturer's directions for your preferred method. Cell-free extract aliquots from several time points or different experiments can be kept frozen at −80 °C, thawed on ice and assayed for protein concentration at the same time.
2. Calculate urease activity as units/mg of total protein for each cell-free extract (*see* **Note 16**).
3. Additionally, urease activity can be normalized to CFU present for each strain. This is important when activity in cocultures is determined and therefore the proportion of each strain needs to be assessed. For a description of how to calculate CFU/mL, *see* **Note 8**.

3.2 Continuous Detection of Live Cell Urease Activity Using Phenol Red

3.2.1 Cultures

1. Inoculate 3 mL of sterile lsLB broth from a single *P. mirabilis* colony growing on an lsLB plate (*see* **Note 1**). Prepare cultures of other species similarly, if testing any. Incubate cultures at 37 °C with shaking at 200 rpm overnight.
2. The next day, inoculate 400 μL of each overnight culture into separate 50-mL conical tubes containing 40 mL of sterile lsLB. If growing polymicrobial cultures, split the inoculum as desired among different strains in additional conical tubes, keeping the total inoculum volume constant.
3. Incubate all tubes at 37 °C at 200 rpm until log phase (1–2 h, $OD_{600} \sim 0.25$–0.5). Incubation time will decrease significantly if medium is prewarmed at 37 °C prior to inoculation.
4. While cultures are growing, prewarm urine in 37 °C water bath and prepare microtiter plate and plate reader for assay.

3.2.2 Preparation of Microtiter Plate

1. Prepare urea–phenol red mix by adding 1 mL of 0.1% phenol red to 4 mL of 5 M urea. Vortex to mix and keep at 4 °C until needed (*see* **Notes 6** and 7).
2. Using a multichannel pipette, transfer 170 μL of the urea–phenol red solution into all eight wells of the last column of the 96-well plate. This is used instead of a reservoir since only 6–8 columns are used in a typical assay setup, but if more columns are needed, place the urea–phenol red solution in a sterile reagent boat.

a

1-2 t=0		3-4 t=30		5-6 t=60							
Pm	CC	Pm	CC	Pm	CC						U/P
Pm	CC	Pm	CC	Pm	CC						U/P
Pm	CC	Pm	CC	Pm	CC						U/P
Pm	CC	Pm	CC	Pm	CC						U/P
Ps	1:1	Ps	1:1	Ps	1:1						U/P
Ps	1:1	Ps	1:1	Ps	1:1						U/P
Ps	1:1	Ps	1:1	Ps	1:1						U/P
Ps	1:1	Ps	1:1	Ps	1:1						U/P

b

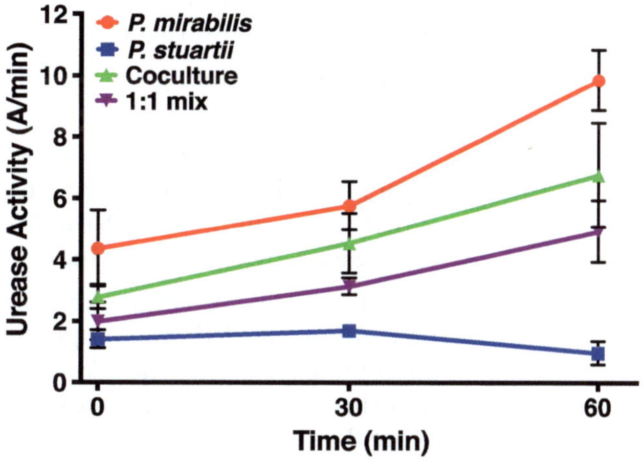

Fig. 2 Sample experiment using the phenol red method to determine urease activity of *P. mirabilis*, *P. stuartii*, and coculture. (**a**) Microplate setup for three time points. *Pm P. mirabilis*, *Ps P. stuartii*, *CC* Coculture, *1:1* a 1:1 mixture of both cultures grown separately, *U/P*: Urea–phenol red. (**b**) Urease activity of single cultures, coculture, and a 1:1 mix. The difference in urease activity between strains grown separately and mixed prior to reaction (1:1 mix) and strains grown together (coculture) can be attributed to polymicrobial interactions

3. Pipet 170 μL of pooled human urine at 37 °C into as many columns as needed to accommodate cultures tested (4 wells/culture). Keep at 37 °C with lid on until culture inoculation.
4. Turn plate reader on and warm up to 37 °C. Select a wavelength of 562 nm and a kinetic read protocol running for 300 s with reads every 5 s (or the minimum time between reads the spectrophotometer allows). Select the wells to read; this will change with every time point. In the setup shown in Fig. 2a for two strains and their coculture, columns 1–2 are read at time = 0, columns 3–4 at time = 30 min, and columns 5–6 are at time = 60 min.

3.2.3 Induction of Urease Activity in Urine

1. After ~1 h incubation at 37 °C, remove 1 mL aliquots of cultures and record optical density at 600 nm. Blank the spectrophotometer with 1 mL of lsLB. Incubate longer if necessary until required optical density is achieved.
2. Once cultures have reached OD_{600} between 0.250 and 0.500, centrifuge at 2500 × *g* for 10 min to precipitate cells. Discard the supernatant.
3. Calculate volume of urine required to resuspend each culture to an $OD_{600} = 0.5$ (*see* **Note 17**).
4. Using a serological pipette, add the calculated volume of filter-sterilized urine to each culture, and mix gently until pellet has been completely resuspended.
5. Start a timer counting up. This is the point of time = 0 of urease induction.
6. Immediately transfer 1 mL of each culture or coculture into separate prelabeled microfuge tubes. If quantifying urease activity of cocultures, *see* **Note 18**. This aliquot will be used for urease activity determination.
7. Transfer a 100 μL aliquot of each culture or coculture into separate prelabeled microfuge tubes. This aliquot will be appropriately diluted and plated at the end of the experiment if normalization of urease activity to cell density is needed (*see* **Note 8**).
8. Continue incubating the urine culture at 37 °C with shaking at 200 rpm while the first time point is assayed until timer indicates next time point.

3.2.4 Urease Assay

1. Centrifuge the culture aliquots corresponding to time = 0 in microfuge tubes at 16,000 × *g* for 2.5 min, to pellet cells.
2. Remove the supernatant using a micropipette or a transfer pipette; do not disturb the pellet.
3. Add 100 μL of 0.9% saline, and mix by pipetting up and down until pellet has been resuspended.
4. Pipet 20 μL of each of the cultures being assayed into the microtiter plate preloaded with urine at 37 °C, with four technical replicates for each culture. *See* Fig. 2a for an example of a plate setup for two strains, coculture of both strains, and mixed cultures as controls.
5. Pipet 10 μL of phenol red–urea solution from the last column into all wells of columns 1–2 corresponding to time = 0 using a multichannel pipette (change tips for each column). Mix well by pipetting up and down.
6. Immediately start the microtiter plate kinetic protocol, making sure only the first two columns are being read.

7. Record kinetic reads over 300 s. An increase in absorbance correlates with the increase in pH generated by urease activity.
8. Repeat **step 6** in Subheading 3.2.3 to **step 7** in Subheading 3.2.4 for the next time point, using columns 3–4 of the microplate in **step 5** of Subheading 3.2.4. Continue in this fashion for all remaining time points desired. Do not stop the timer while collecting samples.
9. Examine the curves and remove initial nonlinear points, typically the first 100 s. Make sure to remove the same reads for all curves. Calculate slopes for the remaining linear portions of the curves.
10. Use linear regression to calculate the average slope for each culture (ΔA/min). Plot the average slope for each culture vs. time (Fig. 2b).

4 Notes

1. Due to the ability of *P. mirabilis* to swarm on regular agar plates, low salt Luria–Bertani (lsLB) agar is used. For consistency, lsLB broth is also used when liquid medium is required. Alternatively, if a different growth medium is desired, swarming can also be prevented by preparing 4% agar plates.
2. Urine may be used fresh or kept frozen in 50 mL aliquots at −20 °C for up to 2 months. If any precipitate is observed upon thawing, centrifuge urine again before use.
3. PEM buffer should be stored at 4 °C and is stable for several weeks. UPEM substrate can be prepared fresh as needed by adding urea from a 5 M stock solution to PEM buffer. The resulting UPEM substrate needs to be brought to room temperature or desired assay temperature prior to adding enzyme source.
4. Phenol is an extremely caustic skin and mucosal irritant. Always wear appropriate skin and eye protection when handling phenol and consult MSDS sheets prior to use. Phenol–nitroprusside solution should be stored at 4 °C in an amber bottle, and it is stable for 4 weeks.
5. Prepare a 10 U/mL stock solution of purified urease in PEM buffer. This solution can be stored at 4 °C for up to a week. Test dilutions ranging from 0.005 to 0.5 U/mL in the final assay volume. After initial test, include two suitable dilutions one order of magnitude apart as positive controls in each plate.
6. Phenol red solution can be kept at 4 °C for up to 4 weeks.

7. Prepare 5 M urea fresh on the day of each assay. Failure to do so will result in higher variability among replicates. Discard remaining urea and urea–phenol red solution after use.
8. This step is important for normalization among biological replicates. If measuring urease activity of polymicrobial cultures, plate on differential agar that will allow for CFU/mL enumeration of each individual strain. An appropriate dilution should yield 30–300 CFU/plate. We used a rough estimate of 10^8 CFU/mL in a $OD_{600} = 1$ culture and calculated a dilution that would yield 30–300 colonies when plating 0.1 mL based on that estimate. We then prepared the estimated dilution and the dilutions one order of magnitude above and below, and plated 0.1 mL of each in lsLB agar or suitable medium. After appropriate incubation, we counted colonies from the plate that yielded 30–300 CFU, and used the following formula to calculate CFU/mL: CFU/mL = No. colonies/(dilution factor × volume plated).
9. Alternatively, freezing can be accomplished in a −80 °C freezer, or in an ethanol–dry ice bath. Cells can be stored at −80 °C for up to a week after the last freeze. If storing, thaw on ice before proceeding to sonication.
10. Keep cells on ice during the entire procedure and allow for 5 min of cooling between sonication cycles. Do not exceed amplitude recommended by manufacturer as microtip can be easily damaged. Always maintain microtip submerged in cell suspension while sonicating.
11. It is very important to keep cell-free lysate at 4 °C or less. Lysates should be assayed for urease activity immediately after being made, as repeated freeze–thaw cycles will result in loss of enzymatic activity. In our hands, cell-free lysates of *P. mirabilis* retained similar levels of urease activity after being frozen up to 4 weeks and after one freeze–thaw cycle, but we saw significant loss of activity on the second freeze–thaw cycle. Cell-free lysates that are not going to be tested for urease activity, but only for total protein concentration, can be kept frozen at −80 °C for several months.
12. For ease of dispensing solutions A, B, and water, we used a repetitive dispenser and disposable tips (Eppendorf).
13. Reactions must be read immediately and without delay between tubes since absorbance will keep increasing over time and the assay will lose linearity after 10 min. We advise against developing more than 14 tubes at once for this reason.
14. This is when the urease reaction will take place. Time and incubation temperature can be modified to obtain suitable absorbance readings. Cultures with very low urease activity (including different bacterial species as well as *P. mirabilis*

mutants) might require longer incubation times or increased temperature to obtain readable results.

15. If measuring urease activity of polymicrobial species in coculture, grow cells together from **step 2** in Subheading 3.1.1. As controls for the coculture, use appropriate mixes of lysates of single species cultures.
16. Once you have calculated a slope from the calibration curve (Fig. 1), convert the absorbance values to micromoles of ammonia generated during the 5 min reaction. Every micromole of ammonia generated in a 5-min period at room temperature represents a unit of urease activity (U). In this way, *urease activity*, or U/mL lysate can be calculated. Once protein concentration of the lysate (mg/mL) has been determined in parallel using a protein concentration kit and following manufacturer's instruction, *urease specific activity*, or U/mg total protein, can be calculated.
17. Cells are incubated in urine for urease induction at the same cell density, equal to $OD_{600} = 0.5$. For this, use $C_1 V_1 = C_2 V_2$ to calculate the volume of urine needed to resuspend each culture. Sample calculation: if $OD_{600} = 0.4$ and culture volume is 39 mL, since the desired final $OD_{600} = 0.5$, the volume of urine required will be 0.4 (C_1) × 39 mL (V_1)/0.5 (C_2) = 31.2 mL of urine (V_2).
18. If quantifying urease activity of cocultures in addition to single strains, an additional control tube needs to be prepared at this point. This control tube consists of a mix of the individually cultured strains present in the coculture just before being assayed (not cocultured).

References

1. Armbruster CE, Mobley HLT, Pearson MM (2018) Pathogenesis of *Proteus mirabilis* infection. EcoSal Plus 8(1). https://doi.org/10.1128/ecosalplus.ESP-0009-2017
2. Sumner JB (1926) The isolation and crystallization of the enzyme urease: preliminary paper. J Biol Chem 69:435–441
3. Dixon NE, Gazzola C, Blakeley RL, Zerner B (1975) Jack bean urease (EC 3.5.1.5). Metalloenzyme. Simple biological role for nickel. J Am Chem Soc 97:4131–4133
4. Schaffer JN, Norsworthy AN, Sun T-T, Pearson MM (2016) *Proteus mirabilis* fimbriae- and urease-dependent clusters assemble in an extracellular niche to initiate bladder stone formation. Proc Natl Acad Sci U S A 113:4494–4499
5. Griffith DP, Musher DM, Itin C (1976) Urease. The primary cause of infection-induced urinary stones. Invest Urol 13:346–350
6. Musher DM, Griffith DP, Yawn D, Rossen RD (1975) Role of urease in pyelonephritis resulting from urinary tract infection with *Proteus*. J Infect Dis 131:177–181
7. Jones BD, Lockatell CV, Johnson DE, Warren JW, Mobley HLT (1990) Construction of a urease-negative mutant of *Proteus mirabilis*: Analysis of virulence in a mouse model of ascending urinary tract infection. Infect Immun 58:1120–1123
8. Johnson DE, Russell RG, Lockatell CV, Zulty JC, Warren J, Mobley HLT (1993) Contribution of *Proteus mirabilis* urease to persistence, urolithiasis, and acute pyelonephritis in a mouse model of ascending urinary-tract infection. Infect Immun 61:2748–2754
9. Stickler DJ (2008) Bacterial biofilms in patients with indwelling urinary catheters. Nat Clin Pract Urol 5:598–608

10. Stickler D, Morris N, Moreno MC, Sabbuba N (1998) Studies on the formation of crystalline bacterial biofilms on urethral catheters. Eur J Clin Microbiol Infect Dis 17:649–652
11. Mobley HLT, Warren JW (1987) Urease-positive bacteriuria and obstruction of long-term urinary catheters. J Clin Microbiol 25:2216–2217
12. Weatherburn MW (1967) Phenol-hypochloride reaction for determination of ammonia. Anal Chem 39:971–974
13. Wakisaka S, Tachiki T, Sung HC, Kumagai H, Tochikura T, Matsui S (1987) A rapid assay method for ammonia using glutamine synthetase from glutamate-producing bacteria. Anal Biochem 163:117–122
14. Okyay TO, Rodrigues DF (2013) High throughput colorimetric assay for rapid urease activity quantification. J Microbiol Methods 95:324–326
15. Deng HH, Hong GL, Lin FL, Liu AL, Xia XH, Chen W (2016) Colorimetric detection of urea, urease, and urease inhibitor based on the peroxidase-like activity of gold nanoparticles. Anal Chim Acta 915:74–80
16. Senior BW, Bradford NC, Simpson DDS (1980) The ureases of *Proteus* strains in relation to virulence for the urinary tract. J Med Microbiol 13:507–512
17. Armbruster CE, Smith SN, Yep A, Mobley HLT (2014) Increased incidence of urolithiasis and bacteremia during *Proteus mirabilis* and *Providencia stuartii* coinfection due to synergistic induction of urease activity. J Infect Dis 209:1524–1532
18. Armbruster CE, Smith SN, Johnson AO, DeOrnellas V, Eaton KA, Yep A, Mody L, Wu W, Mobley HLT (2017) The pathogenic potential of *Proteus mirabilis* is enhanced by other uropathogens during polymicrobial urinary tract infection. Infect Immun 85(2): e00808-16. https://doi.org/10.1128/IAI.00808-16

Chapter 10

Siderophore Detection Using Chrome Azurol S and Cross-Feeding Assays

Stephanie D. Himpsl and Harry L. T. Mobley

Abstract

More than 500 siderophores that bind ferric iron have been characterized and grouped by type based on their chemical structure. The chrome azurol S (CAS) assay is a universal colorimetric method that detects siderophores independent of their structure. In this assay, siderophores scavenge iron from an Fe-CAS-hexadecyltrimethylammonium bromide complex, and subsequent release of the CAS dye results in a color change from blue to orange. Solution-based experiments with CAS result in a quantitative measure of siderophore production, while an observable color change on CAS agar plates can be performed for qualitative detection of siderophores. Cross-feeding assays are another useful method to detect and characterize siderophores produced by bacteria. Under iron-limiting conditions, cross-feeding assays test the ability of an indicator strain to grow when supplied with a specific siderophore (from a test strain) to which it has a cognate receptor required for import into the cell. The cross-feeding assay can be tested with a variety of wild-type strains, siderophore biosynthesis mutants, and siderophore receptor mutants.

Key words Siderophores, Iron chelators, Iron uptake

1 Introduction

Siderophores are small molecular weight molecules synthesized and secreted by bacteria to chelate ferric iron from the environment and subsequently transport into the cell through ferrisiderophore specific outer membrane receptors [1–3]. More than 500 siderophores have been identified and are divided into three main classes based upon their chemical structure and affinity for iron [3]. At physiological pH, catecholates (e.g., enterobactin and salmochelin) [4–6] have the highest affinity, while the hydroxamates (e.g., ferrichrome and desferrioxamine B) [7] have lower affinities for iron. The third class, composed of carboxylates (e.g., citrate and staphyloferrin A) [8, 9], chelate iron with the lowest affinity as compared to catecholates and hydroxamates but are able to chelate iron more efficiently at acidic pH [3]. A fourth class known as "mixed type" (e.g.,

Melanie M. Pearson (ed.), *Proteus mirabilis: Methods and Protocols*, Methods in Molecular Biology, vol. 2021,
https://doi.org/10.1007/978-1-4939-9601-8_10,

yersiniabactin and aerobactin) [10, 11] contain two or more chemical elements found in siderophores of the three main classes.

Previously, *P. mirabilis* was thought to lack siderophores due to its poor ability to chelate iron [12]. Additionally, supernatants of *P. mirabilis* tested negative in colorimetric assays [13] developed to detect the presence of catecholates (Arnow's method [14]) and hydroxamates (Csaky test [15]). Here we describe two siderophore detection methods: a sensitive chemical method [16] that detects siderophores independent of structure in a liquid or agar plate assay that employs chrome azurol S (CAS), and a cross-feeding assay in which bacterial growth on iron-limited media is used as a measure of its ability to import and use a ferrisiderophore. Both methods have been reported to successfully detect previously uncharacterized siderophore molecules in *P. mirabilis* HI4320 [17].

In the CAS assay, chelation of iron from the Fe-CAS-HDTMA complex by a siderophore and subsequent release of the CAS dye results in a color change from blue to orange. To avoid false-positive results, interference by nonsiderophore iron chelators in the CAS assay must be avoided. All extraneous chelators must be excluded throughout the preparation of reagents, solutions, and growth media. Higher concentrations of lower affinity ligands such as phosphate, citrate, and 2,3-dihydroxybenzoic acid (DHBA) must be omitted [16].

To detect siderophores that bacteria can utilize, cross-feeding assays can be performed to exploit siderophore production by a "test" bacterium and subsequently examine the ability of an "indicator" bacterium to take up the siderophore produced by the test strain. These assays are carried out under iron-limiting conditions that restrict growth of the indicator strain unless supplied with a ferrisiderophore that can be taken up using a specific receptor. For the assay to work successfully, the indicator strain examined must be unable to produce the cognate siderophore for which it has a receptor.

2 Materials

For CAS experiments: clean all glassware used to prepare solutions with 6 M HCl and rinse with double distilled water. Prepare all solutions in sterile, purified water. When possible, prepare reagents in polyethylene to eliminate contaminating sources of iron. If possible, store prepared solutions in polyethylene bottles, in the dark, at room temperature, unless otherwise stated (*see* **Note 1**). Filter-sterilize all supplements added to agar and growth media.

2.1 Bacterial Propagation and CAS Substrate Preparation

1. Lysogeny broth (LB): per liter, 10 g of tryptone, 5 g of yeast extract, 0.5 g of NaCl. Autoclave to sterilize (*see* **Note 2**).
2. Antibiotics: add to LB and mix thoroughly. In this protocol, 25 μg/mL of kanamycin and 20 μg/mL of chloramphenicol

were added to LB to propagate siderophore biosynthesis mutants constructed with kanamycin and chloramphenicol resistance markers.

3. 20% casamino acids, chelex-treated (*see* **Note 3**).
4. Neidhardt MOPS defined medium (commercially available, e.g., Teknova) [18] with and without 0.1 mM $FeCl_3 \cdot 6H_2O$.- For culture of *P. mirabilis*, supplement with 1 mL of 1% nicotinic acid, 1 mL of 1 M $MgSO_4 \cdot 7H_2O$, 10 mL of 20% glycerol, and 1 mL of 20% chelex-treated casamino acids per L of medium.
5. 37 °C shaker incubator.
6. Microcentrifuge and superspeed centrifuge.
7. 0.22 μm filter unit with 50 mL collection tube.
8. *P. mirabilis* strains of interest (e.g., wild-type HI4320 (siderophore-producing strain) and HI4320 mutant strains (siderophore-deficient strain)). For the CAS assay protocol, we describe siderophore biosynthesis mutant strains *pbtA* (proteobactin), *nrpR* (yersiniabactin-like), and *nrpR pbtA* (yersiniabactin-like proteobactin double mutant).
9. *E. coli* strains: wild-type positive control (siderophore-producing strain) and mutant strain negative control (siderophore-deficient strain) (*see* **Note 4**).

2.2 CAS Solution Preparation and CAS Assay

1. Iron(III) solution: 1 mM $FeCl_3 \cdot 6H_2O$, 10 mM HCl.
2. pH meter.
3. Fe-CAS-HDTMA solution: prepare a 100 mL glass graduated cylinder with a stir bar. Add 6 mL of 10 mM hexadecyltrimethylammonium bromide (HDTMA) (*see* **Note 5**) and dilute with no more than 50 mL of water. While stirring, slowly add a mixture of 1.5 mL iron(III) solution and 7.5 mL of 2 mM aqueous chrome azurol S (CAS) (*see* **Note 5**).
4. CAS assay solution: place 4.307 g of anhydrous piperazine in a 25 mL glass beaker and add water until dissolved (approximately 10 mL). Carefully add 6.25 mL of 10 N hydrochloric acid to achieve pH = 5.6 (*see* **Note 6**). Pour this buffered solution into the graduated cylinder containing the Fe-CAS-HDTMA solution and add water up to a final volume of 100 mL.
5. CAS shuttle solution: 100 mL CAS assay solution with 4 mM 5-sulfosalicyclic acid. Transfer CAS shuttle solution to a plastic container with lid (recommended) or prepared glass bottle (*see* **Note 1**) and store at room temperature in the dark.
6. 96-well plate, sterile, flat bottom.
7. Plate reader capable of detecting a wavelength of 630 nm.

8. Filtered supernatants of *P. mirabilis* and *E. coli* cultures grown in MOPS buffer with and without 0.1 mM $FeCl_3{\cdot}6H_2O$. Detailed preparation of supernatants is described in Subheading 3.1.
9. Reference and positive control: sterile culture media without and with iron chelator (15 μM deferoxamine (Desferal, Sigma)), respectively (*see* **Note 7**).

2.3 CAS Agar Plate Preparation

1. Iron(III) solution: 1 mM $FeCl_3{\cdot}6H_2O$, 10 mM HCl.
2. 10× MM9 Salts: per liter, 3 g of KH_2PO_4, 0.5 g of NaCl, 10 g of NH_4Cl. Autoclave to sterilize (*see* **Note 8**).
3. Fe-CAS-HDTMA solution: In a 1 L glass bottle, while stirring, dissolve 605 mg of CAS (*see* **Note 5**) in 500 mL of water. Add 100 mL of iron(III) solution. Slowly add a solution of 400 mL of water containing 729 mg of HDTMA (*see* **Note 5**) and stir to dissolve. Remove stir bar, autoclave, and cool to 50 °C.
4. CAS agar plates:
   (a) In a 2 L glass flask containing 750 mL of water, while stirring, dissolve 6 g of NaOH. Stir 30.24 g of 1,4-piperazinediethanesulfonic acid (PIPES) (free acid) into solution. Add 100 mL of 10× MM9 salts and 15 g of Bacto agar. Autoclave and cool to 50 °C.
   (b) To the cooled PIPES-Bacto agar solution, while stirring, add 30 mL of 10% casamino acids (*see* **Note 9**), 10 mL of 20% glucose, 1 mL of 1 M $MgCl_2$, 1 mL of 100 mM $CaCl_2$, 1 mL of 1% nicotinic acid, and 1 mL of 1 M $MgSO_4{\cdot}7H_2O$.
   (c) Carefully, under gentle stirring to avoid bubbles and foaming, add 100 mL of the Fe-CAS-HDTMA solution to the glass wall of the flask containing the PIPES-Bacto agar solution.
   (d) Add approximately 20–25 mL of the blue agar into sterile petri dishes. Allow to solidify overnight at room temperature. To prevent agar plates from drying out, store at 4 °C until ready for use. Use within 3 months.
5. Cultures of *P. mirabilis* and *E. coli* grown in 5 mL of LB with aeration at 37 °C to stationary phase.
6. 30 °C incubator.
7. 20 μL pipette.

2.4 Cross-Feeding Assay

1. Iron-limiting LB agar: LB plus 15 g of agar/L. Autoclave to sterilize and cool to 55 °C. Add deferoxamine (Desferal) to a final concentration of 25 μM (*see* **Note 10**) and 40 μg/mL of X-gal (*see* **Note 11**). Pour into petri dishes, approximately 20–25 mL per plate.

2. Indicator and test strains grown overnight in 5 mL of LB supplemented with antibiotic, if needed, at 37 °C with aeration. For this cross-feeding assay protocol, we describe two indicator strains: proteobactin synthesis/receptor mutant (Pbt^{SR}) and yersiniabactin-related synthesis/receptor mutant (Nrp^{SR}) [17]. To cross-feed *P. mirabilis*, we have used the following *E. coli* test strains: wild-type CFT073, wild-type 536, 536 *entF::kan* (enterobactin synthesis mutant) and 536 *entF ybtS::kan* (enterobactin and yersiniabactin synthesis mutant).
3. 1 μL sterile loop.
4. 37 °C incubator.

3 Methods

3.1 Siderophore Detection Using the CAS Assay

1. Individually inoculate 5 mL of LB (add antibiotics if needed) with a *P. mirabilis* wild-type strain, *P. mirabilis* siderophore biosynthesis mutant, and *E. coli* (siderophore-producing, positive control strain and siderophore-deficient, negative control strain) from single colonies and grow overnight with aeration at 37 °C.
2. Harvest 1.5 mL of overnight culture by centrifugation (5 min, 9600 × *g*, 4 °C). Wash pellet two times with MOPS without 0.1 mM $FeCl_3{\cdot}6H_2O$ and resuspend the pellet in 650 μL of MOPS without 0.1 mM $FeCl_3{\cdot}6H_2O$ (*see* **Note 12**).
3. Inoculate 300 μL of each bacterial resuspension into 30 mL of fresh MOPS buffer with and without 0.1 mM $FeCl_3{\cdot}6H_2O$ (that is, a 1:100 inoculum) and grow to stationary phase for approximately 7 h (*P. mirabilis* strains) and 4 h (*E. coli* strains) to an $OD_{600} \sim 1.0$ with aeration at 37 °C. Place samples on ice for 15 min.
4. Centrifuge (20 min, 4300 × *g*, 4 °C) and carefully collect supernatant from pellet, approximately 25 mL. Pass supernatant through a 0.22 μm filter supplied with a 50 mL conical tube and store at 4 °C in the dark (*see* **Note 13**).
5. Repeat **steps 1–4** for biological replicates.
6. In a flat bottom 96 well plate, mix 150 μL of siderophore-containing supernatant with 150 μL of CAS shuttle solution (*see* **Note 14**). Include a reference of sterile medium that was used for culturing bacteria. Also prepare sterile medium containing 15 μM deferoxamine (Desferal) to serve as a positive control for iron chelation (*see* **Note 15**).
7. Following incubation at room temperature for 1 h, measure absorbance at 630 nm using a plate reader.

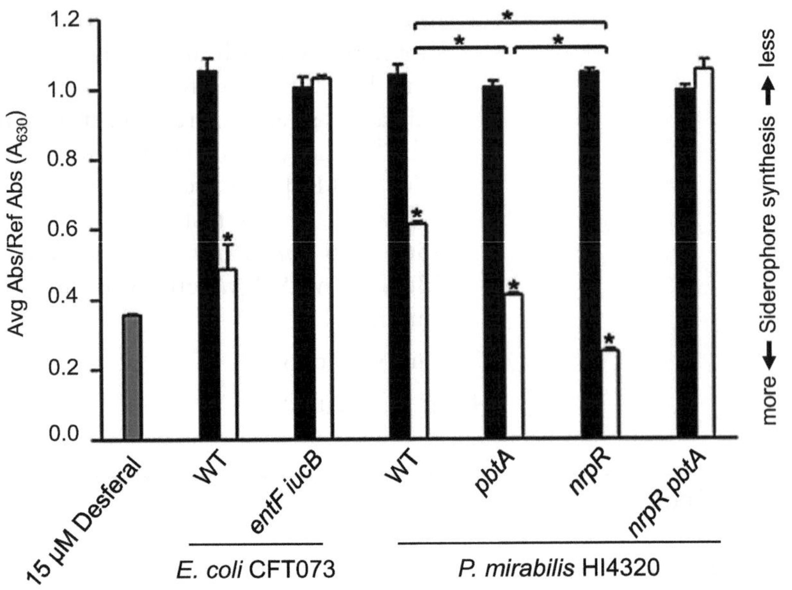

Fig. 1 Siderophore production detected by CAS shuttle solution. Siderophore production measured at 630 nm (A_{630}) using a CAS assay and filtered bacterial supernatants following culture in MOPS defined medium with (black bars) and without (white bars) 0.1 mM $FeCl_3 \cdot 6H_2O$ of *P. mirabilis* HI4320 wild type and *pbtA* (proteobactin), *nrpR* (yersiniabactin-like), and *nrpR pbtA* (double) mutants and *E. coli* CFT073 and *entF iucB* (enterobactin aerobactin) double mutant. *P. mirabilis* supernatants were concentrated 50-fold. Gray bar, positive control of phosphate buffered saline containing 15 μM deferoxamine (Desferal). A lower A_{630} indicates greater siderophore production. Significant differences (∗, $P \leq 0.0022$) were determined using a two-tailed unpaired *t*-test; bars show standard error of the mean. (From: Himpsl, SD, Pearson MM, Arewång CJ, Nusca TD, Sherman DH, Mobley HLT. 2010. Proteobactin and a yersiniabactin-related siderophore mediate iron acquisition in *Proteus mirabilis*. Mol Microbiol. 2010 Oct; 78(1): 138–157)

8. Calculate the absorbance ratio by dividing the average absorbance for the sample by the absorbance of the reference. Calculate the standard error of the mean and perform an unpaired *t*-test on replicates (Fig. 1).

3.2 Siderophore Detection on CAS Agar Plates

1. Inoculate 5 mL of LB (add antibiotics if needed) with a *P. mirabilis* wild-type strain, *P. mirabilis* siderophore biosynthesis mutant, and *E. coli* (siderophore-producing, positive control strain and siderophore-deficient, negative control strain) colony and grow overnight to stationary phase with aeration at 37 °C.
2. Deposit 5 μL of bacterial culture onto CAS agar plates. Once the spot is dried, invert the agar plate and incubate at 30 °C for 18 h (*see* **Note 16**).

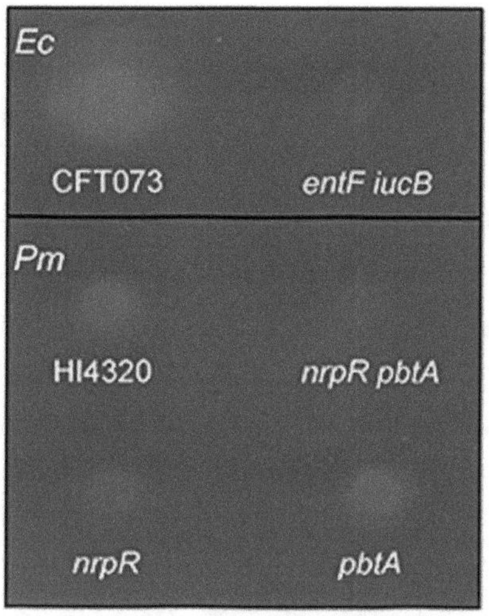

Fig. 2 Siderophore production on CAS agar. Overnight cultures of *E. coli* CFT073 and *entF iucB* (enterobactin aerobactin) double mutant, and *P. mirabilis* HI4320 and *nrpR* (yersiniabactin-like), *pbtA* (proteobactin), and *nrpR pbtA* (double) mutants were grown to stationary phase in LB and spotted in 5 μL volumes onto CAS agar, left to dry, inverted, and incubated at 30 °C for 18 h. A color change of chrome azurol S (CAS) agar from blue to orange indicates siderophore production. (From: Himpsl, SD, Pearson MM, Arewång CJ, Nusca TD, Sherman DH, Mobley HLT. 2010. Proteobactin and a yersiniabactin-related siderophore mediate iron acquisition in *Proteus mirabilis*. Mol Microbiol. 2010 Oct; 78(1): 138–157)

3. In the morning, examine plates for a color change from blue to orange where bacterial culture was deposited (Fig. 2) and compare to positive and negative controls.

3.3 Cross-Feeding Assay

1. Inoculate separate tubes containing 5 mL of LB with a single colony of test and indicator strains and grow overnight at 37 °C with shaking. For example, a properly controlled experiment could include wild-type *P. mirabilis*, *P. mirabilis* proteobactin synthesis/receptor mutant (Pbt^{SR}), *P. mirabilis* yersiniabactin-related synthesis/receptor mutant (Nrp^{SR}), *E. coli* CFT073, *E. coli* 536, and *E. coli* 536 mutant strains *entF::kan* and *entF ybtS::kan*.
2. Using a sterile 1 μL loop, streak cultures onto LB agar supplemented with 25 μM deferoxamine (Desferal), an iron-limited condition in which *P. mirabilis* is unable to grow, and 40 μg/mL of X-gal, used to distinguish growth of the *lac*-positive *E. coli* strains, and *lac*-negative *P. mirabilis* strains. The strain to be tested should be streaked perpendicular to, but not touching, the iron-supplying strain (*see* **Note 17**).
3. Following incubation at 37 °C for 18 h, score agar plates for promotion of growth of the *P. mirabilis* strains (Fig. 3).

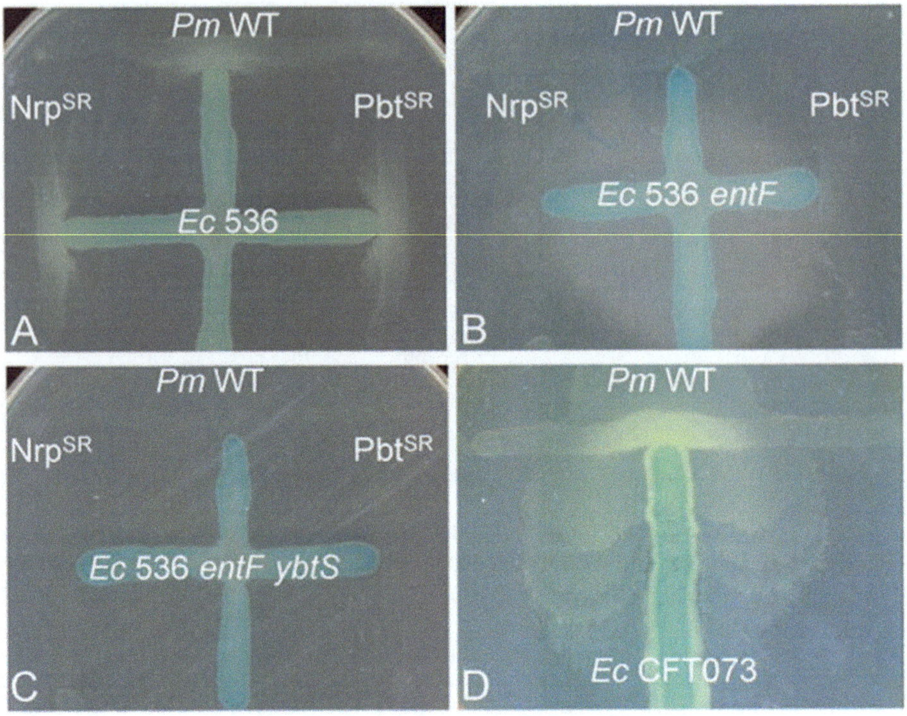

Fig. 3 *P. mirabilis* is unable to utilize yersiniabactin produced by uropathogenic *E. coli* 536. LB medium was supplemented with 25 μM deferoxamine, the minimum concentration of chelator required to completely suppress growth of *P. mirabilis* HI4320. Lactose-fermenting *E. coli* were distinguished from *P. mirabilis* by using the chromogenic indicator X-gal. (**a**) Growth of wild-type *P. mirabilis*, the proteobactin synthesis/receptor mutant (Pbt^{SR}), and the yersiniabactin-related siderophore/receptor mutant (Nrp^{SR}) is restored when cross-fed by *E. coli* 536 that produces both enterobactin and yersiniabactin. (**b** and **c**) Restoration of *P. mirabilis* growth for all strains tested is abolished when cross-fed by either the *E. coli* 536 enterobactin mutant, *entF::kan*, or the enterobactin/yersiniabactin double mutant, *entF ybtS::kan*. (**d**) *P. mirabilis* growth is restored using *E. coli* CFT073, an enterobactin-producing strain incapable of synthesizing yersiniabactin. (From: Himpsl SD, Pearson MM, Arewång CJ, Nusca TD, Sherman DH, Mobley HLT. 2010. Proteobactin and a yersiniabactin-related siderophore mediate iron acquisition in *Proteus mirabilis*. Mol Microbiol. 2010 Oct; 78 (1): 138–157)

4 Notes

1. Increased absorbance readings have been noted with use of CAS assay solutions following long-term storage in glass bottles despite cleaning the glassware with 6 M HCl. Reversible crystallization in the CAS shuttle solution has been reported to take place when stored below 25 °C.
2. Reports have suggested that rich media, such as LB, may hinder the assay [16], but we found no noticeable indication that our LB recipe interfered. A lower salt concentration in LB was used in these experiments to eliminate *P. mirabilis* swarming on

agar. High salt concentrations of 20–100 mM NaCl will cause *P. mirabilis* to swarm on agar [19].

3. In preparation to chelex-treat casamino acids, glassware should be rinsed many times with or bathed in a strong acid such as 5 M HCl, followed by multiple rinses with ddH_2O. Add 5 g of Chelex-100 to 100 mL of casamino acids, mix, and let sit for 1 h at room temperature.
4. For positive and negative controls, *E. coli* CFT073 and *entF::kan iucB::cam* [20], an enterobactin and aerobactin double siderophore biosynthesis mutant, respectively, were successfully used in CAS assays.
5. For agar to result in a blue color, HDTMA and CAS reagents must be purchased from Fluka Chemical.
6. Add additional HCl as needed to achieve pH = 5.6.
7. The naturally occurring siderophore, deferoxamine, has been shown to have a bacteriostatic effect on *P. mirabilis* [21, 22]. Growth of *P. mirabilis* HI4320 was examined in various concentrations of the iron chelator deferoxamine (Desferal) in LB [17]. Slight restriction of *P. mirabilis* HI4320 growth was observed in 15 μM deferoxamine and used in these experiments. The appropriate concentration of deferoxamine to use with other strains should be determined independently.
8. Reduce the salt concentration in 10× MM9 solution from 5.0 g/L to 0.5 g/L to eliminate *P. mirabilis* swarming on CAS agar. Higher salt concentrations of 20–100 mM NaCl will cause *P. mirabilis* to swarm on agar [19]. 10× MM9 solution can also be filter-sterilized instead of sterilization by autoclave.
9. Extract casamino acids with an equal volume of 3% 8-hydroxyquinoline in chloroform to remove contaminating iron. Mix the casamino acids solution and hydroxyquinoline--chloroform solution in a separatory funnel and let sit. Over the course of 3–4 h, continue to mix the solution thoroughly and let sit several times. Allow the solution to separate completely overnight before removing the casamino acids layer. Store in a polyethylene bottle or prepared glassware.
10. Growth of *P. mirabilis* HI4320 was previously examined on various concentrations of deferoxamine (Desferal) in LB agar and was restricted on 25 μM [17]. The appropriate concentration of deferoxamine to use with other strains should be determined independently.
11. The addition of 40 μg/mL of X-gal to the agar distinguishes growth of the LacZ-positive *E. coli* (blue) from LacZ-negative *P. mirabilis* (white).

12. Do not use *P. mirabilis* minimal salts medium (PMSM) [23] to culture *P. mirabilis* for these experiments. PMSM contains sodium citrate which binds iron in subsequent CAS experiments and will give inaccurate results. We found that as little as 0.47 g/L of sodium citrate reproducibly interfered with the assay. Purchase and use of commercially available Neidhardt MOPS defined medium without iron eliminates the possibility of iron contamination in lab preparation of the various components that make up the medium.
13. Supernatants tested from overnight LB cultures have low CAS activity, and may need to be concentrated. We lyophilized *P. mirabilis* wild-type and mutant samples 50-fold from 50 mL volumes of filtered supernatant. Lyophilized samples were subsequently resuspended in 1 mL of phosphate buffered saline (PBS) before testing in the CAS assay.
14. Experiments can be performed with both the CAS assay solution (omission of 4 mM 5-sulfosalicylic acid) and the CAS shuttle solution (addition of 4 mM 5-sulfosalicylic acid). The CAS shuttle solution is best for experiments in which siderophore structure is unknown, as the exchange rate of iron from CAS to the siderophore varies widely and is dependent on the structure of the siderophore. Sulfosalicylic acid expedites iron exchange from CAS to the siderophore. Here we describe use of CAS shuttle solution in the CAS assay experiments for *P. mirabilis* HI4320 since little was known about its siderophore production prior to publication on the subject [17].
15. Lyophilized samples reconstituted with PBS require a reference of PBS alone and PBS containing 15 μM deferoxamine as a positive control.
16. Incubation of plates at 37 °C overnight also resulted in a positive readout of siderophore production; however, color change from blue to orange was slightly less noticeable as compared to 30 °C.
17. In this example, *E. coli* CFT073, *E. coli* 536, and *E. coli* 536 siderophore mutant strains *entF::kan* and *entF ybtS::kan* were inoculated perpendicular to *P. mirabilis* strains in close proximity without contact.

Acknowledgments

We would like to thank Shelly Payne for helpful insight on the CAS agar protocol, Tyler Nusca for chelex-treated casamino acids and lyophilization of samples used in the CAS assay described in this protocol, and Alfredo Torres for generously providing the CFT073 *entF::kan iucB::cam* strain. We also thank Christopher Alteri for insight on designing the cross-feeding assay for *P. mirabilis* and Erin Garcia for construction of *E. coli* 536 mutants *entF::kan* and *entF ybtS::kan*. This work was supported in part by the Public Health Service Grants AI043360 and AI059722 from the National Institutes of Health.

References

1. Braun V (1995) Energy-coupled transport and signal transduction through the gram-negative outer membrane via TonB-ExbB-ExbD-dependent receptor proteins. FEMS Microbiol Rev 16(4):295–307
2. Tuckman M, Osburne MS (1992) In vivo inhibition of TonB-dependent processes by a TonB box consensus pentapeptide. J Bacteriol 174 (1):320–323
3. Miethke M, Marahiel MA (2007) Siderophore-based iron acquisition and pathogen control. Microbiol Mol Biol Rev 71(3):413–451. https://doi.org/10.1128/MMBR.00012-07
4. Pollack JR, Neilands JB (1970) Enterobactin, an iron transport compound from *Salmonella typhimurium*. Biochem Biophys Res Commun 38(5):989–992
5. Hantke K, Nicholson G, Rabsch W, Winkelmann G (2003) Salmochelins, siderophores of *Salmonella enterica* and uropathogenic *Escherichia coli* strains, are recognized by the outer membrane receptor IroN. Proc Natl Acad Sci U S A 100(7):3677–3682. https://doi.org/10.1073/pnas.0737682100
6. Bister B, Bischoff D, Nicholson GJ, Valdebenito M, Schneider K, Winkelmann G, Hantke K, Sussmuth RD (2004) The structure of salmochelins: C-glucosylated enterobactins of *Salmonella enterica*. Biometals 17 (4):471–481
7. Muller G, Matzanke BF, Raymond KN (1984) Iron transport in *Streptomyces pilosus* mediated by ferrichrome siderophores, rhodotorulic acid, and enantio-rhodotorulic acid. J Bacteriol 160(1):313–318
8. Meiwes J, Fiedler HP, Haag H, Zahner H, Konetschny-Rapp S, Jung G (1990) Isolation and characterization of staphyloferrin A, a compound with siderophore activity from *Staphylococcus hyicus* DSM 20459. FEMS Microbiol Lett 55(1–2):201–205
9. Guerinot ML, Meidl EJ, Plessner O (1990) Citrate as a siderophore in *Bradyrhizobium japonicum*. J Bacteriol 172(6):3298–3303
10. Drechsel H, Stephan H, Lotz R, Haag H, Zahner H, Hantke K, Jung G (1995) Structure elucidation of yersiniabactin, a siderophore from highly virulent *Yersinia* strains. Liebigs Ann 10:1727–1733
11. Gibson F, Magrath DI (1969) The isolation and characterization of a hydroxamic acid (aerobactin) formed by *Aerobacter aerogenes* 62-I. Biochim Biophys Acta 192(2):175–184
12. Miles AA, Khimji PL (1975) Enterobacterial chelators of iron: their occurrence, detection, and relation to pathogenicity. J Med Microbiol 8(4):477–490. https://doi.org/10.1099/00222615-8-4-477
13. Evanylo LP, Kadis S, Maudsley JR (1984) Siderophore production by *Proteus mirabilis*. Can J Microbiol 30(8):1046–1051
14. Arnow LE (1937) Colorimetric determination of the components of 3,4-dihydroxyphenylalanine tyrosine mixtures. J Biol Chem 118(2):531–537
15. Csaky TZ (1948) On the estimation of bound hydroxylamine in biological materials. Acta Chem Scand 2(5–6):450–454. https://doi.org/10.3891/acta.chem.scand.02-0450
16. Schwyn B, Neilands JB (1987) Universal chemical assay for the detection and determination of siderophores. Anal Biochem 160 (1):47–56. https://doi.org/10.1016/0003-2697(87)90612-9
17. Himpsl SD, Pearson MM, Arewång CJ, Nusca TD, Sherman DH, Mobley HLT (2010) Proteobactin and a yersiniabactin-related siderophore mediate iron acquisition in *Proteus mirabilis*. Mol Microbiol 78(1):138–157. https://doi.org/10.1111/j.1365-2958.2010.07317.x
18. Neidhardt FC, Bloch PL, Smith DF (1974) Culture medium for enterobacteria. J Bacteriol 119(3):736–747
19. Armbruster CE, Hodges SA, Mobley HLT (2013) Initiation of swarming motility by *Proteus mirabilis* occurs in response to specific cues present in urine and requires excess L-glutamine. J Bacteriol 195(6):1305–1319. https://doi.org/10.1128/JB.02136-12
20. Torres AG, Redford P, Welch RA, Payne SM (2001) TonB-dependent systems of

uropathogenic *Escherichia coli*: aerobactin and heme transport and TonB are required for virulence in the mouse. Infect Immun 69 (10):6179–6185. https://doi.org/10.1128/IAI.69.10.6179-6185.2001

21. van Asbeck BS, Marcelis JH, Marx JJ, Struyvenberg A, van Kats JH, Verhoef J (1983) Inhibition of bacterial multiplication by the iron chelator deferoxamine: potentiating effect of ascorbic acid. Eur J Clin Microbiol 2 (5):426–431
22. Keberle H (1964) The biochemistry of desferrioxamine and its relation to iron metabolism. Ann N Y Acad Sci 119:758–768
23. Belas R, Erskine D, Flaherty D (1991) Transposon mutagenesis in *Proteus mirabilis*. J Bacteriol 173(19):6289–6293

Chapter 11

Using Hemagglutination, Surface Shearing, and Acid Treatment to Study Fimbriae in *Proteus mirabilis*

Stephanie D. Himpsl, Melanie M. Pearson, and Harry L. T. Mobley

Abstract

A critical first step in bacterial virulence and colonization is adherence to mucosal surfaces, often mediated by fimbriae and other protein adhesins. Here are described three short methods for studying these surface proteins and their behaviors, using protocols developed for the opportunistic pathogen *Proteus mirabilis*. Unlike the mannose-binding type 1 fimbriae produced by *Escherichia coli*, most *P. mirabilis* strains produce mannose-resistant/*Proteus*-like (MR/P) fimbriae. Both types of fimbrial production and adhesion can be easily demonstrated by a simple and economical hemagglutination assay which uses a model system of erythrocytes. The second and third fimbrial methods presented here show how to shear surface-exposed proteins and use acid treatment to separate interlocked fimbrial subunits into monomers.

Key words Hemagglutination, Mannose-resistant/Proteus-like fimbriae, Fimbriae, Pili

1 Introduction

Bacterial proteins that project from the surface are typically used to mediate interactions with their surroundings [1]. Proteins in this category include adhesins, such as fimbriae (also known as pili), and flagella. Because they are present on the bacterial surface, the functions of these proteins may be studied using either whole, live bacteria or purified proteins removed from bacteria using shear force.

Adherence by *P. mirabilis* type strain HI4320 is facilitated by 17 different chaperone-usher fimbriae, 14 of which have been reported to have roles in virulence [2], and other nonfimbrial adhesins. One means of assessing fimbrial adherence is a hemagglutination assay, where the bacterial binding of erythrocytes causes the cells to clump and fall to the bottom of the assay surface. Classically, type 1 fimbriae produced by *Escherichia coli* and related species bind mannose, and hemagglutination of erythrocytes by type 1 fimbriae is inhibited by addition of free mannose [3]. Thus, hemagglutination can be differentiated as mannose-

Melanie M. Pearson (ed.), *Proteus mirabilis: Methods and Protocols*, Methods in Molecular Biology, vol. 2021,
https://doi.org/10.1007/978-1-4939-9601-8_11,

sensitive (MSHA) or mannose-resistant (MRHA) by the addition of exogenous mannose as a competitive inhibitor [4]. The most well-studied fimbriae encoded by *P. mirabilis*, mannose-resistant/*Proteus*-like (MR/P) fimbriae, agglutinate erythrocytes independent of D-mannose [5, 6]. Reports also indicate that some *P. mirabilis* isolates produce one or more unknown fimbriae that are able to agglutinate tannic acid-treated bovine erythrocytes, a pattern known as mannose-resistant/*Klebsiella*-like (MR/K) hemagglutination [7].

Because they are generally long and thin, fimbriae and flagella are prone to being broken off in the presence of sufficient shear force. The second protocol described here is a simple method that results in a concentrated preparation of sheared surface proteins suitable for further analysis [6, 8].

Fimbriae are formed from multiple joined units of the major structural protein. In chaperone-usher fimbriae, the preceding subunit contributes a strand that completes an immunoglobulin-like fold in the next subunit (donor-strand complementation) [9]. These structures are very stable, and may not separate into individual subunits, even when subjected to reducing conditions typically used in SDS-PAGE. A brief treatment with acid, described as the third method in this chapter, may allow individual subunits to be analyzed, particularly using immunoblot or mass spectrometry [6, 8, 10].

2 Materials

2.1 Hemagglutination Assay

1. Lysogeny broth (LB): per liter, 10 g of tryptone, 5 g of yeast extract, 0.5 g of NaCl. Autoclave to sterilize.
2. *Proteus mirabilis* strain expressing MR/P fimbriae (*see* **Note 1**).
3. *Escherichia coli*, or other mannose-sensitive control strain (*see* **Note 2**).
4. 37 °C shaking and static incubators.
5. Spectrophotometer and 1.5 mL semi-micro cuvettes.
6. Microcentrifuge and superspeed centrifuge.
7. Phosphate-buffered saline (PBS): per liter, 8 g of NaCl, 0.2 g of KCl, 1.42 g of Na_2HPO_4, 0.24 g of KH_2PO_4, pH 7.4. Autoclave or filter-sterilize.
8. Guinea pig blood, washed and suspended in Alsever's solution (*see* **Notes 3–6**).
9. 100 mM mannose, sterile.
10. Sterile, round bottom, 96-well microtiter plates (*see* **Note 7**).

11. Pedestal mirror: frame to hold 96-well plate with mirror mounted below (optional; this facilitates viewing and photographing the bottom of a microtiter plate).

2.2 Shear Preparation

1. Growth medium of choice, sterilized. *P. mirabilis* is routinely cultured in LB (per liter, 10 g of tryptone, 5 g of yeast extract, 0.5 g of NaCl) (*see* **Note 8**).
2. Sterile Erlenmeyer flasks, or other suitable culture vessel; culture bacteria for the condition you want to study (*see* **Note 9**).
3. Incubators, typically capable of shaking and set to 37 °C with aeration, but this will depend on the specific proteins and conditions you want to study (*see* **Note 9**).
4. 50 mL conical centrifuge tubes.
5. Centrifuges (floor, benchtop, and ultracentrifuge models).
6. Centrifuge bottles or tubes.
7. 0.2 μm filter vacuum sterilization system.
8. Centrifugal concentrators, 10,000 molecular weight cutoff.
9. Phosphate-buffered saline (PBS), sterile: 0.138 M NaCl, 0.0027 M KCl, 0.01 M Na_2HPO_4, 0.0018 M KH_2PO_4; pH 7.4.
10. Glycerol.

2.3 Acid Preparation

1. *P. mirabilis* culture or shear prep.
2. Sterile, distilled water, pH 1.8. Adjust pH with HCl.
3. Heat block or boiling water bath.
4. 4× Tris–Cl/SDS pH 6.8: 6.05 g of Tris base in 40 mL of dH_2O, pH to 6.8 with 1 N HCl. Add dH_2O to 100 mL, and add 0.4 g of SDS. Store at 4 °C.
5. 6× SDS sample buffer: 7 mL of 4× Tris–HCl/SDS pH 6.8, 3 mL of glycerol, 0.93 g of dithiothreitol (DTT), 1.2 mg of bromophenol blue, 1.0 g of SDS. Store in 1 mL aliquots at −20 °C.
6. 15% SDS-PAGE gel and electrophoresis equipment.

3 Methods

3.1 Hemagglutination Assay

1. Culture bacteria overnight in LB, at 37 °C with aeration.
2. Dilute overnight cultures 1:100 into 5 mL of LB.
3. Incubate *P. mirabilis* statically at 37 °C. Passage three times for 48 h each for optimal MR/P fimbrial production. For *E. coli* strains, incubate statically in 5 mL of LB at 37 °C for 72 h (*see* **Note 9**).

4. Measure the OD_{600} of each culture.
5. Standardize cultures to $OD_{600} = 0.8$ with LB (approximately $\geq 10^8$ CFU/mL).
6. Harvest bacteria from 1 mL of the standardized culture by microcentrifugation at 5200 rpm for 5 min at 23 °C.
7. Gently resuspend cell pellets in 0.1 mL of PBS.
8. Prepare the blood by centrifuging 1.5 mL in a 15 mL conical at 3500 × *g* for 5 min at 4 °C (*see* **Note 10**).
9. Pipette off all serum/Alsever's solution.
10. Wash with PBS, using approximately 2–3 times the volume of erythrocytes.
11. Centrifuge at 3500 × *g* for 5 min at 4 °C and pipette off the supernatant.
12. Repeat **steps 3** and **4** two more times.
13. Note the volume of the pellet following the third wash and make a 3% solution (v/v) of erythrocytes in PBS. Keep on ice (*see* **Notes 11** and **12**).
14. Prepare 3% erythrocytes with mannose: aliquot a portion of erythrocytes prepared in **step 13** into a new microcentrifuge tube, and add mannose to a final concentration of 50 mM. Keep on ice.
15. Set up the hemagglutination assay: place 25 μL of PBS into all wells of a round bottom 96-well microtiter plate, except for the first column and the second to last column.
16. Pipette 50 μL of bacteria (from **step 7**) into the first column on the 96 well plate. Each strain to be tested should be placed in a different row (Fig. 1).
17. Serially dilute the bacteria in twofold dilutions: take 25 μL of bacteria from the first column and add it to the next. Mix well.
18. Pipette 25 μL from the second well into the third column; continue this serial dilution across the entire plate, stopping before the last two columns on the plate (*see* **Note 13**).
19. Pipette 25 μL of undiluted bacteria into the second to last column of wells.
20. Add an equal volume (25 μL) of 3% erythrocyte suspension to each well, except for the second to last column.
21. Add 30 μL of erythrocytes mixed with 50 mM mannose to the undiluted bacterial samples in the second to last column (*see* **Note 14**).
22. Gently mix the wells by rocking the plate side to side for a few seconds.
23. Incubate at 23 °C to allow erythrocytes to settle to the bottom of the well, and view at 30 min and 1 h (*see* **Note 15**).

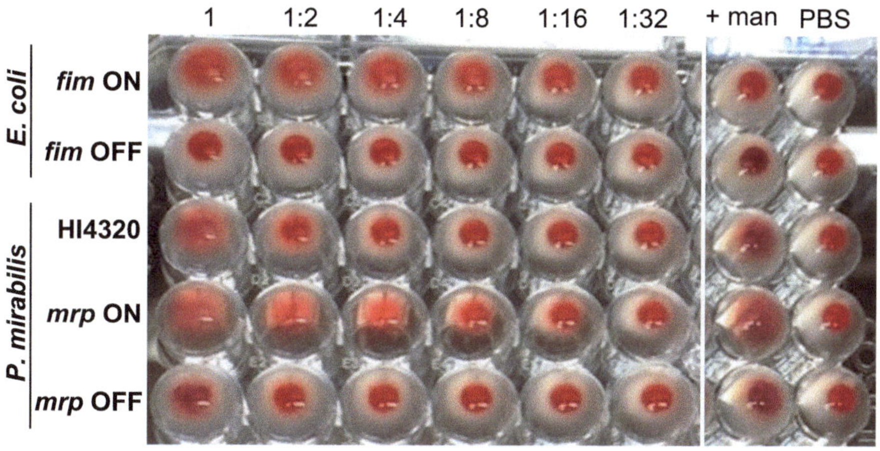

Fig. 1 Hemagglutination results for *E. coli* and *P. mirabilis*. Bacteria were diluted twofold in PBS across ten wells of a 96-well plate (the first six columns are shown). The second to last column of the plate contains undiluted bacteria and erythrocytes with 50 mM mannose (+ man). 3% (vol/vol) guinea pig erythrocytes in PBS were used to detect hemagglutination by mannose-resistant/*Proteus*-like (MR/P) fimbriae in *P. mirabilis* and mannose-sensitive hemagglutination by type 1 fimbriae in *E. coli*

24. Place the microtiter plate on top of a pedestal mirror, and examine erythrocytes through the bottom of the wells.
25. Record the last dilution at which hemagglutination is visible, and note whether or not mannose affects the reaction.

Agglutinated erythrocytes form a diffuse mat while nonagglutinated erythrocytes form a tight button (Fig. 1). Depending on the assay conditions, a positive result can also appear as a clump of particulates [11]. For the MRHA assay, *P. mirabilis* will create a diffuse mat of agglutinated erythrocytes in the presence or absence of mannose, indicating the presence of MR/P fimbriae (*see* **Note 16**). *E. coli* will agglutinate erythrocytes, but in the presence of mannose will create a tight button of nonagglutinated erythrocytes because its type 1 fimbriae are mannose-sensitive (MSHA).

3.2 Shear Preparation

1. Culture bacteria overnight in LB at 37 °C with aeration.
2. Dilute bacteria 1:100 into 250 mL of fresh LB and culture for 8 h at 37 °C with aeration (or as needed to induce the target protein(s); *see* **Notes 9** and **17**).
3. If using a complex growth medium, and removal of peptides and dead bacteria from the medium is desired, pellet cells by centrifugation at 7000 × *g* for 20 min at 4 °C. Gently remove the supernatant, and suspend cells in 100 mL of PBS (*see* **Notes 8** and **18**).
4. Transfer bacteria to a bottle, and tightly fasten the cap. Shake vigorously by hand for 5 min (*see* **Note 19**).

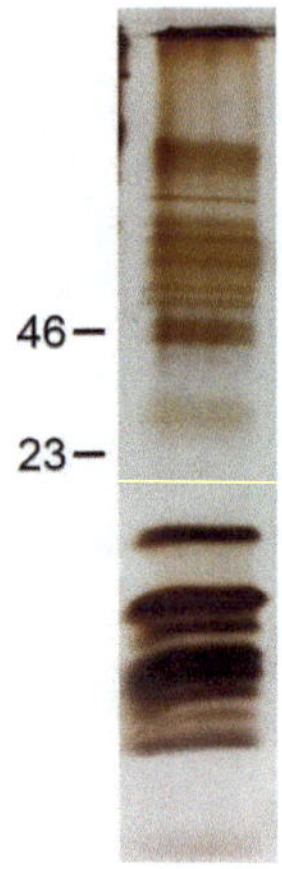

Fig. 2 Shear preparation. Surface proteins were sheared from *P. mirabilis* HI4320 locked on, separated on a 15% SDS-PAGE gel, and visualized using silver stain. A mixture of proteins is present

5. Spin cells down at 7000 × *g* for 20 min at 4 °C. Sheared proteins will remain in the supernatant.
6. Sterilize the supernatant using a 0.2 μm pore vacuum filter (*see* **Note 20**).
7. Add supernatant to a 15 mL 10,000 molecular weight cutoff centrifugal concentration filter (*see* **Note 21**).
8. Spin in a tabletop centrifuge at 1600 × *g* for 20 min at 4 °C.
9. After each spin, discard the flow-through and add more supernatant until the entire sample has been concentrated. Continue concentrating until less than 3 mL of supernatant remains.
10. Ultracentrifuge at 200,000 × *g* for 1 h at 4 °C (*see* **Notes 22** and **23**).
11. Resuspend pellet in PBS (1.5 mL for a 250 mL culture).
12. For preservation, add glycerol to 10% final volume and freeze at −80 °C.
13. Assess results by running samples on SDS-PAGE and visualize proteins using Coomassie blue or silver stain (Fig. 2).

Fimbriae sheared in this way tend to retain their structure, and will not necessarily dissociate into monomers, even after boiling under reducing conditions (such as Laemmli sample buffer). If visualization of individual subunits is desired, mix an aliquot of shear prep 1:1 with pH 1.8 water and proceed with **step 6** of Subheading 3.3. Likewise, a shear preparation can be directly subjected to mass spectrometry for identification of individual proteins. However, fimbriae that are still assembled are resistant to digestion with trypsin, and may appear to be absent from a shear preparation unless first treated with acid (*see* Subheading 3.3).

3.3 Acid Preparation

1. Culture bacteria under a fimbria-inducing condition (*see* **Note 9**).
2. Measure the OD_{600} of the bacterial culture.
3. Adjust the culture density to $OD_{600} = 1.0$ (*see* **Note 24**).
4. Using a microcentrifuge, pellet 1 mL of culture by spinning at 6000 rpm for 6 min; discard the supernatant (*see* **Note 25**).
5. Suspend the bacterial pellet in 100 μL of water, pH 1.8 (*see* **Note 26**).
6. Boil for 10 min (*see* **Note 27**).
7. Add 20 μL of 6× sample buffer (solution will be yellow).
8. Add 1 N NaOH in 0.5 μL increments, until the solution turns purple/blue (*see* **Note 28**).
9. Samples may be visualized by SDS-PAGE (**step 10**), sent for analysis using mass spectrometry (**step 12**), or stored at −20 °C.
10. Load 10 μL of an acid-treated whole cell lysate on 12.5% SDS-PAGE.
11. Visualize separated proteins by immunoblot (Fig. 3) (*see* **Note 29**).
12. Identify acid-treated shear prep components using mass spectrometry (*see* **Notes 17** and **30**).

4 Notes

1. For this protocol, *P. mirabilis* HI4320 and *mrp* locked-on and *mrp* locked-off strains were used [12]. The *mrp* locked strains were generated by mutation of *mrpI*, which encodes a recombinase that dictates the orientation of an invertible element in the *mrp* promoter. These mutants are either locked into constitutively producing MR/P fimbriae, or do not produce MR/P fimbriae at all.
2. For this protocol, *E. coli* CFT073 *fim* locked-on and CFT073 *fim* locked-off strains were used [13]. The *fim* locked strains were generated by mutation of an invertible element in the *fim* promoter, leading to mutants that are either locked into constitutively producing type 1 fimbriae, or do not produce type 1 fimbriae at all. Although most *E. coli* strains produce type 1 fimbriae, not all do.
3. Use whole blood with citrate and EDTA as an anticoagulant, or in Alsever's solution.
4. MR/P fimbriae agglutinate erythrocytes from many different species, including bovine (ox), chicken, guinea pig, horse, sheep, and human [5, 14, 15]. Although guinea pig

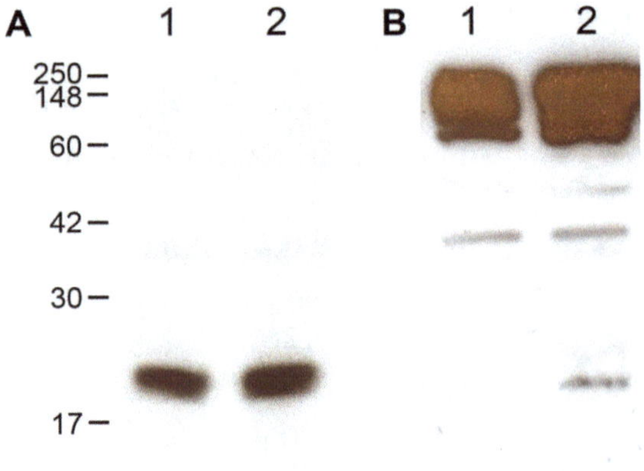

Fig. 3 Acid treatment of fimbriae. Two whole cell lysates of *P. mirabilis* HI4320 were prepared and: (**a**) subjected to acid treatment, or (**b**) not treated with acid. Both sets of lysates were separated by 12.5% SDS-PAGE. MrpA, the major structural subunit of MR/P fimbriae, was detected by immunoblot. In panel (**a**), MrpA runs as a single band, whereas in panel (**b**), MrpA is not uniform and mostly runs near the top of the gel. In lane 2 of panel (**b**), a minor band runs as a monomer. Molecular weight markers are shown on the left, in kilodaltons. Depending on the antibody used, non-acid-treated fimbriae may be detected at the top of the blot, as in panel (**b**), or may not be detected at all

erythrocytes are shown here, chicken erythrocytes have previously been reported to elicit a particularly strong MR/P-dependent agglutination response [14].

5. Mannose-sensitive hemagglutination by *E. coli* results in strong agglutination with erythrocytes of domesticated fowl, guinea pig and horse [16]. Hemagglutination by type 1 fimbriae is inhibited by the addition of mannose [3, 4]. α-methyl mannoside can also be used for this assay, as Gram-negative bacteria are unable to metabolize this mannose analog [3, 4]. To inhibit type 1 fimbrial adherence with α-methyl mannoside, mix 1 mL of 3% guinea pig erythrocytes with 100 μL of 10 mg/mL α-methyl mannoside. Keep this mixture on ice.
6. To examine P fimbrial hemagglutination by *E. coli*, a 3% (vol/vol) solution of human erythrocytes can be used. This reaction is mannose-resistant; hemagglutination occurs in the presence or absence of mannose.
7. Assessment of agglutination or nonagglutination of erythrocytes in a round bottom microtiter plate allows for best interpretation of results. Flat bottom microtiter plates should not be used.
8. Choice of growth medium: rich, complex media like LB result in consistent, robust growth. However, proteins or peptides in the medium may become concentrated along with sheared bacterial proteins. Peptides will interfere with protein

quantification and, depending on size of protein of interest, the ability to view using SDS-PAGE. This may not be a problem, depending on the downstream application. The issue can be avoided using chemically defined media.

9. To maximize fimbrial expression, in general, bacteria may be cultured at 37 °C without aeration for 24–48 h (that is, static culture). Specific culture conditions to induce other fimbriae or different agglutinins will need to be determined empirically. For example, ambient temperature fimbriae (ATF) are produced at room temperature [17].
10. Always keep erythrocytes at 4 °C or on ice. Gently mix by inversion and gently tapping the erythrocyte pellet with a finger.
11. By the end of the third wash, the supernatant should be clear, with no red blood cell lysis. Store blood for no more than 28 days and washed erythrocytes at 4 °C for no more than 2 weeks. The erythrocytes will naturally settle during storage. If the liquid turns pink or red, the cells are lysing and a fresh suspension needs to be prepared. If inconsistent hemagglutination results are observed, prepare a fresh batch of washed erythrocytes.
12. Mannose-resistant *Klebsiella*-like hemagglutination, in which bacteria agglutinate tannic acid-treated bovine erythrocytes in a mannose-resistant manner, has been described for some *P. mirabilis* strains. To obtain tannic acid-treated erythrocytes, mix equal volumes of 3% (vol/vol) washed erythrocytes and 0.01% (w/v) tannic acid in PBS, incubate at 37 °C for 15 min, wash three times and resuspend in PBS to 3% (v/v).
13. Discard 25 μL of bacteria from the final dilution column.
14. This column is used to detect mannose-sensitive hemagglutination, such as that mediated by *E. coli* type 1 fimbriae.
15. If an agglutination reaction has still not occurred, let the plate sit on the bench overnight at room temperature (~23 °C) and check in the morning.
16. Mannose-resistant/*Klebsiella*-like (MR/K) hemagglutination can be detected using tannic acid-treated bovine erythrocytes but not untreated erythrocytes [5]; this reaction is also not inhibited by 50 mM mannose.
17. The 8 h culture time suggested here is a compromise to encourage fimbriation (e.g., stationary phase) but have relatively little cell death and lysis. If contamination with cytoplasmic proteins is not a concern, bacteria may be cultured statically as in Subheading 3.1, **step 3**. This will increase the amount of fimbrial protein in the culture but will also include increased dead cell material and metabolic by-products.

18. If using this alternative wash step, be gentle when transferring cultures to centrifugation bottles. Do not overcentrifuge. Extra or overly vigorous manipulation of bacteria may prematurely shear off the protein(s) of interest, causing them to be lost.

19. This protocol is a modification of an older protocol that used a blender to shear fimbriae [11]. To use this method, add the bacterial suspension to a Waring blender, and blend for 5 min on setting 4. Then, proceed with protein purification (Subheading 3.2, **step 5**).

20. Move quickly between **steps 5** and **6**, because the cell pellet will be loose. Bacteria that are collected with the supernatant will clog the filter.

21. This protocol is written for 15 mL filters, which fit in a typical swinging bucket benchtop centrifuge. Ideally, a concentrator large enough to accommodate the entire sample is used; however, larger centrifugal concentrators may require specialized rotors. If using a larger filter, the sample should still be concentrated to <3 mL. If necessary, transfer sample to a smaller-sized secondary filter to reach the target end volume.

22. Depending on the downstream application, proteins may be sufficiently concentrated that ultracentrifugation is not required. This step can also help get rid of small peptides.

23. Instead of ultracentrifugation, it is also possible to precipitate proteins using trichloroacetic acid (TCA). TCA precipitation can also be done on larger, more dilute samples, but the pellet will be more diffuse and therefore harder to see and resuspend; concentrating first leads to an identifiable solid pellet that is much easier to work with. TCA precipitation is not recommended if bacteria were cultured in a complex, peptide-based medium. However, TCA precipitation will dissociate fimbrial proteins into monomers, and thus, subsequent acid treatment (Subheading 3.3) is not needed. TCA precipitation: in a fume hood, add 50% TCA to the sample, for a final concentration of 10% TCA. Place sample on ice for 10 min to overnight. Spin samples in a microcentrifuge at maximum speed for 15 min at 4 °C, and discard the supernatant. Add 300 μL of cold acetone, tap the pellet gently, and spin at maximum speed for 5 min at 4 °C. Remove the supernatant, and dry the pellet by placing at 37 °C with the tube open. Suspend protein in 500 μL of PBS or 1× SDS loading buffer.

24. Using too much bacteria, such as an undiluted overnight culture, leads to inconsistent results, usually with either incomplete dissociation of fimbriae, or degradation of nonfimbrial proteins.

25. The pellet may be stored at −20 °C at this point until ready to move on to the next steps.
26. Whole cell lysates may alternately be subjected to TCA precipitation (*see* **Note 23**). In this case, make sure that the bacterial pellet is very well suspended, with no clumping. Vortex after addition of TCA. After addition of cold acetone, let pellet sit on ice for 15 min to overnight. Suspend the final pellet in 50 μL of 1× SDS loading buffer.
27. Don't boil samples for too long, or proteins will start to degrade (no more than 15 min). The sample should clear during boiling; if it does not, try using less starting material.
28. If NaOH is added too quickly, the protein will precipitate. The solution should be clear, not cloudy. The amount of NaOH needed to neutralize the acid prep varies, from less than 0.5 μL to 3 μL.
29. Anti-MrpA antibodies used to generate Fig. 3 are from [14]. The polyclonal antibodies were generated by subcutaneous injection of 100 μg of a maltose binding protein (MBP)-MrpA fusion in Freund's complete adjuvant into rabbits, followed by two boosts with 100 μg of protein in incomplete Freund's adjuvant at 4 and 6 weeks after the primary immunization. Antibodies were affinity purified from serum to remove antibodies against MBP and *E. coli* DH5α [14].
30. Sometimes residual peptide carryover from the culture medium can be a problem. Three ways to deal with this are (1) use a chemically defined medium for culturing bacteria; (2) perform the optional wash step in Subheading 3.2, **step 3**; or (3) run the final protein sample slightly into an SDS-PAGE gel and excise the protein band (the shorter peptides will run ahead).

Acknowledgments

We thank Jessica Schaffer for refinement of the shear prep protocol and example of silver-stained sheared proteins. We also thank Xin Li and Chelsea Lane for tailoring the acid treatment protocol to *P. mirabilis*.

References

1. Beachey EH (1980) Receptors and recognition, vol 6. Bacterial Adherence, vol B. Chapman and Hall, London and New York
2. Armbruster CE, Mobley HLT, Pearson MM (2018) Pathogenesis of *Proteus mirabilis* infection. EcoSal Plus 8(1). https://doi.org/10.1128/ecosalplus.ESP-0009-2017
3. Goldhar J (1995) Erythrocytes as target cells for testing bacterial adhesins. Methods Enzymol 253:43–50

4. Duguid JP, Gillies RR (1957) Fimbriae and adhesive properties in dysentery bacilli. J Pathol Bacteriol 74(2):397–411. https://doi.org/10.1002/path.1700740218
5. Old DC, Adegbola RA (1982) Haemagglutinins and fimbriae of *Morganella*, *Proteus* and *Providencia*. J Med Microbiol 15(4):551
6. Bahrani FK, Johnson DE, Robbins D, Mobley HLT (1991) *Proteus mirabilis* flagella and MR/P fimbriae: isolation, purification, N-terminal analysis, and serum antibody response following experimental urinary tract infection. Infect Immun 59(10):3574–3580
7. Duguid JP (1959) Fimbriae and adhesive properties in *Klebsiella* strains. J Gen Microbiol 21:271–286. https://doi.org/10.1099/00221287-21-1-271
8. Kuan L, Schaffer J, Zouzias CD, Pearson MM (2014) Characterization of 17 chaperone-usher fimbriae encoded by *Proteus mirabilis* reveals strong conservation. J Med Microbiol 63(7):911–922. https://doi.org/10.1099/jmm.0.069971-0
9. Busch A, Waksman G (2012) Chaperone-usher pathways: diversity and pilus assembly mechanism. Philos Trans R Soc Lond Ser B Biol Sci 367(1592):1112–1122. https://doi.org/10.1098/rstb.2011.0206
10. Brinton CC Jr (1965) The structure, function, synthesis and genetic control of bacterial pili and a molecular model for DNA and RNA transport in gram negative bacteria. Trans N Y Acad Sci 27(8):1003–1054
11. Li X, Zhao H, Geymonat L, Bahrani F, Johnson DE, Mobley HLT (1997) *Proteus mirabilis* mannose-resistant, *Proteus*-like fimbriae: MrpG is located at the fimbrial tip and is required for fimbrial assembly. Infect Immun 65 (4):1327–1334
12. Li X, Lockatell CV, Johnson DE, Mobley HLT (2002) Identification of MrpI as the sole recombinase that regulates the phase variation of MR/P fimbria, a bladder colonization factor of uropathogenic *Proteus mirabilis*. Mol Microbiol 45(3):865–874
13. Gunther IV NW, Snyder JA, Lockatell V, Blomfield I, Johnson DE, Mobley HLT (2002) Assessment of virulence of uropathogenic *Escherichia coli* type 1 fimbrial mutants in which the invertible element is phase-locked on or off. Infect Immun 70(7):3344
14. Li X, Johnson DE, Mobley HLT (1999) Requirement of MrpH for mannose-resistant *Proteus*-like fimbria-mediated hemagglutination by *Proteus mirabilis*. Infect Immun 67 (6):2822–2833
15. Bahrani FK, Mobley HLT (1993) *Proteus mirabilis* MR/P fimbriae: molecular cloning, expression, and nucleotide sequence of the major fimbrial subunit gene. J Bacteriol 175 (2):457–464
16. Duguid JP, Clegg S, Wilson MI (1979) The fimbrial and non-fimbrial haemagglutinins of *Escherichia coli*. J Med Microbiol 12 (2):213–227. https://doi.org/10.1099/00222615-12-2-213
17. Massad G, Bahrani FK, Mobley HLT (1994) *Proteus mirabilis* fimbriae: identification, isolation, and characterization of a new ambient-temperature fimbria. Infect Immun 62 (5):1989–1994

Chapter 12

Phase Variation of the *mrp* Fimbrial Promoter

Melanie M. Pearson

Abstract

Mannose-resistant *Proteus*-like (MR/P) fimbriae, produced by the uropathogen *Proteus mirabilis*, are major contributors to urinary tract infection. Expression of *mrp* genes is controlled by an invertible element in the *mrp* operon promoter, and promoter orientation is controlled by a single recombinase, MrpI, which determines whether the element is in the on or off orientation. Detailed here is a simple assay to determine the orientation of the invertible element in a population of *P. mirabilis*, a rapid screen to detect element-locked *mrpI* mutants, and quantification of mixtures of on and off bacteria.

Key words Fimbriae, Pili, DNA inversion, Phase variation, Recombinase, MR/P, Invertible element

1 Introduction

One method that bacteria use to control gene expression is an invertible element (IE) within the promoter [1]. The IE is flanked by inverted repeats, and orientation of the element is mediated by one or more recombinases. *Proteus mirabilis* encodes an IE in the promoter of the *mrp* fimbrial operon which contains the transcriptional +1 site (Fig. 1) [2, 3]. Transcription only occurs if the IE is pointing toward the first gene of the operon, *mrpA* ("on orientation").

In *P. mirabilis*, a single recombinase, MrpI, controls the *mrp* IE [4]. Therefore, null mutation of *mrpI* causes the IE to be locked in either the on or off orientation, which results in bacteria that either constitutively produce mannose-resistant/*Proteus*-like (MR/P) fimbriae or do not produce MR/P fimbriae at all [4]. Other bacterial IEs may be more complicated; for example, the *Escherichia coli* type 1 fimbrial operon IE is controlled by two main recombinases, FimB and FimE [5], and some *E. coli* isolates may encode additional recombinases [6]. In cases such as the type 1 fimbria, isolation of IE-locked mutants may be more difficult, but the IE assay described below can be easily adapted to screen IE orientation [7].

Melanie M. Pearson (ed.), *Proteus mirabilis: Methods and Protocols*, Methods in Molecular Biology, vol. 2021,
https://doi.org/10.1007/978-1-4939-9601-8_12,

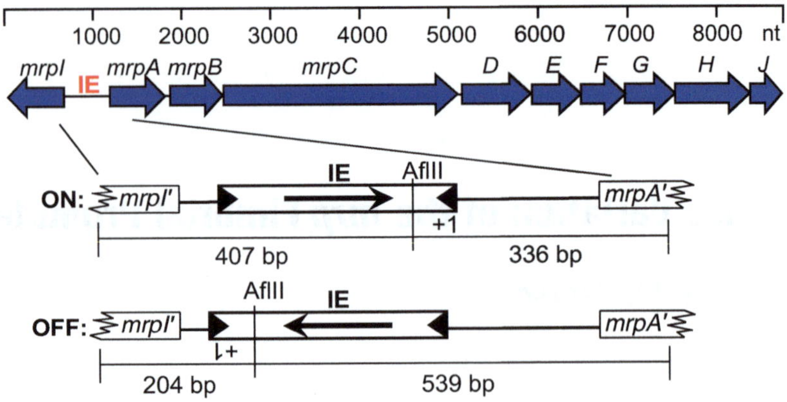

Fig. 1 The *mrp* invertible element. Top, the genes of the *mrp* locus are shown. *mrpA-J* comprise an operon that encodes all components of the MR/P fimbria, plus a transcriptional regulator, MrpJ. Divergently transcribed is *mrpI*, which encodes a recombinase that controls the orientation of an invertible element (IE) in the *mrp* promoter. Bottom, IE schematic. In the forward (on) orientation, the transcriptional start site (+1) points toward *mrpA* and allows transcription of the *mrpA-J* operon. In the off orientation, *mrp* transcription is halted because the transcriptional start site is displaced and in the reverse orientation from the *mrpA-J* operon. An AflII restriction site, located in the IE, can be used to distinguish IE orientation using a PCR-amplified fragment of the *mrp* locus

MR/P fimbriae are a major contributor to urinary tract infection (UTI) [8]. Genes in the *mrp* operon are among the most highly expressed of the entire *P. mirabilis* genome during experimental infection in mice [9]. Correspondingly, bacteria isolated from the bladder overwhelmingly have the *mrp* IE in the on orientation. Bacteria isolated from the kidneys may be either on or off, suggesting that MR/P fimbriae specifically contribute to bladder infection [3, 4].

MR/P fimbriae also contribute to biofilm formation in vitro, and the ability to phase vary is key to robust biofilm development. A locked-off strain does not form much biofilm on coverslips, while a locked-on strain produces a very short but dense biofilm. In contrast, wild-type *P. mirabilis* produces a thick, well-developed biofilm [10]. In most laboratory conditions, including culture on agar plates and aerated broth cultures, the *mrp* IE is in the off orientation [11, 12]. Because of its contribution to virulence and biofilm formation, it can be useful to assess the proportion of *mrp*-expressing bacteria in a given population using an IE assay.

This straightforward IE assay presented here provides information on the IE orientation of a population of *P. mirabilis*, including the relative proportion of on or off bacteria. Bacteria are cultured, adjusted to the same density, and lysed so the IE can be amplified by PCR. The amplicon is digested with a restriction enzyme that cuts within the IE (Fig. 1). Orientation of the element, or proportion of the population in each orientation, is then visible by gel electrophoresis. Use of *mrpI* mutants with the IE locked on or off is strongly recommended as controls. A second, PCR-only assay that is not quantitative but is more amenable to high-throughput screening is also presented.

2 Materials

1. *P. mirabilis* strain to be tested.
2. Control bacteria: MR/P locked-on and locked-off *P. mirabilis* (*see* **Notes 1** and **2**).
3. Heat block that can accommodate 1.5 mL tubes.
4. PCR reagents (*Taq* DNA polymerase and buffer, dNTPs) and thermocycler.
5. Primers Pm IE P1 (5′-GCA TCA ATA AAG GGT TGT GTT TT-3′) and Pm IE P2 (5′-GTA ATT GAG CAA GGA GCA TCA AT-3′). Working stock is 10 μM; store at −20 °C (*see* **Note 3**).
6. AflII restriction enzyme and buffer (store at −20 °C) (*see* **Note 4**).
7. 37 °C incubator.
8. Agarose gel electrophoresis supplies and gel documentation system.
9. Primer IE-flip (only needed for nonquantitative PCR-only assay) (5′-ATT TGA CTT AAG TTA GAA GCC GG-3′). Working stock is 10 μM; store at −20 °C (*see* **Note 3**).

3 Methods

3.1 Quantitative IE Assay

1. Culture *P. mirabilis* under the desired test condition(s) (*see* **Note 5**).
2. Adjust the OD_{600} of the culture to 0.4 (*see* **Note 6**).
3. Incubate 50 μL of the culture in a microcentrifuge tube at 100 °C for 10 min ("boil prep"). Store at −20 °C until ready to use (*see* **Note 7**).
4. Set up a PCR reaction (per 50 μL): 34.5 μL of water, 4 μL of dNTPs, 5 μL of 10× buffer, 2 μL of primer Pm IE P1, 2 μL of primer Pm IE P2, 2 μL of boil prep, and 0.5 μL of *Taq* DNA polymerase. Cycle at 94 °C for 2 min, (94 °C for 45 s, 57 °C for 45 s, 72 °C for 1 min) × 30 cycles, 72 °C for 5 min, hold at 4 °C.

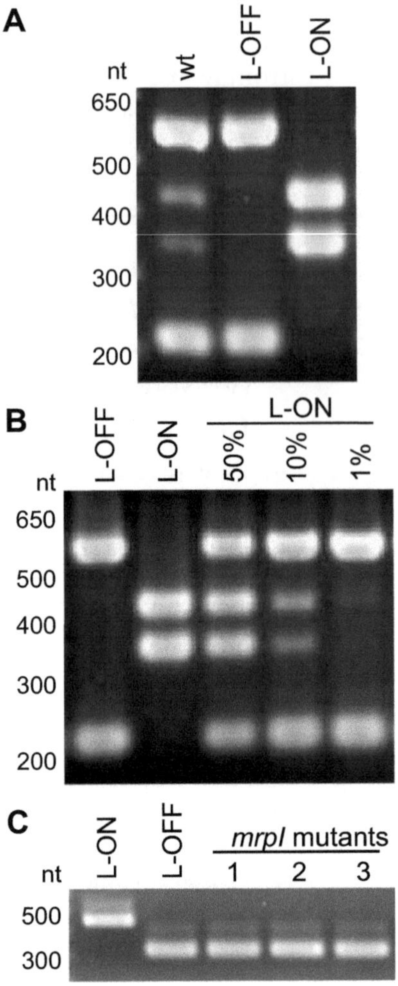

Fig. 2 Invertible element assays. (**a**) IE assay with *mrp* locked-off (L-OFF) and locked-on (L-ON) controls. The wild-type (wt) sample contains a mixture of on and off bacteria. Bacteria were cultured in 125 mL flasks containing 60 mL of LB at 37 °C with aeration for 24 h. (**b**) Controlled mixtures of L-OFF and L-ON bacteria provide standards for estimating on and off populations in a test sample. (**c**) A PCR-only IE assay provides high throughput evaluation of samples but does not quantify the amount of on or off bacteria in a population

5. Digest DNA (*see* **Notes 8** and **9**): mix 15 μL of the PCR product, 2 μL of 10× CutSmart buffer, 2 μL of water, and 1 μL of AflII. Incubate at 37 °C for 2 h.
6. Run 20 μL of the digest on a 2% agarose gel. IE "on" bacteria will result in two bands of 407 and 336 bp, while IE "off" bacteria will result in two bands of 204 and 539 bp. A mixture of on and off bacteria results in four bands. An example of IE assay results is shown in Fig. 2a.

7. If quantification of the ratio of on–off is desired, prepare cultures of *mrp* locked-on and locked-off bacteria as in **steps 1** and **2**. Mix the cultures at the following amounts: 0%, 2%, 5%, 10%, 20%, 30%, 40%, 50%, 60%, 70%, 80%, 90%, 95%, 98%, 100% locked-on. Proceed with the IE assay **steps 3–6**.
8. The proportion of on–off may be quickly assessed by a visual comparison against the standards (Fig. 2b), or the standards may be quantified using densitometry via a gel documentation system.

3.2 PCR-Based Rapid IE Screen

1. Culture *P. mirabilis* under the desired test condition(s) (*see* **Notes 5** and **10**).
2. Add 50 μL of culture to a 1.5 mL tube. Alternately, add 50 μL of water to a 1.5 mL tube, touch a pipet tip to colony or biofilm-cultured bacteria, and add bacteria to water.
3. Incubate at 95 °C for 10 min.
4. Set up PCR reaction (per 40 μL): 30.2 μL of water, 3.2 μL of dNTPs, 4 μL of 10× buffer, 0.4 μL of primer Pm IE P1, 0.4 μL of primer Pm IE P2, 0.4 μL of primer IE-flip, 1 μL of the boil prep, and 0.4 μL of *Taq* DNA polymerase. Cycle at 94 °C for 1 min, (94 °C for 15 s, 56 °C for 30 s, 72 °C for 30 s) × 30 cycles, 72 °C for 3 min, hold at 4 °C (*see* **Note 11**).
5. Check 10 μL of the PCR product on an agarose gel; make sure to include L-ON and L-OFF controls (Fig. 2c).

4 Notes

1. MrpI is the sole recombinase that controls the invertible element [4]. Thus, *mrpI* mutants have an *mrp* invertible element that is locked in either the on orientation or off orientation. These locked mutants serve as controls for the IE assay that will aid assessment of the test population. The mutants are particularly needed if quantification of on–off in the target population is desired.
2. Mutation of the recombinase works well for the *mrp* IE because its orientation is only controlled by MrpI. To produce IE-locked mutants as controls for other IE assays, for example, type 1 fimbriae in *Escherichia coli*, it may be necessary to knock out multiple recombinases [13] or mutate the IE itself [14].
3. These primers were designed based off of the HI4320 genome. If using another strain, ensure these sequences match the target strain, and if necessary, modify. The IE is very AT-rich, and therefore the IE-flip primer must be carefully designed.

4. This assay was optimized for *P. mirabilis* strain HI4320. The presence of the AflII site within the IE must be confirmed in other strains. If there is not an AflII site in the target strain's IE, seek an alternative restriction enzyme that cuts once within the IE, but not elsewhere in the target amplicon.
5. Conditions that favor *mrp* IE on or off are outlined in the introduction. Briefly, conditions that favor *mrp* on include bladders or urine of infected mice, culture in 5% oxygen, or serial passaging of 48 h static cultures in LB medium [9, 11, 15]. Conditions that favor *mrp* off include logarithmic phase cultures, aerated cultures, and agar-grown colonies [11, 12].
6. If culture is below OD_{600} 0.4, use an equivalent amount of bacteria to equal OD_{600} 0.4 in a 50 μL volume. Spin down this larger volume in a microcentrifuge at 10,000 × *g* for 5 min and resuspend in 50 μL of LB. Proceed with boil preparation as usual.
7. Some strains of *P. mirabilis* have higher nuclease activities than others. Using the boil prep immediately in the PCR reaction can prevent degradation of the template. Alternatively, the boil prep may be centrifuged in a microcentrifuge at full speed for 1 min, and the supernatant removed to a new tube for storage at −20 °C.
8. Some sources of AflII may require addition of BSA to the reaction. In that case, substitute 2 μL of 10× BSA for water.
9. No purification of the PCR product is required before proceeding with the restriction digest.
10. The rapid PCR screen is primarily useful for identifying *mrp* locked-on or locked-off mutants. Because the PCR reactions for the on and off orientations are not equally efficient, this assay is not quantitative for the proportion of "on" and "off" bacteria in a population.
11. Although the Pm IE P1 and Pm IE P2 primers theoretically could yield a second PCR product, only one PCR product is obtained due to the higher efficiency of the shorter product with the IE-flip primer coupled with shorter extension times.

Acknowledgments

I gratefully acknowledge Hui Zhao and Xin Li, who originally developed the quantitative IE protocol, and M. Chelsea Lane and Amy Simms, who refined it.

References

1. Blomfield IC (2001) The regulation of pap and type 1 fimbriation in *Escherichia coli*. Adv Microb Physiol 45:1–49
2. Bode NJ, Debnath I, Kuan L, Schulfer A, Ty M, Pearson MM (2015) Transcriptional analysis of the MrpJ network: modulation of diverse virulence-associated genes and direct regulation of *mrp* fimbrial and *flhDC* flagellar operons in *Proteus mirabilis*. Infect Immun 83 (6):2542–2556. https://doi.org/10.1128/IAI.02978-14
3. Zhao H, Li X, Johnson DE, Blomfield I, Mobley HLT (1997) In vivo phase variation of MR/P fimbrial gene expression in *Proteus mirabilis* infecting the urinary tract. Mol Microbiol 23(5):1009–1019
4. Li X, Lockatell CV, Johnson DE, Mobley HLT (2002) Identification of MrpI as the sole recombinase that regulates the phase variation of MR/P fimbria, a bladder colonization factor of uropathogenic *Proteus mirabilis*. Mol Microbiol 45(3):865–874
5. Schwan WR (2011) Regulation of *fim* genes in uropathogenic *Escherichia coli*. World J Clin Infect Dis 1(1):17–25. https://doi.org/10.5495/wjcid.v1.i1.17
6. Sarkar S, Roberts LW, Phan MD, Tan L, Lo AW, Peters KM, Paterson DL, Upton M, Ulett GC, Beatson SA, Totsika M, Schembri MA (2016) Comprehensive analysis of type 1 fimbriae regulation in *fimB*-null strains from the multidrug resistant *Escherichia coli* ST131 clone. Mol Microbiol 101(6):1069–1087. https://doi.org/10.1111/mmi.13442
7. Lim JK, Gunther NW, Zhao H, Johnson DE, Keay SK, Mobley HLT (1998) In vivo phase variation of *Escherichia coli* type 1 fimbrial genes in women with urinary tract infection. Infect Immun 66(7):3303
8. Schaffer JN, Pearson MM (2015) *Proteus mirabilis* and urinary tract infections. Microbiol Spectr 3(5). https://doi.org/10.1128/microbiolspec.UTI-0017-2013
9. Pearson MM, Yep A, Smith SN, Mobley HLT (2011) Transcriptome of *Proteus mirabilis* in the murine urinary tract: virulence and nitrogen assimilation gene expression. Infect Immun 79(7):2619–2631. https://doi.org/10.1128/IAI.05152-11
10. Jansen AM, Lockatell V, Johnson DE, Mobley HLT (2004) Mannose-resistant *Proteus*-like fimbriae are produced by most *Proteus mirabilis* strains infecting the urinary tract, dictate the in vivo localization of bacteria, and contribute to biofilm formation. Infect Immun 72 (12):7294–7305
11. Lane MC, Li X, Pearson MM, Simms AN, Mobley HLT (2009) Oxygen-limiting conditions enrich for fimbriate cells of uropathogenic *Proteus mirabilis* and *Escherichia coli*. J Bacteriol 191(5):1382–1392
12. Massad G, Bahrani FK, Mobley HLT (1994) *Proteus mirabilis* fimbriae: identification, isolation, and characterization of a new ambient-temperature fimbria. Infect Immun 62 (5):1989–1994
13. Bryan A, Roesch P, Davis L, Moritz R, Pellett S, Welch RA (2006) Regulation of type 1 fimbriae by unlinked FimB- and FimE-like recombinases in uropathogenic *Escherichia coli* strain CFT073. Infect Immun 74 (2):1072–1083
14. Gunther IV NW, Snyder JA, Lockatell V, Blomfield I, Johnson DE, Mobley HLT (2002) Assessment of virulence of uropathogenic *Escherichia coli* type 1 fimbrial mutants in which the invertible element is phase-locked on or off. Infect Immun 70(7):3344
15. Bahrani FK, Johnson DE, Robbins D, Mobley HLT (1991) *Proteus mirabilis* flagella and MR/P fimbriae: isolation, purification, N-terminal analysis, and serum antibody response following experimental urinary tract infection. Infect Immun 59(10):3574–3580

Chapter 13

Adherence of *Proteus mirabilis* to Uroepithelial Cells

María José González, Victoria Iribarnegaray, Paola Scavone, and Pablo Zunino

Abstract

Bacterial adherence to eukaryotic cells is mediated by different adhesins that can act at different stages in bacteria–host interaction. Abundant evidence has suggested that adherence is critical for infection by bacterial pathogens. *Proteus mirabilis* is an opportunistic pathogen which frequently infects the human urinary tract, particularly in patients with indwelling urinary catheters. Sequencing of the genome of this pathogen has revealed the existence of a remarkable amount of complete fimbrial operons. In this chapter, we describe in vitro adherence assays of *P. mirabilis* to uroepithelial cells, which can provide relevant results to assess virulence of uropathogenic strains.

Key words *Proteus mirabilis*, Uropathogen, Urinary tract infection, Uroepithelial cells, Adherence, Virulence

1 Introduction

Bacterial adherence to eukaryotic cells involves the participation of different adhesins that can act at the same time or during different stages in bacteria–host interaction events. Both pathogenic and commensal bacteria have to colonize specific environments and overcome natural clearing mechanisms [1].

In the case of pathogenic bacteria, adherence to mucosal surfaces is a crucial stage in the establishment and further progress of the infection. This mechanism, mediated by different factors, must get over different physiological defenses depending on the organs involved like mucus layers, ciliary movements, peristalsis, fluid flow or the presence of native microbiota. Also, complex immunological systems associated with mucosae provide efficient defensive mechanisms to the host against pathogens [2].

Several pathogenic bacteria display fimbriae (also called pili), homopolymeric or heteropolymeric fibers that protrude from the bacterial surface and mediate adhesion to epithelial cell receptors, collaborating in the subsequent successful colonization of the host

Melanie M. Pearson (ed.), *Proteus mirabilis: Methods and Protocols*, Methods in Molecular Biology, vol. 2021, https://doi.org/10.1007/978-1-4939-9601-8_13,

[3]. These virulence factors also mediate interbacterial aggregation and biofilm formation helping also, in some cases, in bacterial adhesion to inert surfaces [4, 5]. However, there is a variety of other nonfimbrial factors that can collaborate in the adhesion of bacteria to the host cells like flagella, outer membrane proteins, autotransporters, capsule, or LPS, among others [2, 6].

Proteus mirabilis is an opportunistic pathogen which frequently infects the human urinary tract of patients with indwelling urinary catheters. It is a ubiquitous bacterium that can be found in different environments and is a commensal of the gastrointestinal tracts of humans and animals [7]. *P. mirabilis* pathogenesis is multifactorial, and several virulence attributes are required for the colonization of the urinary tract [8]. Adhesion to host cells is considered a central pathogenicity mechanism of uropathogens to establish a successful infection [9].

The first *P. mirabilis* sequenced genome revealed that this pathogen has 17 chaperone-usher fimbrial operons, suggesting that it can synthesize more fimbriae than any other bacterial species sequenced to date [10]. Also, *P. mirabilis* encodes a variety of structures potentially related to virulence like autotransporters or even type IV fimbriae [9, 10]. Therefore, it can be assumed that adherence is crucial in the context of *P. mirabilis* urinary tract infections. In this chapter, different in vitro assays designed to assess adherence of *P. mirabilis* to uroepithelial cells are described. Classical methods to assess *P. mirabilis* adherence are usually based on the use of cultured human cells (particularly the bladder carcinoma T24 cell line) or exfoliated uroepithelial cells from urine.

The use of cultured cells for bacterial adhesion assays is extensive. The control of environmental parameters, cell homogeneity, or ease of storage and preservation are advantages associated with cell cultures. On the other hand, these methods have different disadvantages like the possibility of contamination, the need of special equipment and technical expertise, or the chance of genetic variation due to very rapid growth rates, among others [11].

Exfoliated cells may be heterogeneous and influenced by the host physiological status, but they can be easily obtained and manipulated. Different authors have reported in early studies that these assays helped to correlate adherence and infection of the urinary tract [12].

2 Materials

Solutions need to be prepared using ultrapure water. Analytical grade reagents must be used throughout the whole procedure. Carefully read the materials safety data sheet for the proper handling of the products. Follow all existing waste disposal regulations.

2.1 Adherence to Cultured Cells

1. Human bladder T24 cell line (HTB-4TM, Biosafety Level 1) (*see* **Note 1**) [13]. This transitional cell line derived from human urinary bladder carcinoma can be purchased from certified banks (e.g., American Type Culture Collection, ATCC). Cells must be stored in liquid nitrogen vapor phase (storage at −80 °C can result in loss of viability).
2. D-MEM (Dulbecco's Modified Eagle's medium) (*see* **Note 2**) with 2 mM L-glutamine (filter-sterilized) and 1 mM sodium pyruvate. These factors can be added, but ready-to-use media, including all necessary additives, can also be purchased.
3. Fetal bovine serum (FBS).
4. Standard CO_2 cell incubator (5% CO_2, 37 °C).
5. Phosphate-buffered saline (PBS), per liter: 0.41 g of potassium phosphate monobasic (KH_2PO_4), 1.7 g of sodium phosphate dibasic (Na_2HPO_4) and 8.2 g of sodium chloride (NaCl), pH 7.4. Sterilize by autoclaving at 121 °C for 15 min.
6. Trypsin–EDTA solution: 0.025% trypsin and 0.01% EDTA in PBS. Sterilize by filtration using 0.2 μm membranes.
7. Trypan blue dye: 0.4% solution (w/v) trypan blue in PBS, pH 7.4. Sterilize by filtration using 0.2 μm membranes.
8. Neubauer chamber (hemacytometer).
9. 25 cm^2 cell culture flask.
10. Sterile 6-well culture plates and sterile glass coverslips, placed at the bottom of the wells.
11. Luria Bertani (LB) broth: 1% w/v tryptone, 0.5% w/v yeast extract, 0.5% w/v NaCl. Mix thoroughly and sterilize by autoclaving at 121 °C for 15 min.
12. Standard 37 °C incubator for bacterial growth.
13. 4% Paraformaldehyde (PFA) solution in PBS: this is a standard reagent for cell fixation, but it can cause serious damage if managed improperly. Therefore, handling indications provided by the manufacturer must be carefully followed.
14. PBS—Ca/Mg, per liter: 0.1 g of calcium chloride ($CaCl_2$), 0.1 g of magnesium chloride ($MgCl_2{\cdot}6H_2O$), 0.2 g of potassium chloride (KCl), 0.2 g of potassium phosphate monobasic (KH_2PO_4), 2.16 g of sodium phosphate dibasic (Na_2HPO_4), and 8.0 g of sodium chloride (NaCl), pH 7.4. Sterilize by filtration using 0.2 μm membranes.
15. Methylene blue dye (Loeffler formula): Prepare a saturated alcoholic methylene blue solution (1.6 g of the dye in 100 mL of ethanol 95%). Add 30 mL of this solution to 100 mL of 0.01% (w/v) KOH.
16. Optical microscope, equipped with a 100× oil immersion lens.

2.2 Adherence to Desquamated Cells from Urine

1. Pooled urine from three healthy female donors.
2. Centrifuge.
3. Refrigerated centrifuge.
4. Buffered saline with gelatin (BSG): 0.15 M NaCl, 0.004 M KH_2PO_4, 0.1% gelatin, pH 6.2.
5. Glass microscope slides.
6. Methanol (stored at −20 °C).
7. Methylene blue dye (Loeffler formula): Prepare a saturated alcoholic methylene blue solution (1.6 g of the dye in 100 mL of ethanol 95%). Add 30 mL of this solution to 100 mL of 0.01% (w/v) KOH.
8. Optical microscope, equipped with a 100× oil immersion lens.

3 Methods

Media and solutions used for cultured cell manipulation must be prewarmed at 37 °C, unless specific directions are provided.

3.1 Cell Growth and Maintenance

1. Thaw the T24 cell stock vial at 37 °C.
2. Once the cellular stock is thawed, add 5 mL of D-MEM.
3. Rapidly suspend the cells and centrifuge for 5 min at 100 × *g* (20 °C). Discard the supernatant, which contains cryoprotective substances.
4. Add 5 mL of D-MEM supplemented with 10% FBS (D-MEM/FBS) and gently resuspend the cells.
5. Transfer the cell suspension to a 25 cm^2 bottle for culture at 37 °C and 5% CO_2 and incubate until 40–70% confluency (monitored by optic microscopy with low magnification, e.g., 10× objective) is achieved. The cap must remain vented to allow gas exchange.
6. Once desired cell semiconfluency is reached, discard the media and detach the cell monolayer by adding 1 mL of trypsin–EDTA solution for 3 min under incubation conditions (*see* **Note 3**).
7. Once the cells are detached from the plastic substrate, add 4 mL of D-MEM/FBS medium to inactivate trypsin and resuspend the cells.
8. Analyze cell viability by trypan blue vital staining. This technique is based on the principle that viable cells do not take up the dye, while nonviable cells do; therefore, noncolored cells are considered viable while blue cells are considered nonviable. Cell viability should be at least 95% for healthy cultures. Add 0.1 mL of a trypan blue stock solution to 0.1 mL of cells and count using a Neubauer chamber (hemacytometer, usually

used for blood cell counting) to estimate cell concentration and viability [11] (*see* **Note 4**).

9. Wash the cells with PBS that has been prewarmed to 37 °C.
10. Before the addition of the cells to a 6-well culture plate, place a sterile coverslip in the bottom of each well using sterile forceps.
11. Taking into account the concentration assessed after counting, suspend cells in D-MEM/FBS and transfer the cells to 6-well culture plates (1×10^5 cells per well in 2 mL of medium) (*see* **Note 5**).
12. Incubate the cells at 37 °C and 5% CO_2 for 24 h.

3.2 Preparation of *P. mirabilis* Suspensions

1. Subculture *P. mirabilis* in 10 mL of LB broth and incubate statically at 37 °C for 48 h in a 50 mL centrifuge tube (*see* **Note 6**).
2. Centrifuge the cultures at $8000 \times g$ for 15 min at 4 °C.
3. Discard the supernatant and resuspend the bacteria contained in the pellet in 10 mL of PBS.
4. Repeat **step 2** and resuspend bacteria in D-MEM (without FBS) to generate a 2×10^8 colony-forming unit (CFU)/mL suspension (*see* **Note 7**).
5. Use D-MEM to dilute bacteria to 1×10^6 CFU/mL.

3.3 Evaluation of Bacterial Adherence to Cultured Cells

1. Remove culture medium from the T24 cells.
2. Add 2 mL of 1×10^6 CFU/mL bacteria per well of T24 cells. To one well of each plate, do not add bacteria and instead, add medium only as a negative control for the assay.
3. Incubate the plate at 37 °C in 5% CO_2 for the desired time (e.g., 5, 10, and 15 min) (*see* **Note 8**).
4. After incubation, wash the cell monolayer with PBS at 4 °C three times to remove the nonadherent bacteria. Remove the final wash.
5. Fix the coverslips at the bottom of the wells in 1 mL of 4% PFA for 20 min on ice.
6. Remove the PFA and wash the coverslips twice with 1 mL of PBS—Ca/Mg.
7. Use forceps to pick up each coverslip and blot dry.
8. Stain the coverslip with Loeffler methylene blue, by flooding with the dye for 5 min.
9. Wash, blot dry and mount the coverslip on a glass slide.
10. Assess bacterial adherence to the cells by direct counting under an optical microscope using a 100× oil immersion lens (Fig. 1). Record 20 fields and count only bacteria adhered to the cells (*see* **Note 9**). Compare the results of attached bacteria using appropriate statistical tests.

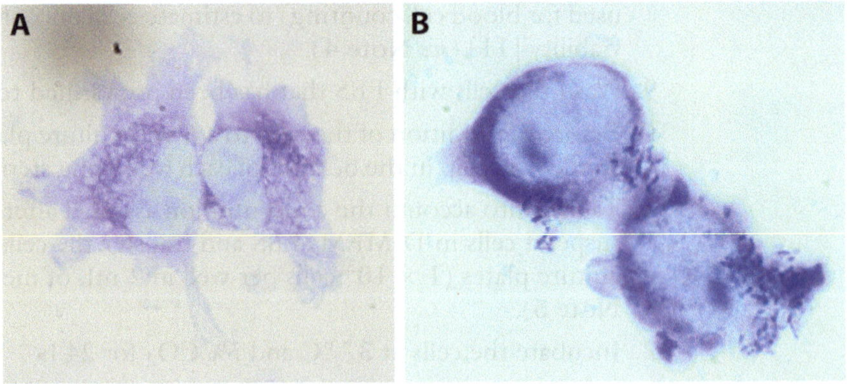

Fig. 1 Adhesion of *P. mirabilis* to T24 cultured cells. (**a**) Cells without attached bacteria. (**b**) Cells with attached bacteria after incubation for 10 min. A ×100 objective was used for cell visualization

3.4 Adherence to Desquamated Cells from Urine

1. Centrifuge the pooled urine of at least three healthy women at 400 × *g* for 10 min to recover the cell sediment [14] (*see* **Note 10**).
2. Remove the supernatant and resuspend the sediment in 30 mL of BSG.
3. Centrifuge at 500 × *g* for 10 min at 20 °C.
4. Repeat **steps 2** and **3** twice more, so three washes are made under the same conditions.
5. Discard the supernatant of the last centrifugation and resuspend the cell pellet in 1 mL of BSG.
6. Assess the cell quality and concentration using a Neubauer chamber (*see* Subheading 3.1, **step 9**).
7. Generate a cell suspension containing 4×10^5 desquamated uroepithelial cells/mL in BSG.
8. Transfer and extend 100 μL of the cellular suspension (without added bacteria) on a glass slide and air dry. Fix with methanol for 5 min, wash, and blot-dry. Stain with Loeffler methylene blue by flooding with the dye for 5 min. Wash and blot dry. Then, examine under an optical microscope using a 100× oil immersion lens.
9. Centrifuge 1 mL of the cell suspension prepared in **step 4** and remove the supernatant.
10. In parallel, prepare bacterial suspensions containing 1×10^9 CFU/mL of *P. mirabilis* bacteria in BSG.
11. Add 1 mL of the bacterial suspension to the tube containing the uroepithelial cell pellet prepared in **step 7**. Resuspend and incubate for 1 h at 37 °C with gentle agitation.
12. After incubation, centrifuge the tubes at 400 × *g* for 10 min, and discard the supernatant. Resuspend in 1 mL of BSG and

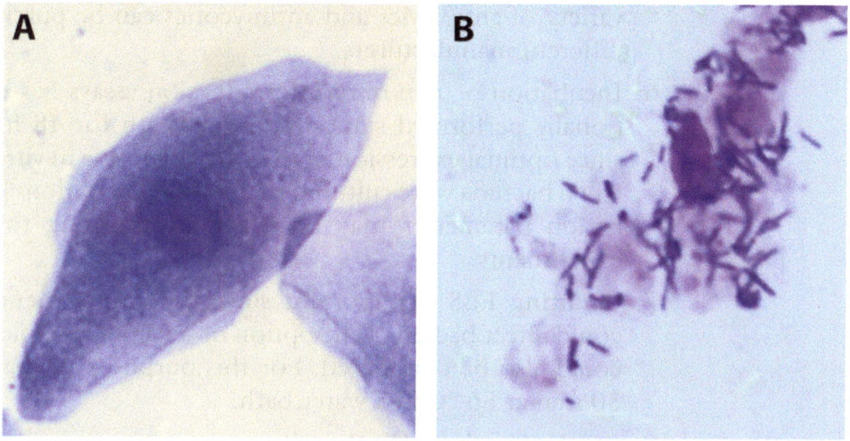

Fig. 2 Adhesion of *P. mirabilis* to desquamated uroepithelial cells. (**a**) Cells without attached bacteria. (**b**) Cells with attached bacteria after incubation for 1 h. A ×100 objective was used for cell visualization

repeat the procedure, so the cells are washed five times to remove nonadhered bacteria.

13. After the last wash, resuspend the cell pellet in 1 mL of BSG and prepare smears for light microscopy as described in Subheading 3.3, **steps 5–10** (Fig. 2). Count bacteria attached to 50 uroepithelial cells. Express the results as the number of adhered bacteria per cell [15].

4 Notes

1. The T24 cell line, derived from human bladder uroepithelial carcinoma, is the most frequently used cell line for adherence assays [6]. However, other studies have been performed using alternative cell lines (e.g., Vero cells [6] or EJ/28 [16]).
2. D-MEM (Dulbecco's Modified Eagle's medium) is widely used to support the growth of a wide range of mammalian cells. Other media may also be used, like McCoy's 5a medium supplemented with 2 mM glutamine and 10% FBS for cell growth.
3. After incubating for 3 min, the bottle can be gently tapped a few times to help cell detachment. Take into account that the use of high trypsin concentrations for long periods may severely damage or even kill the cells.
4. When T24 cell 100% confluency is reached, a cell yield of about 3×10^6 cells can be expected.
5. Antibiotics and antimycotics can be added to the cell culture medium in the first steps of the experiments but not when the cells are finally cultured for the bacterial adhesion assay since these agents could affect the viability of added bacteria. A wide

variety of antibiotics and antimycotics can be purchased from different manufacturers.

6. Incubation of *P. mirabilis* for adhesion assays has been traditionally performed statically in LB broth for 48 h at 37 °C, since optimal expression of the MR/P fimbria in vitro occurred when bacteria were cultured under these conditions [17]. Incubation parameters may be changed according to particular experiments.
7. Omitting FBS is important since the complement in serum could affect bacteria. The option of serum decomplementation could also be considered. For this purpose, incubate FBS for 30 min at 56 °C in a water bath.
8. Incubation times for the adherence assay cannot be long since *P. mirabilis* cytotoxins exert a strong deleterious effect on the cell monolayer.
9. Since *P. mirabilis* highly adheres to abiotic surfaces, microscopy-based counting methods are chosen rather than the tests based on cell detachment and lysis and bacterial plate-counting. To use the plate-counting method, after the last wash, add 1 mL of PBS with 0.5% (v/v) Triton X-100 per well and incubate for 15 min at 37 °C in 5% CO_2 to lyse eukaryotic cells. Dilute bacteria in LB, plate on LB agar (without NaCl, to avoid *P. mirabilis* swarming), and incubate overnight at 37 °C. Determine the number of remaining attached bacteria by plate counts of viable bacteria.
10. Regarding the adherence assays based on urine exfoliated cells, variations can be performed, like using scraped vaginal cells [18].

References

1. Rendón MA, Saldaña Z, Erdem AL, Monteiro-Neto V, Vázquez A, Kaper JB, Puente JL, Girón JA (2007) Commensal and pathogenic *Escherichia coli* use a common pilus adherence factor for epithelial cell colonization. Proc Natl Acad Sci U S A 104(25):10637–10642
2. Brumell JH, Finlay BB (2000) Bacterial adherence, colonization, and invasion of mucosal surfaces. In: Brogden K, Roth J, Stanton T, Bolin C, Minion F, Wannemuehler M (eds) Virulence mechanisms of bacterial pathogens, 3rd edn. ASM Press, Washington, DC, pp 3–17
3. Jonson AB, Normak S, Rhen M (2005) Fimbriae, pili, flagella and bacterial virulence. In: Russell W, Herwarld H (eds) Concepts in bacterial virulence, vol 12. Karger Publishers, Basel, pp 67–89
4. Klemm P, Schembri M (2004) Type 1 fimbriae, curli, and antigen 43: adhesion, colonization, and biofilm formation. EcoSal Plus 1(1). https://doi.org/10.1128/ecosalplus.8.3.2.6
5. Scavone P, Iribarnegaray V, Caetano AL, Schlapp G, Härtel S, Zunino P (2016) Fimbriae have distinguishable roles in *Proteus mirabilis* biofilm formation. Pathog Dis 74(5):pii: ftw033. https://doi.org/10.1093/femspd/ftw033
6. Scavone P, Villar S, Umpiérrez A, Zunino P (2015) Role of *Proteus mirabilis* MR/P fimbriae and flagella in adhesion, cytotoxicity and genotoxicity induction in T24 and Vero cells. Pathog Dis 73:pii: ftv017. https://doi.org/10.1093/femspd/ftv017
7. Armbruster CE, Mobley HLT (2012) Merging mythology and morphology: the multifaceted

lifestyle of *Proteus mirabilis*. Nat Rev Microbiol 10(11):743–754

8. Nielubowicz GR, Mobley HLT (2010) Host-pathogen interactions in urinary tract infection. Nat Rev Urol 7(8):430–441
9. Alamuri P, Löwer M, Hiss JA, Himpsl SD, Schneider G, Mobley HLT (2010) Adhesion, invasion, and agglutination mediated by two trimeric autotransporters in the human uropathogen *Proteus mirabilis*. Infect Immun 78 (11):4882–4894
10. Pearson MM, Sebaihia M, Churcher C, Quail MA, Seshasayee AS, Luscombe NM, Abdellah Z, Arrosmith C, Atkin B, Chillingworth T, Hauser H, Jagels K, Moule S, Mungall K, Norbertczak H, Rabbinowitsch E, Walker D, Whithead S, Thomson NR, Rather PN, Parkhill J, Mobley HL (2008) Complete genome sequence of uropathogenic *Proteus mirabilis*, a master of both adherence and motility. J Bacteriol 190 (11):4027–4037
11. Freshney RI (1987) Culture of animal cells: a manual of basic technique, 2nd edn. Alan R. Liss, New York
12. Bruce AW, Chan RC, Pinkerton D, Morales A, Chadwick P (1983) Adherence of gram-negative uropathogens to human uroepithelial cells. J Urol 130(2):293–298
13. Shih C, Weinberg RA (1982) Isolation of a transforming sequence from a human bladder carcinoma cell line. Cell 29(1):161–169
14. Wray S, Hull S, Cook J, Barrish J, Hull R (1986) Identification and characterization of urinary cell adhesin from a uropathogenic isolate of *Proteus mirabilis*. Infect Immun 54 (1):43–49
15. Zunino P, Sosa V, Allen AG, Preston A, Schlapp G, Maskell DJ (2003) *Proteus mirabilis* fimbriae (PMF) are important for both bladder and kidney colonization in mice. Microbiology 149(11):3231–3237
16. Latta RK, Schur MJ, Tolson DL, Altman E (1998) The effect of growth conditions on *in vitro* adherence, invasion, and NAF expression by *Proteus mirabilis* 7570. Can J Microbiol 44(9):896–904
17. Bahrani FK, Mobley HLT (1994) *Proteus mirabilis* MR/P fimbrial operon: genetic organization, nucleotide sequence, and conditions for expression. J Bacteriol 176(11):3412–3419
18. Szabo RJ, Shortliffe LM, Stamey TA (1987) Adherence of *Escherichia coli* and *Proteus mirabilis* to human transitional cells. J Urol 137 (4):793–797

Chapter 14

An In Vitro Bladder Model for Studying Catheter-Associated Urinary Tract Infection and Associated Analysis of Biofilms

Jonathan Nzakizwanayo, Harriet Pelling, Scarlet Milo, and Brian V. Jones

Abstract

Urethral catheters are among the most widely used medical devices, applied to manage a wide range of conditions in hospital, community, and care home settings. In long-term catheterized individuals, infection with *Proteus mirabilis* frequently complicates the care of patients owing to formation of extensive crystalline biofilms. Here we describe the use of an in vitro bladder model of the catheterized urinary tract and associated analyses to study *P. mirabilis* crystalline biofilm formation. The model originally described by Stickler et al. (1999, 310:494–501, Methods Enzymol) replicates a complete sterile closed drainage system as used in clinical practice, and permits formation of biofilms directly on catheters under conditions representative of those encountered in vivo. Models may be used to replicate either established infection or early stage colonization, and we describe a range of associated methods for quantification and visualization of biofilms formed on catheters. These methods are also easily adapted to study catheter-associated biofilm formation by other urinary tract pathogens.

Key words *Proteus mirabilis*, Catheter-associated urinary tract infection, Urethral catheters, Bladder models, Catheter blockage, Crystalline biofilms

1 Introduction

Urinary catheters are the most commonly used medical devices and are employed in a wide range of healthcare settings, ranging from short-term use in hospitalized patients to long-term management of incontinence in community care or nursing home residents [1–6]. It has been estimated that over 100 million urethral catheters are fitted each year, with many used for long-term bladder management where catheters are typically in place for 4–12 weeks between changes [4]. Although catheterization provides significant benefits to many patients, these are often undermined by the acquisition of a catheter associated urinary tract infection (CAUTI) and, given the scale of catheter use, it is perhaps unsurprising that CAUTI are the most common infections in many healthcare systems. Patients undergoing long-term catheterization are at most

Melanie M. Pearson (ed.), *Proteus mirabilis: Methods and Protocols*, Methods in Molecular Biology, vol. 2021,
https://doi.org/10.1007/978-1-4939-9601-8_14,

risk of acquiring a CAUTI, with those catheterized for a period greater than 28 days almost certain to become bacteriuric [7]. Although in many cases colonization may be asymptomatic, the presence of urease-producing species can lead to clinical complications that pose a serious risk to patient health.

Proteus mirabilis is a particularly problematic pathogen in this regard, associated with up to 45% of CAUTIs, and causing particular problems in the care of long-term catheterized patients [8–10]. A central feature and major complication of *P. mirabilis* CAUTI is the ability of this organism to form extensive crystalline biofilms that encrust catheters and block urine flow [9, 11–14]. Once catheters become blocked, infected urine will accumulate and often reflux to the kidneys leading to pyelonephritis, septicemia, and endotoxic shock [4, 6, 9]. Outside the hospital setting, blockage is often not detected until more serious clinical symptoms arise [4]. Around 50% of long-term catheterized patients will experience blockage, most commonly related to *P. mirabilis* infection, and for many this becomes a chronic recurring problem [2, 15].

Along with a strong ability to attach to catheter surfaces and form biofilms, *P. mirabilis* also produces a highly potent urease enzyme, which drives the mineralization of biofilms and catheter encrustation [9, 11, 12]. Hydrolysis of urea by the *P. mirabilis* urease enzyme generates ammonia which increases urinary pH to highly alkaline levels (~pH 8–9) [9, 11, 12]. Under these conditions, calcium and magnesium phosphates precipitate and form microcrystals that become trapped in the growing biofilm matrix. Within the biofilm, crystal growth is enhanced and stabilized by the highly alkaline microenvironment, and the attraction of calcium and magnesium ions to the biofilm matrix [9, 13, 14, 16–18]. Eventually this leads to the calcification and mineralization of the biofilm, and formation of extensive crystalline biofilm structures which occlude urine flow.

Here we describe the setup and operation of an in vitro model of the catheterized urinary tract, and associated methods to evaluate the formation of *P. mirabilis* crystalline biofilms on urinary catheters. This "bladder model" was originally developed and described by Stickler et al. [19], and protocols for model setup and operation are adapted from their original work. This model simulates a complete closed-drainage system as used in clinical practice and permits the use of human urine or artificial urine medium. The "bladder" consists of a double-walled glass vessel which is maintained at 37 °C. The model can be used to simulate a range of infection scenarios from initial low level colonization to more established infection [13, 14, 19–21].

The model has proven to be a highly valuable resource in the study of bacterial biofilm formation and CAUTI. Examples include the application of this model to understand the basic mechanisms of *P. mirabilis* biofilm formation [13, 14, 22–26], to determine the ability of commercially available catheters to resist colonization and encrustation [27], to develop novel early warning systems for catheter blockage [28, 29], and to test novel approaches for the control and prevention of catheter encrustation and blockage [20, 21, 29–32]. However, it should also be noted that the model and protocols presented here may also be readily adapted to study biofilm formation by other uropathogens in this system.

2 Materials

2.1 In Vitro Bladder Models

2.1.1 Preparation of Artificial Urine Stock Solutions

1. Artificial urine Solution 1: 3.25 g of magnesium chloride hexahydrate, 23 g of sodium chloride, 11.5 g of sodium sulfate, 3.25 g of trisodium citrate, 0.1 g of disodium oxalate, 14 g of potassium dihydrogen phosphate, 8 g of potassium chloride, 5 g of ammonium chloride, 25 g of gelatin, 5 g of tryptone soya broth. Initially dissolve compounds in 850 mL of deionized water, adjust the pH to ~5.7 using NaOH or HCl as appropriate, adding dropwise to prevent precipitation, then bring to a final volume of 1 L by addition of deionized water. Adjusting the pH to ~5.7 at this stage is important to ensure that the final pH of 1× strength artificial urine will be appropriate once reconstituted and diluted for use in models. This solution should be sterilized by autoclaving, and then be stored for up to three months at room temperature (*see* **Note 1**).
2. Artificial urine Solution 2: Dissolve 2.45 g of calcium chloride and 125 g of urea in 400 mL of sterile deionized water. These components cannot be autoclaved and must be sterilized by filtration through 0.45 μm pore size filters. Once sterilized, this solution can be stored for up to three months at room temperature (*see* **Notes 1** and **2**).

2.1.2 Bladder Model Assembly

1. Double-walled glass vessel representing the "bladder." This is custom-made glassware (Fig. 1, *see* **Note 3**).
2. Sterile Foley catheter (size 14 Ch).
3. Urine collection bag and bag stand (e.g., Bard, UK).
4. 5 L glass aspirator bottle.
5. Circulating water bath.
6. Single hole rubber stoppers (40 mm diameter).
7. Borosilicate glass tubing, 1 mm bore.
8. Peristaltic pump (e.g., Watson-Marlow, 323S/D).
9. Peristaltic pump tubing (1.6 mm bore, 1.6 mm wall thickness, silicone, for Watson-Marlow 323S/D).
10. All silicone tubing (0.1 cm bore and 0.5 cm bore).

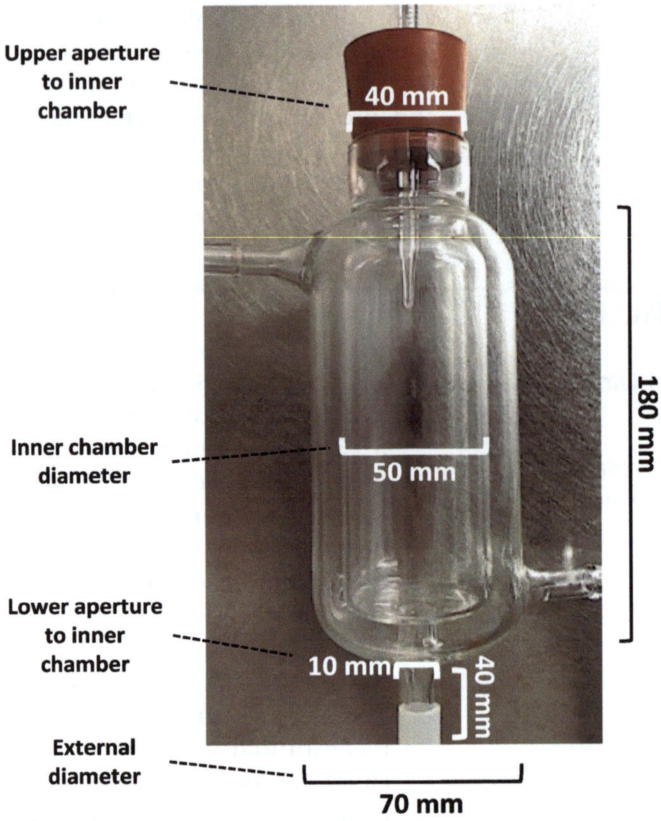

Fig. 1 Image of double-walled glass vessels acting as the "bladder" in models. The image indicates the key features and their dimensions. The capacity of the inner chamber is ~200 mL

11. Tubing connectors.
12. Cable ties.
13. Sterile artificial urine media.
14. Tubing clips or gate clamps.
15. Retort stands.

2.1.3 Inoculation and Operation of the Bladder Model

1. Inoculum: exponential phase batch culture of the test organism growing in suitable growth medium (e.g., *Proteus mirabilis* cultured in tryptone soya broth).
2. Solidified agar medium appropriate for enumeration of viable cells of organisms being tested. For *P. mirabilis*, this medium must also suppress swarming to permit development of single colonies for enumeration (e.g., MacConkey agar without salt).
3. Small volume (~200 mL) of artificial urine media for preparation of inoculum.
4. 70% (v/v) Industrial methylated spirit (IMS).

5. Sterile phosphate-buffered saline (PBS; 10 mM phosphate, 150 mM sodium chloride, pH 7.4).
6. Centrifuge tubes (50 mL).
7. Sterile universal tubes.
8. Sterile colony spreader.

2.2 Quantification of Biofilms

2.2.1 Measurement of Crystalline Biofilm Calcium

1. Centrifuge tubes (10 mL).
2. Scalpel.
3. Solutions of calcium standards 0, 25, 50, 75, and 100 ppm (*see* **Note 4**).
4. Ammonium oxalate and oxalic acid solution: 95% ammonium oxalate and 5% oxalic acid [vol/vol]; prepare from 0.1 M stock solutions.
5. 0.05 M perchloric acid.
6. Vortex mixer.

2.2.2 Measurement of Biofilms by Crystal Violet Staining

1. Centrifuge tubes (50 mL).
2. Scalpel.
3. Deionized water.
4. 0.5% w/v crystal violet solution.
5. Dimethylsufoxide (DMSO).
6. Microcentrifuge tubes (1.5 mL).
7. Vortex mixer.
8. Spectrophotometer.
9. Semi-micro cuvettes.

2.3 Scanning Electron Microscopy

2.3.1 Environmental Scanning Electron Microscopy of Catheter Cross Sections

1. Fully hydrated catheter samples freshly harvested from bladder model experiments.
2. Scalpel.
3. Aluminum stubs.
4. Tissue-Tek CRYO-OCT compound mountant (Agar Scientific) or comparable mounting compound.
5. Environmental scanning electron microscope (e.g., Zeiss evoLS15) equipped with Peltier cooling stage.

2.3.2 Scanning Electron Microscopy of Catheter Cross Sections

1. Fully hydrated catheter samples freshly harvested from bladder model experiments.
2. Scalpel.
3. Aluminum stubs.
4. Leit adhesive carbon mounting tabs.
5. Desiccator (nonvacuum, typically with silica desiccant).

6. Sputter coater (e.g., Agar Scientific AGB7431).
7. Scanning electron microscope (e.g., JEOL JSM-6480LV; Zeiss evoLS15).

3 Methods

3.1 In Vitro Bladder Models

3.1.1 Preparation of Artificial Urine

1. Combine Solution 1 and Solution 2 in sterile glass aspirators (which will supply bladder models), and add a further 3.6 L of sterile deionized water to make 5 L of 1× strength artificial urine (*see* **Note 5**). The final pH of 1× artificial urine media should be ~6.1 once reconstituted, and can be stored at room temperature before use in models (*see* **Note 6**).
2. Retain a small volume of 1× strength artificial urine media (~200 mL) for later use in preparing inocula for models.

3.1.2 In Vitro Bladder Model Assembly

1. An overview of the bladder model assembly is shown in Fig. 2. Sterilize all components by autoclaving prior to model assembly. Conduct the main stages of model set up aseptically in a laminar flow cabinet, during which a closed drainage system is established. It is usually convenient to sterilize components in autoclave bags, but the ends of tubing should also be covered with aluminum foil and secured with autoclave tape. This ensures the tubing will remain sealed and sterile during handling and setup of the model in the laboratory.
2. The key component of the model is a bespoke double walled glass vessel which represents the "bladder" (Fig. 1, *see* **Note 3**). The outer chamber surrounds the inner vessel and has its own inlet and outlet, through which water at 37 °C may be circulated to maintain the inner vessel at body temperature (Fig. 2, *see* **Note 7**). The inner vessel holds media/cultures, and the balloon and tip of catheters during model operation. This has a large diameter inlet at the top (40 mm) that can accommodate a standard rubber closure (through which media is supplied), and a narrow outlet at the bottom through which media is drained. A short 10 cm section of 0.5 cm bore silicone tubing is attached to the narrow outlet of the inner vessel to simulate the urethra.
3. Closures for the inner vessel upper inlet consist of standard rubber stoppers with single holes, through which borosilicate glass tubes are inserted so that short ~2 cm sections emerge from the distal side of the stopper that will be inserted into the "bladder." The glass tubing is extended with silicone tubes that will supply media from aspirators via a peristaltic pump.

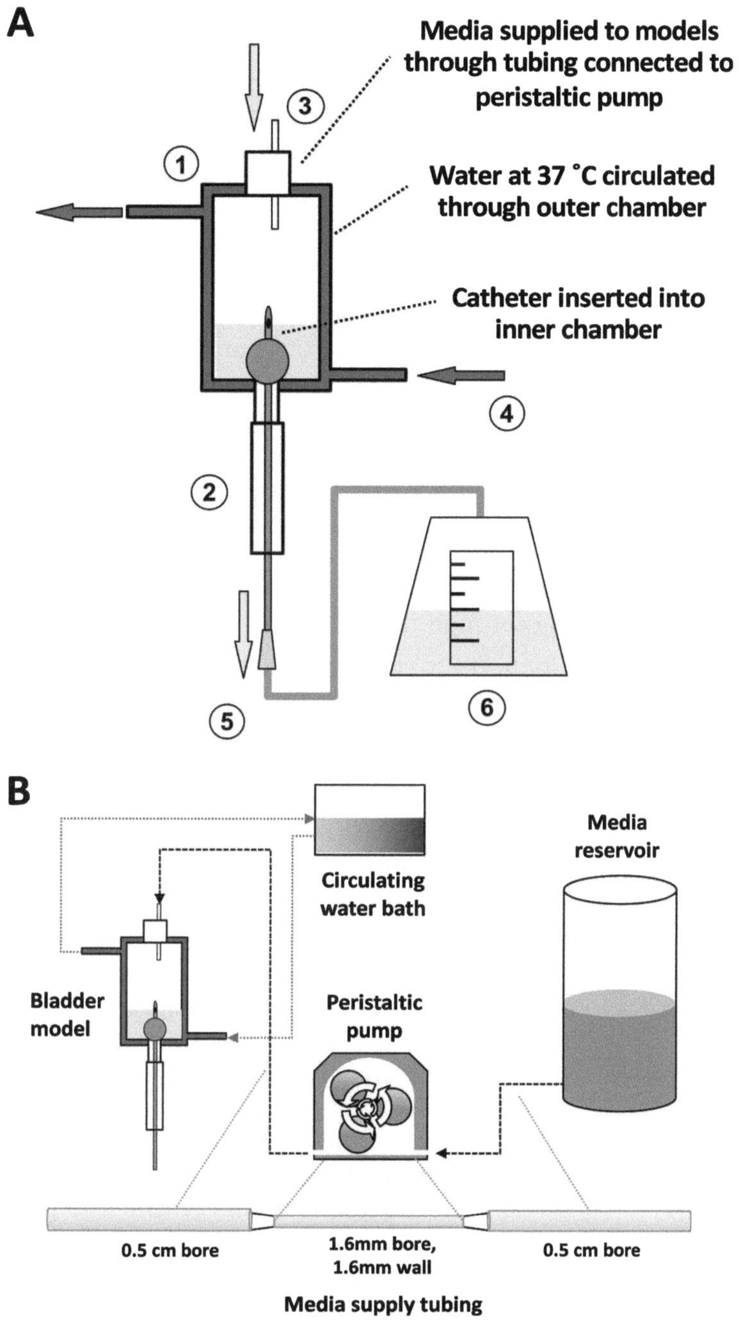

Fig. 2 Illustration of main components and connections in the in vitro bladder model system. According to specifications originally described by Stickler et al. [19]. (**a**) Bladder model, closed drainage system, and media supply: (1) Double walled glass vessel representing the bladder. (2) Foley catheter inserted into the model and connected to drainage bag to form sterile closed drainage system. (3) Sterile urine/artificial urine supplied to "bladder" via a peristaltic pump at a constant flow rate (*see* **Note 11**). (4) Water at 37 °C circulated through outer bladder model chamber to maintain constant temperature (*see* **Note 7**). (5, 6) Drainage tubing and bag collects urine outflow. Diagram is adapted from Stickler et al. [19] and Nzakizwanayo et al. [20]. (**b**) Schematic showing key features of media supply tubing, and flow of media and water through different parts of the model. Media flow is governed by the peristaltic pump, whilst water flow through the bladder model outer chamber is driven by the circulating water bath. Dimensions of tubing associated with the peristaltic pump supplying media should be adjusted to reflect the unit used, and this example relates to use of a Watson-Marlow 323S/D unit (*see* **Note 8**)

4. Media supply tubing consists of 0.1 cm bore silicone tubing of sufficient length to connect media reservoirs with the bladder model, without becoming stretched. The exact length will be determined by the specific set up in your laboratory. Media supply tubing should also contain short sections where the main tubing is connected to thick walled, smaller bore tubing that will pass through the peristaltic pump. This ensures pumps are able to efficiently supply media at the desired flow rate. These short sections of "pump tubing" are connected to the main media supply tubing using tubing connectors and secured by cable ties (Fig. 2 and *see* **Note 8**).
5. Insert a standard Foley catheter into the inner chamber through the narrow outlet at the base and inflate the retention balloon with 10 mL of sterile water (*see* **Note 9**). The balloon is used to create a watertight seal between the bottom of the vessel and the catheter so that media can only escape by draining through the catheter eye-hole.
6. Connect the catheter to a drainage bag to establish a complete sterile closed drainage system. The rubber stoppers with attached media supply tubing are inserted into the large inlet aperture of the inner vessel, but the sterile foil seal at the distal end of media supply tubing is left intact. At this point the main sterile components of the model are assembled and the integrity of the system established. Models may then be carefully transferred to the laboratory bench for final stages of assembly.
7. Hold bladder models in retort stands at similar heights, sufficient to allow the bottom of each catheter to be taped to the bench without stretching. Catheters should be held taut with no slack to ensure balloons retain seals in the inner vessel, but should not be stretched (*see* **Note 10**). Catheters should also be free from kinks and bends that could alter urine flow. Media supply tubing can be connected to aspirators containing sterile artificial urine, and appropriate sections run through peristaltic pumps, but the supply should remain clamped off at this stage using standard clips or gate clamps.
8. Once individual models are set up on the bench and all supply tubing connected, connect the outer chambers of models to a circulating water bath set at 37 °C. Three to four models can be connected in series, by using short sections of 0.5 cm bore silicone tubing to connect the outlet of the first model to the inlet of the next model. The outlet from the last model in the series should return to the water bath (*see* **Note 7**).
9. Once water supply to outer vessels has been established, activate medium supply to the inner vessel. During model operation, peristaltic pumps should supply artificial urine media at a constant flow rate of ~0.7 mL/min (*see* **Note 11**). However,

on activation, the supply rate may be increased so that a residual volume of urine (~ 30 mL) collects in the bladder below the catheter eye-hole level, before flowing through the drainage tube to the collection bag. Pause the flow, and then empty any residual media in drainage bags, before resuming flow at normal speed (*see* **Notes 12** and **13**).

10. Allow uninoculated models to run for 30 min to confirm the urine flow is constant and the system is running as an intact closed drainage system without leakage (*see* **Note 8**). After 30 min, measure the volume of urine that has collected in drainage bags, to verify the rate of urine flow is as expected and similar between individual models.

3.1.3 Bladder Model Operation

1. Remove 10 mL of artificial urine medium from the bladder model inner chamber using a sterile pipette. This can be retained in a sterile universal bottle for later use in **step 4** below. The stopper should be removed for the minimal duration required and replaced immediately (*see* **Note 14**).
2. To initiate growth of biofilms in the bladder model system, inoculate models with bacterial cultures directly into the inner chamber of the bladder model.
3. Culture the *P. mirabilis* strains to be used in bladder models in rich media such as tryptone soya both, at 37 °C with aeration, until mid-exponential phase of growth (OD_{600} ~1.0). For each model, 10 mL of culture will be required for inoculation when models simulating established infection are to be run, where 10^{10} CFU is inoculated into models (*see* **Note 15**).
4. Harvest cells by centrifugation at 4500 × *g* for 10 min at room temperature, discard the supernatant and resuspend cells in 10 mL of artificial urine (it is convenient to use the aliquot of artificial urine taken from models in **step 1** above). For most strains this will provide an initial inoculum of 10^{10} CFU.
5. Introduce the 10 mL cell suspension directly into the bladder model vessel inner chamber, and gently mix with the residual artificial urine media. Remove a 5 mL aliquot of the established cell suspension from the bladder and use to measure the pH and enumerate levels of viable cells at the start of experiments.
6. Allow bacterial cultures to establish in the bladder models for 1 h, before activating bladder models by initiating urine flow, at a constant rate of ~0.7 mL per min. During this time, process aliquots taken from models after inoculation to enumerate viable cells by serial dilution in PBS and plating on solid media. Record the pH of this aliquot (*see* **Notes 16** and **17**).
7. Several options are available in terms of the duration bladder models are allowed to run for. For most experiments with *P. mirabilis*, it is typical to run models until catheters become

blocked and urine begins to accumulate in the "bladder" (past the level of the catheter eye-hole and tip) (e.g., [14, 20, 21]). In this scenario, drainage bags should be emptied regularly and the length of time models have been running noted on each occasion. When it is apparent that the catheters have become blocked, the final volume of urine in the drainage bag can be measured and used to calculate the time of blockage based on the constant flow rate in the system and the activation time. When models are used to simulate an established infection with inocula of ~10^{10} CFU, most strains of *P. mirabilis* will block standard 14 Ch catheters in ~18–24 h.

8. Alternatively, bladder models can be run for a set period of time. This mode of operation is most useful when it is necessary to directly compare biofilms formed by two distinct strains. A period of ~10 h is typically sufficient for most wild-type strains of *P. mirabilis* to form well-developed biofilms on catheters that may be evaluated by approaches described in sections below.
9. At the end of the experiments, deactivate flow and clamp off media supply to prevent leakage of any artificial urine remaining in reservoirs. Collect 5 mL aliquots of the residual urine from the "bladder" in each model, and use to measure pH and enumerate viable cells at the end of experiments (*see* **Notes 16** and **17**).
10. Remove and discard the remaining cultures in the bladder model inner chamber. Once bladders are empty, carefully deflate the catheter balloon, and remove the catheter through the base of the model. If required, catheters can be retained and the biofilms formed on these subjected to further analyses to visualize the architecture and/or quantify the amount of encrustation or biofilm formed (as described below).

3.2 Quantification of Biofilms Formed on Catheter Surfaces

The amount of crystalline biofilm or encrustation formed on catheters can be evaluated by quantifying the amount of calcium on sections of catheters using flame photometry [13, 14, 20, 21].

3.2.1 Measurement of Crystalline Biofilm/Encrustation

1. Section catheters removed from bladder model experiments using a scalpel according to the schematic in Fig. 3. Retain 1 cm sections of interest and further cut longitudinally to expose the catheter lumen and biofilm.
2. Place the sections in separate 10 mL centrifuge tubes and submerge in 2 mL of ammonium oxalate and oxalic acid solution. Vigorously mix immersed sections by vortexing for 3 min to disrupt biofilms and then incubate at room temperature for 30 min to dissolve calcium (*see* **Note 18**).

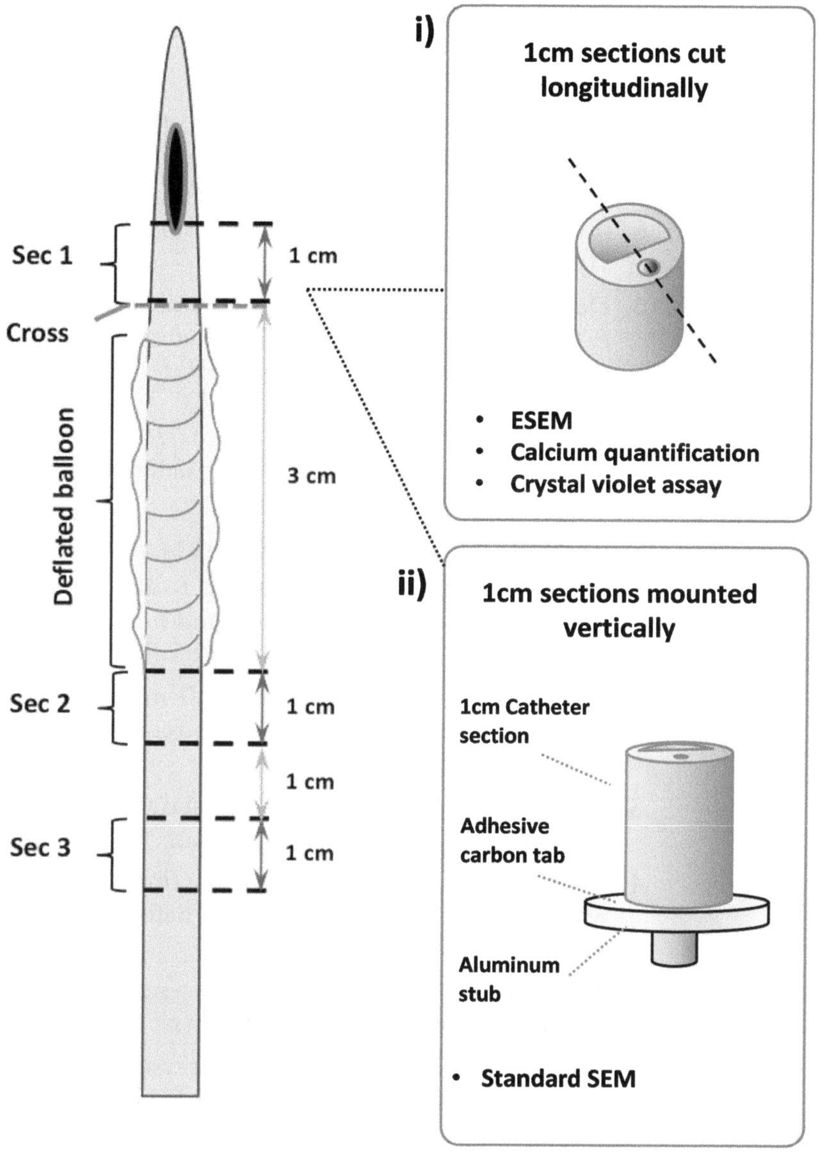

Fig. 3 Sectioning of catheters for imaging by electron microscopy or use in assay to quantify biofilm formation: Adapted from Holling et al. [14] and Nzakizwanayo et al. [20]. *See* also **Note 18**. Catheters obtained from bladder model experiments may be dissected as shown in the main illustration. 1 cm sections for analysis may be obtained from points over the proximal 8 cm starting at the eye-hole where crystalline biofilm formation is greatest and typically of most relevance to blockage. In general, sections from the area containing the deflated balloon are avoided due to difficulties in processing reproducibly, and retention of excess moisture. (1) For use in assays intended to quantify amounts of crystalline biofilm formation, biomass, or environmental scanning electron microscopy (ESEM) of biofilms, target sections should be further cut longitudinally as shown. This exposes the lumen and biofilm. (2) For scanning electron microscopy of catheter lumen cross sections, to evaluate biofilm thickness, catheter sections can be directly mounted on stubs for sputter coating and imaging

3. Remove the washed catheter sections from the tubes, centrifuge the remaining mixture (3000 × *g* for 10 min), and discard the supernatant.
4. Resuspend the resulting pellets in 5 mL of 0.05 M perchloric acid, mix thoroughly, centrifuge (3000 × *g* for 2 min), and recover supernatants to fresh tubes.
5. Calibrate a flame photometer using calcium standards at 100, 75, 50, and 25 ppm.
6. Determine the amount of calcium dissolved in retained solutions (indicative of amount of crystalline biofilm formed) using the flame photometer (*see* **Note 19**).

3.2.2 Quantification of Biofilms Using Crystal Violet Staining

Biofilm formation may also be evaluated by using an adapted version of the crystal violet assay originally described by O'Toole and Kolter [33].

1. Section catheters removed from bladder models into 1 cm cross sections using a scalpel according to the schematic in Fig. 3. Retain 1 cm sections of interest and further cut longitudinally to expose the catheter lumen and biofilm.
2. Place the sections into separate 50 mL centrifuge tubes and wash gently in deionized water to remove unbound cells and debris.
3. Submerge sections in 1 mL of crystal violet solution and incubate for 15 min at room temperature.
4. Thoroughly wash the sections three times in deionized water to remove excess crystal violet stain before transferring to fresh 1.5 mL microcentrifuge tubes.
5. Add 1 mL of DMSO to the microcentrifuge tubes and release the crystal violet dye by vortexing (*see* **Note 18**).
6. Transfer the DMSO with released crystal violet dye to a cuvette and measure absorbance at 595 nm using a spectrophotometer.

3.3 Scanning Electron Microscopy for Analysis of Biofilms In Situ on Urinary Catheters

Note 20 provides a brief overview of the advantages and disadvantages of standard scanning electron microscopy and environmental scanning electron microscopy for analysis of biofilms.

3.3.1 Environmental Scanning Electron Microscopy

1. Use catheters removed from bladder models after blockage, or after models have been run for a set time to allow biofilm growth, immediately. Section catheters using a scalpel according to the scheme shown in Fig. 3.
2. Retain the 1 cm sections to be viewed (sections 1–3 in Fig. 3) and further section longitudinally through the lumen using a

scalpel, to expose the crystalline biofilm. During sectioning, catheter sections must be handled with care to avoid disrupting the biofilm while preparing sections.

3. Directly fix longitudinally sectioned catheters to aluminum stubs using Tissue-Tek CRYO-OCT compound mountant, with the exposed catheter lumen and biofilm facing directly upward for imaging.
4. Place stubs onto a Peltier cool stage, prechilled to a temperature of approximately 1.5 °C, within the chamber of an environmental scanning electron microscope.
5. Directly view biofilms on catheter sections using the following parameters: a 100 μm aperture, a 500 μm lower EDS EP aperture, a chamber pressure ~570 pa with 85% humidity, an accelerating voltage of 20 kV EHT, and a 5 quadrant back scatter detector (5Q-BSD).

3.3.2 Scanning Electron Microscopy of Catheter Cross Sections

1. Use catheters removed from bladder models, after blockage or after models have been run for a set time to allow biofilm growth, immediately. Dissect sections of 1 cm length from the region immediately below the eye-hole (Fig. 3). Other regions of the catheter may also be viewed, but the highest levels of crystalline biofilm formation typically occur in the region around the eye-hole.
2. Directly mount catheter sections onto an aluminum specimen split mount vice holder using adhesive carbon tape. Mount sections upright, so that the cross section of the catheter lumen is facing directly upwards and available for imaging (Fig. 3).
3. Care must be taken during sectioning and mounting of catheter sections to avoid disrupting the crystalline biofilm.
4. Store mounted samples overnight in a desiccator at room temperature to remove excess moisture.
5. Coat catheter sections with gold or platinum using an automatic sputter coater to reduce charging effects and thermal damage, and enhance secondary electron emission. Evacuate the sputter chamber prior to coating to a base pressure of approximately 10^{-6} Pa before backfilling with argon as processing gas. Perform coating under standard conditions, using a 20 mA sputter current, and a 30 mm working distance for a duration of 60 s.
6. Perform imaging of catheter cross sections using a scanning electron microscope operated at 5–10 kV. Use secondary electron imaging for inspection of the sample topography.

4 Notes

1. Generally, we find it is best to prepare all media for use in bladder models fresh for each experiment, and would advise this approach. However, once sterilized, in our experience Solutions 1 and 2 can be stored at room temperature for 2–3 months if necessary. Cloudiness or precipitation is an indication that a solution is no longer suitable for use and should be discarded if this occurs. *See* also **Note 5** below.
2. For preparation of Artificial Urine Solution 2, use glass beakers, measuring cylinders, and magnetic stir bars which are all prewashed with 70% IMS and oven-dried in order to facilitate sterilization. Transfer preweighed reagents into the beaker placed on a hotplate with an integrated magnetic stirrer (heat at 40 °C, to counteract extremely endothermic solvation) then add sterile deionized water, and mix for 5 min. Do not heat/autoclave or refrigerate, or allow significant evaporation. Filter-sterilize through a 0.45 μm pore size membrane filter.
3. The double walled glass vessels used to represent the bladder are not commercially available as standard glassware. We have found that the best approach to procure these is to commission them from specialist suppliers of bespoke glassware. Many universities provide a scientific glassblowing service, often associated with chemistry departments, and in many cases these will also supply external organizations. We have found these to be ideal for procuring bladder models.
4. Although it is possible to purchase calcium standards commercially, these are straightforward to prepare in house. Prepare a stock solution of 1000 ppm calcium standard by transferring 2.4973 g of calcium carbonate powder ($CaCO_3$) into a large beaker then add 25 mL of 1 M HCl carefully. Allow the powder to dissolve then dilute to 1 L with deionized water, and transfer the solution into a borosilicate glass bottle. Standards for each experiment at required concentrations can be prepared from this stock by dilution in deionized water. The 1000 ppm stock solution can be stored at room temperature for a maximum of 6 months.
5. To make the final 1× strength Artificial Urine for each model, the full final volume of Solution 1 and Solution 2 prepared as described in Subheading 2.1.1, **items 1** and **2**, should be mixed with an additional 3.6 L of sterile deionized water. Although the volume of Solution 1 and 2 prepared can be varied and larger volume aliquots made, it is strongly recommended to prepare these solutions exactly as described in Subheading 2.1.1, and prepare individual aliquots at stated volumes for each model. This practice greatly simplifies the generation of

distinct reservoirs of 1× strength artificial urine for each model to be run, and reduces the chances of contamination entering multiple models.

6. It is advisable to carry out the reconstitution of 1× strength artificial urine media under aseptic conditions (in a laminar flow cabinet, or Class 2 safety cabinet). Although we would recommend making fresh media for each experiment, once reconstituted in aspirators and sealed, the 1× artificial urine can be stored at room temperature for 3–4 weeks in our experience. Any cloudiness or precipitation in this media indicates it is no longer suitable for use and should be discarded.

7. The integrated pump in a circulating water bath is sufficient to drive the flow of water through the outer chambers of the bladder models. Standard water baths can also be easily adapted by using separate external circulation units that may be attached to the baths, and these are available from many suppliers of standard laboratory equipment. However, it is advised that no more than four bladder models be connected in series. This avoids the potential for temperature differences to develop across the models that may influence results. Water should be supplied through the lower inlet to the outer chamber of each model, so that the chamber is filled from the bottom. This avoids the potential for air to become trapped in the outer chamber which can prevent efficient and even heating of models.

8. Manufacturers will provide specifications regarding the properties of tubing that should be used with specific models of peristaltic pump, and you should use tubing recommended for your specific pump. The connections between these sections of "pump tubing," and the wider bore standard media supply tubing, are potential points of leakage in the system and should be checked carefully on model setup. When connecting tubing it is important not to over-tighten cable ties as this can damage tubing. Successive rounds of autoclaving may also weaken these joints. If leaks are detected when models are activated, it is usually possible to use additional cable ties on connectors to rectify, and continue with the experiment as planned. Disinfecting the area with 70% IMS can help reduce the risk of contamination when manipulating tubing with a leak.

9. The way in which individual catheters inflate can vary. Even catheters of the same type from the same manufacturer can display variations in the exact size and shape of the balloon, and/or the final angle of the catheter tip in relation to the bottom of the bladder model chamber (this is often not 90°). This is also a possible source of variation between results obtained from replicate experiments. For individual catheters

that appear particularly different in final tip angle compared to other catheters used, it is usually best to replace these at this stage with a new catheter that displays great conformity. However, the variation in catheter characteristics is a feature of the models in general, and it is not possible or desirable to eliminate this entirely.

10. It is advisable to place retort stands and other components in autoclavable plastic trays to capture any spillages that may occur during operation.
11. The peristaltic pump will supply media at a constant rate throughout model operation. The flow rate used in models is intended to reflect the urinary output of catheterized individuals, and is based on original work by Stickler et al. [19]. We have found flow rates of 0.5–0.75 mL/min to generate comparable results in terms of *P. mirabilis* biofilm formation and time to blockage. We recommend operating models using these parameters, though there is potential to vary this aspect of the model operation if necessary to model a particular infection scenario.
12. It is useful to observe the inflow of urine into the chamber in relation to the catheters during this initial set up stage. It is best to avoid the media dropping directly onto the catheter tip itself as this can influence results. If this is occurring, the stopper sealing the bladder model can be angled slightly, without compromising the integrity of the system, to ensure media instead drops into the residual volume of urine collecting below catheter eye-holes.
13. The exact volume of residual urine that collects in a model will be influenced by the characteristics of the catheter in that model. *See* also **Note 9** above.
14. The external surfaces of the bladder model and stopper may be disinfected with 70% IMS prior to opening, to help prevent contamination of the system.
15. An initial inoculum of 10^{10} CFU can be used to simulate established infection (the most common mode of model operation, e.g., [13, 14, 21]). However, these parameters can be adjusted as required to simulate distinct infection scenarios, such as early stage colonization where cell numbers are initially much lower (e.g., ~10^3 CFU [20]). To do this, adjust the amount of cells resuspended in artificial urine. However, the final volume of the inoculum should still be 10 mL to reflect the volume removed from models and ensure flow through the catheter will commence without significant delay when models are activated.
16. It is important to record pH levels of the AU medium at start and end of bladder model experiments to help evaluate the performance of the model and the results. The change in pH to

alkaline conditions is a key factor in precipitation of magnesium and calcium phosphates from the urine, and subsequent formation of crystals (and crystalline biofilms). Measuring pH at the start of models confirms that addition of the inoculum itself has not altered pH significantly, and sets the baseline. For bladder models infected with *P. mirabilis*, urease activity results in the elevation of artificial urine media to pH 8–9, and this should correspond with formation of crystalline biofilms and blockage. In models infected with *P. mirabilis*, blockage (indicated by accumulation of urine above the catheter eye-hole in models) in the absence of any pH elevation should be treated with caution. It is unlikely to be a genuine result, and more probably due to kinks in catheters or drainage tubing, or occasionally air locks in the system. Enumerating viable cells at the start and end of models ensures inocula are comparable between strains and replicates, and evaluates survival of cells during the operation of the model. In terms of survival during operation, the interpretation of viable count data will depend on the aim of the experiment, the parameters under which the model is run, and what exactly is being tested. For standard models with wild-type *P. mirabilis* strains replicating an established infection and operated until blockage, it is typical to see cell numbers retained at levels similar to the original inocula, with no statistically significant reduction at the end of experiments. In other experiment types (such as evaluation of antimicrobial intervention, mutants that may have defects in biofilm formation or survival in the urinary tract, or use of other uropathogens), changes in cell numbers during model operation may be encountered or expected, and a main outcome of the experiment.

17. At the end of experiments and during the operation of models, the medium supply reservoir should also be checked regularly. This should remain clear throughout the experiment and free from any signs of contamination. If the medium in aspirators supplying a model becomes cloudy, it is likely that this is due to microbial growth from contamination. In this case the results from that particular model should be disregarded.
18. The disruption of biofilms is often a critical step in approaches to measure biofilm formation. In flame photometry, the emphasis is on measurement of calcium present in *P. mirabilis* crystalline biofilms, and the oxalate and oxalic acid solution is effective in dissolving this directly from crystalline biofilms. However, other approaches to remove biofilms for measurement for other analyses should first be subject to optimization and validation. In particular, methods that seek to measure biofilm formation through enumeration of recovered biofilm-associated bacteria by plate counts should be treated with

caution. Even once efficient biofilm removal has been verified, recovered cells will require further validated steps to disaggregate and ensure single cells can be enumerated. The physiological status of cells in biofilms, and entry into viable but non cultivatable states, may also be an issue for such methods, which can undermine accuracy. If possible, it is advisable to use methods which seek to evaluate the amount of biofilm formed in situ. However, no measure of biofilm formation is completely reliable and without limitation, and it is best to use a mixture of qualitative and quantitative approaches together, including direct visualization.

19. It is important to note that this method measures the mineral precipitate incorporated into crystalline biofilms rather than biomass. In this case, the amount of crystalline biofilm formation as a whole is inferred from this measurement.

20. For the evaluation of catheter biofilms, standard scanning electron microscopy (SEM) and environmental scanning electron microscopy (ESEM) have distinct features that make them best suited to visualize biofilms in different ways (*see* also Holling et al. [13] for more detailed discussion):

    Standard SEM: This requires more laborious and specialist sample preparation that involves the dehydration and fixation of samples, and usually also critical point drying under high pressure. These steps make it difficult to image the biofilm in situ, and likely also introduce artifacts that can make it difficult to draw meaningful conclusions regarding features of biofilm architecture. However, the minimal approach to sample prep described in this article permits imaging of biofilms in cross section and provides a useful and reasonably accessible way to measure thickness in conjunction with other methods of biofilm evaluation. Furthermore, standard SEMs are more likely to be available than the more specialist ESEMs.

    ESEM: In contrast to standard SEM, ESEM permits the visualization of fully hydrated biofilms in situ on catheter surfaces. Sample preparation relates solely to dissection of the catheter to expose the biofilm and this method can be used to more meaningfully compare major architectural features of biofilms (though such data can still be highly subjective). The lack of sample processing also permits samples visualized by ESEM to be taken directly into other analyses of biofilm formation, such as calcium quantification or measurements of biomass. This permits visual data to be more directly linked to other measures of biofilm formation. A trade-off is the relatively lower magnification that can be used with ESEM compared to standard SEM; however, our experience is that this is still very useful for evaluation of biofilm formation.

References

1. Weinstein RA, Darouiche RO (2001) Device-associated infections: a macroproblem that starts with microadherence. Clin Infect Dis 33 (9):1567–1572. https://doi.org/10.1086/323130
2. Getliffe K, Newton T (2006) Catheter-associated urinary tract infection in primary and community health care. Age Ageing 35 (5):477–481. https://doi.org/10.1093/ageing/afl052
3. Loveday HP, Wilson JA, Pratt RJ, Golsorkhi M, Tingle A, Bak A, Browne J, Prieto J, Wilcox M, UK Dept of Health (2014) epic3: national evidence-based guidelines for preventing healthcare-associated infections in NHS hospitals in England. J Hosp Infect 86 Suppl 1:S1–70. https://doi.org/10.1016/S0195-6701(13)60012-2
4. Stickler DJ (2014) Clinical complications of urinary catheters caused by crystalline biofilms: something needs to be done. J Intern Med 276 (2):120–129. https://doi.org/10.1111/joim.12220
5. McNulty C, Freeman E, Smith G, Gunn K, Foy C, Tompkins D, Brady A, Cartwright K (2003) Prevalence of urinary catheterization in UK nursing homes. J Hosp Infect 55 (2):119–123. https://doi.org/10.1016/S0195-6701(03)00224-X
6. Jacobsen SM, Stickler DJ, Mobley HLT, Shirtliff ME (2008) Complicated catheter-associated urinary tract infections due to *Escherichia coli* and *Proteus mirabilis*. Clin Microbiol Rev 21(1):26–59. https://doi.org/10.1128/CMR.00019-07
7. Stamm WE (1991) Catheter-associated urinary tract infections: epidemiology, pathogenesis, and prevention. Am J Med 91(3B):65S–71S
8. Mobley HLT (1996) Virulence of *Proteus mirabilis*. In: Mobley HLT, Warren JW (eds) Urinary tract infections: molecular pathogenesis and clinical management. ASM Press, Washington, DC, pp 245–265
9. Stickler DJ (2008) Bacterial biofilms in patients with indwelling urinary catheters. Nat Clin Pract Urol 5:598. https://doi.org/10.1038/ncpuro1231
10. Warren JW (1991) The catheter and urinary tract infection. Med Clin North Am 75 (2):481–493
11. Griffith DP, Musher DM, Itin C (1976) Urease. The primary cause of infection-induced urinary stones. Investig Urol 13(5):346–350
12. Jones BD, Mobley HLT (1987) Genetic and biochemical diversity of ureases of *Proteus*, *Providencia*, and *Morganella* species isolated from urinary tract infection. Infect Immun 55 (9):2198–2203
13. Holling N, Dedi C, Jones CE, Hawthorne JA, Hanlon GW, Salvage JP, Patel BA, Barnes LM, Jones BV (2014) Evaluation of environmental scanning electron microscopy for analysis of *Proteus mirabilis* crystalline biofilms in situ on urinary catheters. FEMS Microbiol Lett 355 (1):20–27. https://doi.org/10.1111/1574-6968.12451
14. Holling N, Lednor D, Tsang S, Bissell A, Campbell L, Nzakizwanayo J, Dedi C, Hawthorne JA, Hanlon G, Ogilvie LA, Salvage JP, Patel BA, Barnes LM, Jones BV (2014) Elucidating the genetic basis of crystalline biofilm formation in *Proteus mirabilis*. Infect Immun 82(4):1616–1626. https://doi.org/10.1128/IAI.01652-13
15. Kunin CM (1997) Urinary tract infections. Detection, prevention, and management, 5th edn. Williams & Wilkins, Baltimore, pp 226–278
16. Clapham L, McLean RJC, Nickel JC, Downey J, Costerton JW (1990) The influence of bacteria on struvite crystal habit and its importance in urinary stone formation. J Cryst Growth 104(2):475–484. https://doi.org/10.1016/0022-0248(90)90150-J
17. Dumanski AJ, Hedelin H, Edin-Liljegren A, Beauchemin D, McLean RJ (1994) Unique ability of the *Proteus mirabilis* capsule to enhance mineral growth in infectious urinary calculi. Infect Immun 62(7):2998–3003
18. Cox AJ, Hukins DWL (1989) Morphology of mineral deposits on encrusted urinary catheters investigated by scanning electron microscopy. J Urol 142(5):1347–1350. https://doi.org/10.1016/S0022-5347(17)39095-X
19. Stickler DJ, Morris NS, Winters C (1999) Simple physical model to study formation and physiology of biofilms on urethral catheters. Methods Enzymol 310:494–501
20. Nzakizwanayo J, Hanin A, Alves DR, McCutcheon B, Dedi C, Salvage J, Knox K, Stewart B, Metcalfe A, Clark J, Gilmore BF, Gahan CGM, Jenkins ATA, Jones BV (2016) Bacteriophage can prevent encrustation and blockage of urinary catheters by *Proteus mirabilis*. Antimicrob Agents Chemother 60 (3):1530–1536. https://doi.org/10.1128/AAC.02685-15
21. Nzakizwanayo J, Scavone P, Jamshidi S, Hawthorne JA, Pelling H, Dedi C, Salvage JP, Hind CK, Guppy FM, Barnes LM, Patel BA, Rahman KM, Sutton MJ, Jones BV (2017) Fluoxetine and thioridazine inhibit efflux and attenuate crystalline biofilm formation by

Proteus mirabilis. Sci Rep 7(1):12222. https://doi.org/10.1038/s41598-017-12445-w

22. Drew SJ (2014) Clefting syndromes. Atlas Oral Maxillofac Surg Clin North Am 22 (2):175–181. https://doi.org/10.1016/j.cxom.2014.05.001
23. Jones BV, Mahenthiralingam E, Sabbuba NA, Stickler DJ (2005) Role of swarming in the formation of crystalline *Proteus mirabilis* biofilms on urinary catheters. J Med Microbiol 54 (Pt 9):807–813
24. Stickler DJ, Morgan SD (2008) Observations on the development of the crystalline bacterial biofilms that encrust and block Foley catheters. J Hosp Infect 69(4):350–360. https://doi.org/10.1016/j.jhin.2008.04.031
25. Macleod SM, Stickler DJ (2007) Species interactions in mixed-community crystalline biofilms on urinary catheters. J Med Microbiol 56.(Pt 11:1549–1557. https://doi.org/10.1099/jmm.0.47395-0
26. Stickler D, Morris N, Moreno MC, Sabbuba N (1998) Studies on the formation of crystalline bacterial biofilms on urethral catheters. Eur J Clin Microbiol Infect Dis 17(9):649–652
27. Morris NS, Stickler DJ, Winters C (1997) Which indwelling urethral catheters resist encrustation by *Proteus mirabilis* biofilms? Br J Urol 80(1):58–63
28. Milo S, Thet NT, Liu D, Nzakizwanayo J, Jones BV, Jenkins ATA (2016) An in-situ infection detection sensor coating for urinary catheters. Biosens Bioelectron 81:166–172. https://doi.org/10.1016/j.bios.2016.02.059
29. Stickler DJ, Jones SM, Adusei GO, Waters MG (2006) A sensor to detect the early stages in the development of crystalline *Proteus mirabilis* biofilm on indwelling bladder catheters. J Clin Microbiol 44(4):1540–1542. https://doi.org/10.1128/JCM.44.4.1540-1542.2006
30. Morris NS, Stickler DJ (1998) The effect of urease inhibitors on the encrustation of urethral catheters. Urol Res 26(4):275–279
31. Stickler DJ, Jones GL, Russell AD (2003) Control of encrustation and blockage of Foley catheters. Lancet 361(9367):1435–1437
32. Milo S, Hathaway H, Nzakizwanayo J, Alves DR, Esteban PP, Jones BV, Jenkins ATA (2017) Prevention of encrustation and blockage of urinary catheters by *Proteus mirabilis* via pH-triggered release of bacteriophage. J Mater Chem B 5(27):5403–5411. https://doi.org/10.1039/C7TB01302G
33. O'Toole GA, Kolter R (1998) Flagellar and twitching motility are necessary for *Pseudomonas aeruginosa* biofilm development. Mol Microbiol 30(2):295–304

Chapter 15

Independent Transurethral Urinary Tract Inoculation in a Murine Model of Ascending Infection with *Proteus mirabilis*

Sara N. Smith

Abstract

Live animal modeling enables more in-depth realistic methods for studying the pathogenesis of urinary tract infections. Initiating a transurethral urinary tract infection in the female mouse is a challenging endeavor. However, when done with consistency and care, this infection model yields irreplaceable data. The methods necessary to ensure successful mouse transurethral inoculation and colonization in a single strain infection model with *Proteus mirabilis* are presented here.

Key words *Proteus mirabilis*, Urinary tract infection, Independent challenge, Mouse model

1 Introduction

The introduction of an in vivo murine model of urinary tract infection [1] changed the dynamics of urinary tract research capabilities drastically. Utilizing a working mouse model of infection that closely mimics human infection has revolutionized our ability to tease apart distinct niches within the pathogenesis process.

Initially, 50 μL of *Escherichia coli* was introduced into the urinary bladder using a handheld syringe system [1]. The importance of broadening the scope of urinary tract infections to include *Proteus mirabilis* in a mouse model of infection was elucidated with the successful colonization of the murine urinary tract with a urease-negative mutant [2]. Although this model revolutionized the way we examine ascending urinary tract infections, there was concern about inadvertently causing vesicoureteral reflux as a result of the pressure exerted by the human hand while inoculating mice. The abnormal introduction of bacteria directly into the kidney(s), bypassing normal urinary tract progression, was a valid concern. However, with the introduction of an automatically controlled syringe pump system, this fear was minimized. This instrument

Melanie M. Pearson (ed.), *Proteus mirabilis: Methods and Protocols*, Methods in Molecular Biology, vol. 2021,
https://doi.org/10.1007/978-1-4939-9601-8_15,

enhanced the ability to minimize the risk of vesicoureteral reflux and also added consistency to the infection model. Inocula are now able to be given efficiently at a constant flow rate with consistent pressure not only between every group of any given experiment, but each individual mouse. The use of various anesthetic agents has also increased the effectiveness of the ascending model by allowing the inocula to remain in the bladder while the mouse is anesthetized. We employ the use of injectable ketamine mixed with xylazine. Ketamine acts as a dissociative anesthetic and, when mixed appropriately with xylazine, not only offers analgesia [3] but also provides anesthesia lasting 30–45 min [4], which allows bacterial adherence and natural infection progression.

Thus, through continued modification and technological improvement, we have optimized this procedure to successfully inoculate female mice with *P. mirabilis* and induce an ascending urinary tract infection. This method of infection can also be modified to include the introduction of more than one strain at a time (*see* Chapter 16 on cochallenge) as well as other experimental conditions (*see* Chapter 18 on vaccines).

2 Materials

2.1 Prior to Infection

For smooth experimental implementation, pour agar plates one week before the experiment and pre-sterilize all supplies. Methods of sterilization vary depending on the item but include ethylene oxide for precut urinary catheters, filter sterilization for anesthesia and autoclave sterilization for silicone catheter segments, PBS, metal weigh boats, surgical instruments, beakers and flasks. Prior to transurethral instillation of bacteria into mice, always perform in vitro growth curves to identify any in vitro growth defects that may confound experimental results.

1. Black or blue permanent marker (e.g., Sharpie) (*see* **Note 1**).
2. Spectrophotometer.
3. Spectrophotometer cuvettes, 1.5 mL semi-micro.
4. 37 °C incubator plus 37 °C shaker incubator.
5. Sterile phosphate-buffered saline (PBS), pH 7.4.
6. Lysogeny broth (LB) (per liter): 10 g of tryptone, 5 g of yeast extract, 0.5 g of NaCl. Mix well and sterilize by autoclaving.
7. LB agar: LB plus 15 g of agar/L; autoclave to sterilize, cool to 55 °C, and pour into petri dishes, approximately 20–25 mL/plate (*see* **Note 2**).
8. Sterile flask(s) or culture tubes, one per bacterial strain to be tested.

9. Female CBA/J mice, age 5–6 weeks, should be ordered 2 weeks prior to the experimental date (*see* **Note 3**). This allows them proper time to both arrive at the facility and acclimate for 7 days to a new environment. This period of adjustment is necessary due to the impact widespread stress has on physiological systems during, and directly after, transportation [5].

2.2 Day of Infection

1. Sterile white petrolatum and mineral oil eye ointment.
2. Anesthetic stock solution (per 5 mL): 1 mL of 100 mg/mL ketamine, 0.05 mL of 100 mg/mL xylazine, 3.95 mL of 0.9% sterile NaCl. Mix all ingredients together and filter-sterilize with a 0.22 μm PEM membrane filter unit. Every 100 μL of diluted anesthesia is equal to 0.02 mL of ketamine. This stock solution has a shelf life of 7 days (*see* **Note 4**).
3. Sterile 1 mL disposable Luer-lock tip syringes with clear barrel.
4. 26 G × ½ (0.45 mm × 13 mm) needles.
5. 0.22 μm PEM membrane filter unit.
6. Automated syringe pump (e.g., Harvard Apparatus Pump 11 Elite infuse-only single syringe pump).
7. Absorbent underpad (polyethylene, 750 mL/m^2 absorbency).
8. 20″ Sterile Extension Set (approx. priming volume 2.0 mL, latex-free, non-DEHP, kink resistant). This is tubing that connects the infusion pump to the catheter.
9. Sterile 25 mL single use pipette basin (gamma-irradiated).
10. Sterile dissecting dressing forceps (autoclavable, latex free, stainless steel, straight, serrated tip).
11. Sterile 10 mL disposable Luer-lock tip syringes with clear barrel.
12. Polyethylene tubing: I.D. 0.28 mm (0.011″), O.D. 0.61 mm (0.024″), 30.5 m (100′). Cut tubing into 14 mm catheter segments and sterilize using ethylene oxide gas.
13. 10% povidone iodine solution.
14. Sterile polystyrene 100 mm × 15 mm Petri dishes.
15. Tissue wipes (e.g., Kimwipes).
16. 1.5 mL microcentrifuge tubes (autoclavable, polypropylene).
17. Polyester-tipped applicators (6 in., sterile, plastic shaft, single use).
18. 30 G × ½ (0.3 mm × 13 mm) needles.

2.3 Necropsy

1. CO_2 or isoflurane with drop jar.
2. Dissecting blunt-pointed forceps (autoclavable, latex free, chrome-plated steel, straight, serrated, blunt point tip).

3. Sharp-pointed dissecting scissors (autoclavable, surgical grade, stainless steel, sharp/sharp tip).
4. Blunt-end dissection scissors (autoclavable, surgical grade, stainless steel, straight, blunt/blunt tip).
5. Infrared loop sterilizer.
6. 100% ethanol in a squirt bottle.
7. Aluminum weighing dishes (20 mL capacity, 52 mm diameter, fluted sides).
8. Fine analytical balance.
9. Sterile 5 mL clear polystyrene culture tubes.
10. Rotor-stator mechanical laboratory homogenizer (5000–28,000 rpm, capacity 0.03 mL to 10 L).
11. Sawtooth rotor-stator homogenizer probe (stainless steel, 7 mm). Sterilize by enclosing in a sealable sterilization pouch and steam autoclave.
12. Sterile 500 mL plastic beakers.
13. 1 L of sterile ddH_2O.
14. Automated spiral plater (*see* **Note 5**).
15. Bacterial colony counter.
16. Sealable sterilization pouches.

3 Methods

All biohazardous work should be completed in a Class II biological safety cabinet, and all animal procedures should be institutionally reviewed and approved. Sterile technique should be used for all protocols using live bacteria.

3.1 Inoculum Preparation

1. Inoculate *P. mirabilis* from a fresh plate or frozen stock into 6 mL of LB and incubate with shaking at 37 °C overnight. Add antibiotics to the culture if needed.
2. Dilute 5 mL of the overnight culture into 45 mL of fresh, sterile LB.
3. Take the OD_{600} reading using the spectrometer (should be approximately 0.450) (*see* **Note 6**).
4. Dilute with fresh, sterile LB to a culture density of 2×10^8 CFU/mL (approximately $OD_{600} = 0.2$; *see* **Note 6**) in a final volume of 10 mL.
5. Pipette 1 mL of the diluted culture into a sterile microcentrifuge tube to use for quantifying the input.
6. Dilute input 1:10,000 (10^{-4} dilution) and use this for plating. Plate on LB agar, with antibiotic if necessary.
7. After the plates dry, incubate them at 37 °C overnight.

3.2 Infection Procedure

1. Cut a piece of absorbent underpad approximately 15″ × 20″ and place in the safety cabinet with the infusion pump positioned in the middle.
2. Place one sterile, empty petri dish on either side. In the petri dish on the left, place enough povidone iodine to cover at least half of the plate. Place 1 applicator inside so it can start to soak up the iodine. Leave the other petri dish empty.
3. Vortex or mix the diluted bacteria by inverting the tube. Pour the contents into a sterile basin.
4. Using the 10 mL syringe, fill the contents of the basin to the 10 mL marker and then ensure any air bubbles are removed by flicking the tip.
5. Place one end of the extension set onto the end of the 10 mL syringe and place a 30 G needle onto the other end. Insert a 14 mm sterilized catheter onto the needle (Fig. 1). Push the inoculum through the extension set and catheter until a small amount of inoculum dribbles out (*see* **Note 7**). Place this set up inside the sterile, empty petri dish on the right.
6. Place the syringe into the infusion pump and set to the appropriate volume (50 μL per mouse unless otherwise indicated). Set the infuse rate to 0.2 mL/min, ensuring a slow inoculum delivery. Run the apparatus until that volume of liquid consistently dribbles out, typically up to five times, prior to the transurethral inoculation.
7. Weigh each mouse and dose appropriately intraperitoneally (IP) with stock anesthetic solution. Wait until the mice are fully anesthetized prior to attempting to infect them (*see* **Note 8**).

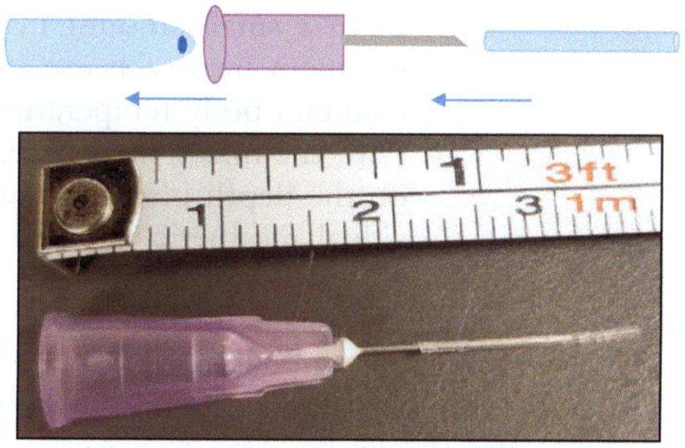

Fig. 1 Placement of a polyethylene catheter onto a 30 G × ½ (0.3 mm × 13 mm) needle. The gas-sterilized, precut catheter segment should be 14 mm in length and extend 4–5 mm past the bevel of the needle

8. Perform a pedal reflex test by gently grasping the feet and applying pressure. If the mouse does not pull her foot away, she has gotten an adequate dose of anesthetic (*see* **Note 9**).
9. Lay the mouse dorsally in front of the apparatus with her head pointing toward the left.
10. Gently blot the periurethral region with the iodine-soaked applicator, then carefully blot the area again with a tissue wipe carefully to remove any excess iodine (*see* **Note 10**).
11. Using your nondominant hand, gently grasp the periurethral area with the blunt-ended forceps. Grasp close enough to the vulva so that it spreads apart and has a tent-like appearance. Gently insert the catheter in the middle of the tent. This is where the urethral opening lies. Sometimes, small redirection of the catheter is necessary either from side to side or top to bottom to find the orifice. Apply gentle pressure and the catheter should enter the urethra smoothly (*see* **Note 11**).
12. If you meet a large amount of resistance or the catheter cannot be inserted all the way to the hub, the catheter is not in the urethra. If this happens, gently remove the catheter and try again. Sometimes, there will be a very small amount of pressure needed to get from the urethra to the bladder, but initially, there should be little, if any, resistance (*see* **Notes 12–15**).
13. Keeping the catheter in place, push the run button on the apparatus and wait until it has completed. After completion, wait an additional 5 s prior to slowly removing the urinary catheter. While removing it, pause a minimum of three times prior to full removal and let the catheter rest for 1 s at each pause. This gives the inoculum time to fully exit the catheter and stabilize itself in the bladder (*see* **Note 16**).
14. Place the mouse on a paper towel dorsally so the inoculum stays in place as long as possible.
15. Place a small amount (1 mm or less) of sterile ophthalmic ointment into each eye (*see* **Note 17**). Use a heat source to maintain body temperature as the anesthesia wears off. Typically, the mice remain anesthetized for 20–30 min depending on the depth of onset anesthesia and variances between individual anesthetic reactions.

3.3 Necropsy

1. The day before the experimental endpoint (typically 4–7 days post-inoculation), label all agar plates with mouse number, organ, and dilution information.
2. Fill culture tubes with 3 mL of sterile PBS. One tube is needed for each organ (bladders, kidneys, or spleen). Store tubes at 4 °C overnight.

3. Turn on the infrared loop sterilizer in the biosafety cabinet 10 min prior to the necropsy to allow temperature to reach 815 °C (1500 °F).
4. If collecting urine for CFUs, label a sterile microcentrifuge tube (*see* **Notes 18** and **19**). Place a microcentrifuge tube rack onto your work surface. Place the tube on the lower portion and position the mouse on the upper so her periurethral region is hovering over the tube. The mouse can grasp the rack while you gently massage the lower abdominal region, encouraging her to release urine into the tube. Some contamination from the periurethral region will be noted if the urine catch is not clean so ensure no fecal matter or hair is in the sample. Repeat for your cohort of mice and place the urine at 4 °C in between collections (*see* **Note 20**).
5. Place a 15″ × 20″ piece of absorbent underpad in the biological safety cabinet along with the sterilized dissecting instruments and autoclaved weigh boats (*see* **Note 21**). The balance should also be ready for use. Place a small beaker with sterile water near the infrared loop sterilizer and have a tissue wipe on hand to clean the instruments in between uses.
6. Humanely euthanize mice by cage as directed in your approved animal use protocol. We use isoflurane and a drop jar (*see* **Note 22**), as exposure to this method allows the mice to become anesthetized until breathing ceases with minimal side effects (Fig. 2).
7. Place the mouse horizontally on the absorbent underpad dorsally with her head pointing to the left. Using the squirt bottle with 100% ethanol, apply a small amount to the abdominal area and smooth the hair. Wetting it down not only keeps the hair out of the way, but also reduces the potential for contamination.
8. Using the blunt-ended dissecting scissors, make a small tent in the diaphragm region of the mouse and make a superficial 5–10 mm incision. Gently grasp both sides of the incision and pull apart in a smooth, steady motion. The entire first layer of hair and flesh should easily be removed from the necropsy area, leaving the organs intact and the sublayer of fascia exposed (*see* **Note 23**, Fig. 3).
9. Retrieve the urinary bladder first and place it gently into a sterile weight boat. It is located in the distal abdominal cavity nestled in between the adipose tissue. Record the weight and repeat for all other organs. The spleen is located on the left side of the mouse, nestled near the liver in close proximity to the left kidney (that is, it will be on the right side from the dissector's perspective). The kidneys are located on either side of the mouse in the lower abdominal wall close to the iliocostal muscle. Note any gross abnormalities such as bladder or kidney

Fig. 2 Proper drop jar setup. The drop jar must be used inside a fume hood. A physical barrier must be present that separates the mouse from the anesthesia soaked cotton balls as direct contact with liquid isoflurane may be irritating to footpads. A mesh screen is ideal as it allows free flowing vapors and can be easily disinfected after each use. Ensure all surfaces that come into direct contact with the mice are protected to prevent undue injury during euthanasia. A secondary method of euthanasia is required after use of an anesthetic as a euthanasia agent; policies may vary by institution

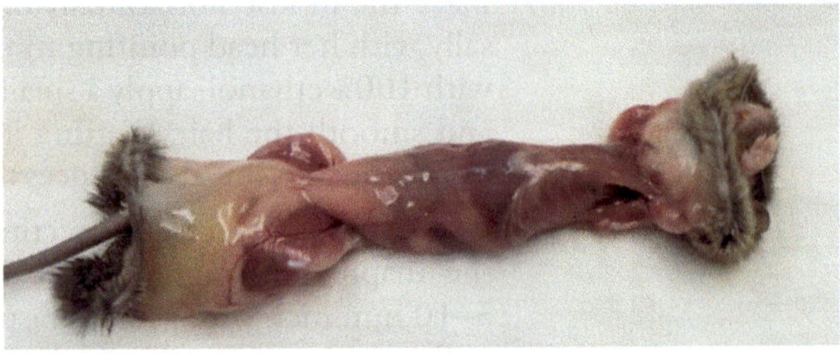

Fig. 3 Reflection of skin. Following euthanasia, place the mouse in dorsal recumbency, apply a small amount of 100% ethanol onto the ventral abdomen, smooth down the hair and make a small incision into the ventral skin, ensuring only the first layer of epidermis is penetrated. Next, reflect the skin over the head and hind limbs by gently pulling on either side of the incision, leaving the abdominal wall intact. It is important to pull the skin back completely to leave a sterile field

stones, which may appear as lighter areas within the tissue or external nodules on the tissue itself (*see* **Note 24**).

10. Place each organ in a sterile culture tube prefilled with 3 mL of sterile PBS. Vortex or tap each tube gently so the organ falls into the bottom.
11. Using the maximum rpm setting, homogenize bladders for 60 s, kidneys for 30–45 s, and spleens for 15 s, ensuring sterility is maintained throughout the process. Keep organs at 4 °C if you have a large number in between groups of mice.
12. After each organ is homogenized, clean the homogenizer by placing the tip into a plastic beaker of ethanol and running for 5–10 s. Then, repeat using a plastic beaker of sterile water (*see* **Note 25**).
13. Measure each urine volume obtained and bring up to 150 μL with sterile PBS (e.g., if 20 μL of urine was collected, add 130 μL of sterile PBS to the tube). Record both the amount of urine you collected and the amount of PBS added for calculation of dilution factor later (*see* **Note 26**).
14. Transfer 1 mL of each homogenate to sterile 1.5 mL tubes. Use sterile PBS to make 1:100 (that is, 10^{-2}) dilutions of the bladder and kidney homogenates and urine samples.
15. Use the automated plater to spiral plate 0 and −2 dilutions for the bladders, kidneys, and urine, and 0 for the spleen.
16. Allow plates to dry, then incubate all plates at 37 °C overnight.

3.4 Data Formatting

1. The next day, count each plate using a colony counter, or count by hand if needed. Save or export the data as a spreadsheet. Properly label the Plate ID if antibiotics were used, as you will need to know this for data calculations later. The limit of detection is 100 CFU, so any zeros should be graphed as 100 for statistical measurement purposes.
2. To calculate CFU/g of tissue, multiply the raw colony counts by 3 (the volume of PBS the organs were homogenized in, in mL). Divide that number by the weight of the organ (in grams). To calculate CFU/mL of urine, divide 150 μL by the amount of urine collected (in μL), then multiply by the raw count (*see* **Notes 27** and **28**).
3. Graph the data. Use the two-tailed Mann–Whitney test for statistical analysis.

4 Notes

1. Use an identification method to differentiate the individual mice (i.e., ear tagging, tail marking, or ear notching). We mark the tails with permanent marker and ensure the marks

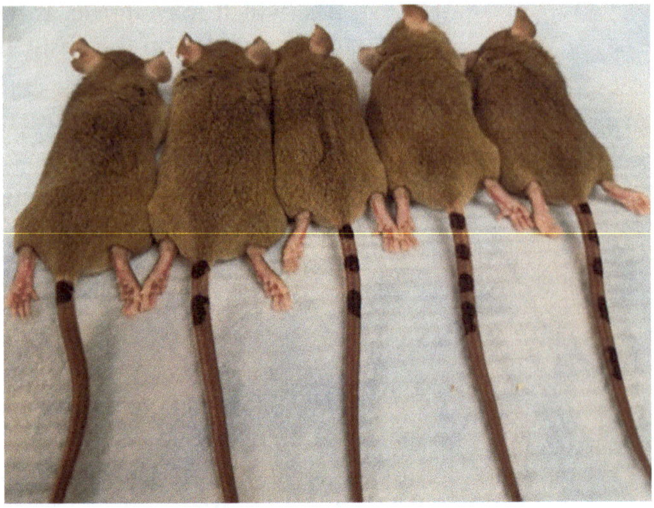

Fig. 4 Identification of mice using a marker. Mice are labeled with 1–5 marks on their tails for short-term, individual identification purposes. Due to grooming, the marker sometimes fades and touch-up is necessary. Long-term studies may require a more permanent identification method such as ear punches (note mice 1 and 2 on the left)

remain by touching up the preexisting marks every other day until completion of the experiment (Fig. 4).

2. Higher concentrations of NaCl, common in many LB recipes, cause *P. mirabilis* isolates to swarm, making isolation of individual colonies impossible.
3. Typical mouse cohorts for urinary tract infections are a minimum of six to ten mice. This allows an adequate number of infected mice for statistical purposes as well as human and host error (i.e., unable to properly infect the mouse, the mouse has an anatomical abnormality, unexpected death, etc.).
4. Ketamine–HCl is a controlled substance requiring careful documentation (e.g., in the USA, it is a Schedule III drug). In order to obtain ketamine, licensing may be required. Individual institutions may have varying policies regarding its distribution and storage.
5. CFU can also be manually plated without the use of a spiral plater. In this case, individually spread-plate 100 μL serial dilutions. Alternatively, drip plating can be used for multiple dilutions on one plate [6].
6. On our spectrophotometer, $OD_{600} = 0.20$ is equivalent to 2×10^8 CFU/mL $= 1 \times 10^7$ CFU/mouse. The correct OD_{600} to obtain this number of CFU may vary depending on the machine used and the bacterial strain, and should be determined independently by others following this protocol.

7. The priming volume needed for the 20″ extension set is 2.0 mL. When preparing inoculum, take this volume into account. We aim for a total volume of 10 mL to ensure there is enough, but the least amount needed for any transurethral inoculation is 4 mL.
8. It is extremely important that the mice are appropriately dosed with anesthesia. As they wake up, their bladders contract, which can lead to premature expelling of inoculum if they are not under anesthesia for an appropriate amount of time. However, if given too much, they may have tremors which also cause bladder contraction.
9. If the mice are having trouble falling asleep, sometimes covering their eyes with a paper towel or cupping them with your gloved hand allows them to relax more quickly.
10. Only a small amount of povidone iodine is necessary to aseptically clean the periurethral region. Blot the excess; do not rub it off or wipe it, as you risk entry of the antiseptic into the urinary tract.
11. Depending on the size of the mouse that is being inoculated, the distance from the urethral opening to the bladder may be difficult to gauge. It is important to be as consistent as possible with where the catheter is being placed. You can use sterile scissors to cut the catheter if necessary. If the catheter is too long, it may be difficult to gently guide into the urethra. If the catheter is too short, there is the potential for rupture of the urethra and/or injury to the delicate tissues of the urogenital region.
12. Female mice sometimes have genital abnormalities which inhibit normal urethral catheterization. The most common abnormality I have encountered is a small lump of skin I call a "nub" which lies in the way of the urethra. There is no simple answer to infecting these mice successfully. Gently probe underneath the "nub" where the urethral opening should be. Do not force the catheter; simply move it in different directions until the opening is found. Always mark these mice specifically so if the infection is not successful they may be excluded from the data set.
13. The typical weight of 7–8-week-old mice is 20 g. However, it is occasionally necessary to use older mice that are heavier for some experiments. Larger mice tend to be more difficult to infect due to the larger amount of abdominal adipose tissue on either side of and/or surrounding the urethra.
14. Due to the close proximity of the urethra and the vagina, it is common to mistake the vagina for the urethra. One way to ensure the catheter is in the urethra is by advancing the catheter gently all the way to the hub of the needle. It should easily glide

in. If, however, the catheter is in the vagina instead, you will not be able to insert the catheter fully. Do not force it. Simply remove it and try again. If you believe the catheter is in the urethra and proceed with the inoculation but it is in the vagina instead, the inoculum will start to dribble out within a few seconds.

15. The tissue around the urogenital region is delicate. Even with careful manipulation, the area is easily irritated and becomes swollen, making inoculation difficult. If the catheter does not easily glide into the urethra, gently try to find the opening. If, with repeated manipulation, the region becomes increasingly inflamed, it will become impossible for a successful infection to take place.
16. Once the catheter has been inserted into the bladder, and the inoculum has been given, let the catheter remain in place for 5 s prior to removal. This allows the full volume to enter the bladder. When pulling the catheter out, do so slowly so the inoculum stays in place and does not follow the catheter back out.
17. Since Ketamine is a dissociative anesthetic which does not allow the mice to close their eyes, a paper towel placed lightly over the mice while they are waking up, as well as protective lubrication, prevents corneal drying and damage. The paper towel also helps maintain body heat during the recovery process.
18. Throughout the infection procedure and necropsy it is common for your gloves to accumulate mouse hair and debris. Instead of using alcohol to clean them off, change gloves to minimize the risk of accidentally killing the bacteria or getting alcohol onto an organ.
19. If a cage is soiled, change the bedding a minimum of 20 min before starting an experiment. This allows for less chance of secondary contamination while the mice are recovering from anesthesia. Conversely, once a urinary tract infection has been established, mice tend to urinate more frequently but when the cages are freshly changed, it excites them, causing them to release urine. If urine is needed for an experiment, allow for extra time so the mice calm down prior to collection. Typically, it takes a minimum of 20 min for urine to accumulate in the bladder again once it has been emptied and multiple collections may be needed.
20. When collecting urine, it is important to remain consistent in the collection process. Too much manual bladder manipulation may lead to irritation and inflammation. Unless required for experimental design (RNA studies, urine being collected at intermediate time points throughout the course of an experiment, etc.), the maximum number of urine collections we will

perform is 3, spaced 30 min apart. Too much manual manipulation may cause excessive inflammation and discomfort. Even if enough urine has been collected from one mouse, manipulate the entire cohort the same number of times as the control.

21. Aluminum weighing dishes can be cleaned with water and autoclaved for multiple uses.
22. A drop jar consists of a plexiglass or glass container with isoflurane-soaked cotton balls in the bottom. The mice need to be prevented from coming into direct contact with the liquid inhalant, generally by the use of a mesh grid covering the cotton.
23. Removal of the entire first layer of skin and hair from the mouse prior to organ removal leads to less chance of hair contamination.
24. During organ removal it is important that sterility is maintained. The intestinal tract will need to be gently moved out of the way to gain access to the kidneys. Ensure you do not accidentally puncture it while removing the kidneys and avoid touching it excessively with your instruments during the procedure.
25. Use plastic beakers to avoid possible shatter hazards from the homogenizer. Change the alcohol and sterile water every 20 mice, or as necessary, if the liquids start retaining organ remnants or become turbid.
26. Urine is typically brought up to 150 μL using PBS. This is to ensure an adequate amount of urine is present for scenarios where one sample will be spread onto multiple plates. If more than two plates are being used with one urine sample, you may need to bring up all the samples to more than 150 μL. Ensure consistency within the experiment and dilute every urine sample within a cohort the same to reduce mistakes during data analysis.
27. CFU/g of tissue calculation example: If original output is 1.80×10^6 CFU and the organ weight is 0.2356 g: $(1.80e6 \times 3)/0.2356 = 2.29 \times 10^7$ CFU/g.
28. CFU/mL of urine calculation example: If CFU output is 1.88×10^4, the amount of urine collected is 20 μL, and 130 μL of PBS is added to bring the volume to 150 μL: $(150/20) \times 1.88e4 = 1.41 \times 10^5$ CFU/mL of urine.

References

1. Hagberg L, Engberg I, Freter R, Lam J, Olling S, Svanborg Edén C (1983) Ascending, unobstructed urinary tract infection in mice caused by pyelonephritogenic *Escherichia coli* of human origin. Infect Immun 40(1):273–283
2. Jones BD, Lockatell CV, Johnson DE, Warren JW, Mobley HLT (1990) Construction of a urease-negative mutant of *Proteus mirabilis*: analysis of virulence in a mouse model of ascending urinary tract infection. Infect Immun 58 (4):1120–1123

3. Mathisen LC, Skjelbred P, Skoglund LA, Oye I (1995) Effect of ketamine, an NMDA receptor inhibitor, in acute and chronic orofacial pain. Pain 61(2):215–220
4. Nowland MH, Lebowsky R (2017) Guidelines on anesthesia and analgesia in mice. https://az.research.umich.edu/animalcare/guidelines/guidelines-anesthesia-and-analgesia-mice. Accessed 18 Dec 2018
5. Obernier JA, Baldwin RL (2006) Establishing an appropriate period of acclimatization following transportation of laboratory animals. ILAR J 47(4):364–369
6. Thomas P, Sekhar AC, Upreti R, Mujawar MM, Pasha SS (2015) Optimization of single plate-serial dilution spotting (SP-SDS) with sample anchoring as an assured method for bacterial and yeast cfu enumeration and single colony isolation from diverse samples. Biotechnol Rep (Amst) 8:45–55. https://doi.org/10.1016/j.btre.2015.08.003

Chapter 16

Cochallenge Inoculation with *Proteus mirabilis* in a Murine Transurethral Urinary Tract Model of Ascending Infection

Sara N. Smith

Abstract

Determining virulence factors which contribute to the ability of *Proteus mirabilis* to establish itself as a prominent cause of urinary tract infections is key in developing targets for potential therapies and treatments. Constructing mutants and testing them in a head-to-head in vivo challenge with the parental strain or other mutants enables determination of what, if any, competitive advantage may exist. Here we describe the methods necessary to ensure successful mouse transurethral inoculation and colonization in a cochallenge infection.

Key words *Proteus mirabilis*, Urinary tract infection, Mutants, Mouse model, Cochallenge, Competitive index

1 Introduction

Incorporation of a modified in vivo murine transurethral urinary tract infection model is an invaluable tool for determining virulence factors which contribute to *Proteus mirabilis* pathogenicity [1]. Independent infections (*see* Chapter 15) determine the overall capability of a given strain or mutant to cause infection. In contrast, a coinfection using the wild-type parent strain in a 1:1 ratio with a mutant allows a more targeted approach to evaluate a specific virulence factor's contribution to the infection process [2–5]. In general, cochallenge experiments are more sensitive than independent challenges and can detect subtle fitness defects. In addition, cochallenges can be used to compete different species; for example, one recent polymicrobial study confirmed having two or more species present may enhance infection severity [6]. However, both in vivo experimental methods are useful in tandem because secreted toxins, cross-feeding and other metabolic processes may be altered in the presence of more than one competing strain of bacteria. For example, utilizing a cross-feeding assay along with cochallenge

Melanie M. Pearson (ed.), *Proteus mirabilis: Methods and Protocols*, Methods in Molecular Biology, vol. 2021,
https://doi.org/10.1007/978-1-4939-9601-8_16,

experimentation has been successful in determining a model of iron acquisition in *P. mirabilis* [7] (*see* Chapter 10).

Competing a mutant in vivo against the parental strain initiates a battle for the limited nutrients available in the host urinary tract environment. If the mutant strain is outcompeted in vivo in a specific organ, we conclude that expression of the missing virulence factor leads to a competitive advantage. With these data, we can narrow our search to target specific genes and pathways that contribute to overall virulence. Cochallenge studies with isogenic mutants have successfully demonstrated the importance of zinc uptake, specific metabolic pathways, fimbriae, and toxins that *P. mirabilis* uses to establish infection [8, 9]. However, successful transurethral inoculation is challenging. Levels of each strain to be tested must be carefully balanced, and then actual infectious dose quantified. In contrast with independent challenges, because mutant and parental bacteria are competing for resources within the same mouse, matched-pair statistical analysis is used to assess mutant fitness [5, 10]. Incorporation of quantified input and output for both mutant and wild type allows for the calculation of a competitive index [11].

This optimized procedure of introducing a urinary tract infection with competing strains of *P. mirabilis* is an invaluable asset. The ability to successfully incorporate in vivo cochallenge data along with in vitro experiments enables us to delve deeper into the underlying mechanisms contributing to virulence.

2 Materials

2.1 Prior to Infection

For smooth experimental implementation, pour agar plates one week before the experiment and pre-sterilize all supplies. Methods of sterilization vary depending on the item but include ethylene oxide for pre-cut urinary catheters, filter sterilization for anesthesia and autoclave sterilization for silicone catheter segments, PBS, metal weigh boats, surgical instruments, beakers, and flasks. Prior to transurethral instillation of bacteria into mice, always perform in vitro growth curves to identify any in vitro growth defects that may confound experimental interpretation.

1. Black or blue permanent marker (e.g., Sharpie) (*see* **Note 1**).
2. Spectrophotometer.
3. Spectrophotometer cuvettes, 1.5 mL semi-micro.
4. 37 °C incubator plus 37 °C shaker incubator.
5. Sterile phosphate buffered saline, pH 7.4.
6. Lysogeny broth (LB) (per liter): 10 g of tryptone, 5 g of yeast extract, 0.5 g of NaCl. Mix well and sterilize by autoclaving.

7. LB agar: LB plus 15 g of agar/L; autoclave to sterilize, cool to 55 °C, and pour into petri dishes, approximately 20–25 mL/plate (*see* **Note 2**).
8. Sterile culture tubes, one per bacterial strain to be tested.
9. Female CBA/J mice, age 5–6 weeks, should be ordered 2 weeks prior to the experimental date (*see* **Note 3**). This allows them proper time to both arrive at the facility and acclimate for 7 days to a new environment. This period of adjustment is necessary due to the impact widespread stress has on physiological systems during and directly after, transportation [12].
10. Bacterial strains to be tested (*see* **Note 4**).

2.2 Day of Infection

1. Sterile white petrolatum and mineral oil eye ointment.
2. Anesthetic stock solution (per 5 mL): 1 mL of 100 mg/mL ketamine, 0.05 mL of 100 mg/mL xylazine, 3.95 mL of 0.9% sterile NaCl. Mix all ingredients together and filter sterilize with a 0.22 μm PEM membrane filter unit. Every 100 μL of diluted anesthesia is equal to 0.02 mL of ketamine. This stock solution has a shelf life of 7 days (*see* **Note 5**).
3. Sterile 1 mL disposable Luer-lock tip syringes with clear barrel.
4. 26 G × ½ (0.45 mm × 13 mm) needles.
5. 0.22 μm PEM membrane filter unit.
6. Automated syringe pump (e.g., Harvard Apparatus Pump 11 Elite infuse-only single syringe pump).
7. Absorbent underpad (polyethylene, 750 mL/m^2 absorbency).
8. 20″ Sterile Extension Set (approx. priming volume 2.0 mL, latex-free, non-DEHP, kink resistant). This is the tubing that connects the infusion pump to the catheter.
9. Sterile 25 mL single use pipet basin (gamma-irradiated).
10. Sterile dissecting dressing forceps (autoclavable, latex free, stainless steel, straight, serrated tip).
11. Sterile 10 mL disposable Luer-lock tip syringes with clear barrel.
12. Polyethylene tubing: I.D. 0.28 mm (0.011″), O.D. 0.61 mm (0.024″), 30.5 m (100′). Cut tubing into 14 mm catheter segments and sterilize using ethylene oxide gas.
13. 10% povidone iodine solution.
14. Sterile polystyrene 100 mm × 15 mm petri dishes.
15. Tissue wipes (e.g., Kimwipes).
16. 1.5 mL microcentrifuge tubes (autoclavable, polypropylene).
17. Polyester-tipped applicators (6 in., sterile, plastic shaft, single use).
18. 30 G × ½ (0.3 mm × 13 mm) needles.

2.3 Necropsy

1. Isoflurane with drop jar.
2. Dissecting blunt-pointed forceps (autoclavable, latex free, chrome-plated steel, straight, serrated, blunt point tip).
3. Sharp-pointed dissecting scissors (autoclavable, surgical grade, stainless steel, sharp/sharp tip).
4. Blunt-end dissection scissors (autoclavable, surgical grade, stainless steel, straight, blunt/blunt tip).
5. Infrared loop sterilizer.
6. 100% ethanol in a squirt bottle.
7. Aluminum weighing dishes (20 mL capacity, 52 mm diameter, fluted sides).
8. Fine analytical balance.
9. Sterile 5 mL clear polystyrene culture tubes.
10. Rotor-stator mechanical laboratory homogenizer (5000–28,000 rpm, capacity 0.03 mL–10 L; e.g., Omni International).
11. Sawtooth rotor-stator homogenizer probe (stainless steel, 7 mm). Sterilize by enclosing in a sealable pouch and autoclaving.
12. Sterile 500 mL plastic beakers.
13. 1 L sterile ddH_2O.
14. Automated spiral plater (*see* **Note 6**).
15. Bacterial colony counter.
16. Sealable sterilization pouches.
17. Graphing software (e.g., GraphPad Prism).

3 Methods

All biohazardous work should be completed in a Class II biological safety cabinet and all animal procedures should be institutionally reviewed and approved. Sterile technique should be used for all protocols using live bacteria.

3.1 Inoculum Preparation

1. Inoculate cultures from a fresh plate or frozen stock of *P. mirabilis* and the mutant strain into separate culture tubes containing 5 mL of LB, and incubate overnight with shaking at 37 °C. Add antibiotics to the culture(s) if needed (*see* Table 1).
2. Dilute 5 mL of each overnight culture into 45 mL of fresh, sterile LB.
3. Take the OD_{600} reading using the spectrometer (should be approximately 0.450 (*see* **Note 7**).
4. Dilute each culture with fresh, sterile LB to a culture density of 2×10^8 CFU/mL (approximately $OD_{600} = 0.2$; *see* **Note 7**) in

Table 1
Antibiotic use with *P. mirabilis*

Antibiotic	Concentration of stock solution (mg/mL)	Final concentration (μg/mL)
Ampicillin	100	100
Kanamycin	50	25
Chloramphenicol	20	20

Antibiotic stock solutions are prepared at 1000–2000× concentration, filter-sterilized, portioned into 1 mL aliquots, and stored at −20 °C. Chloramphenicol is prepared in ethanol

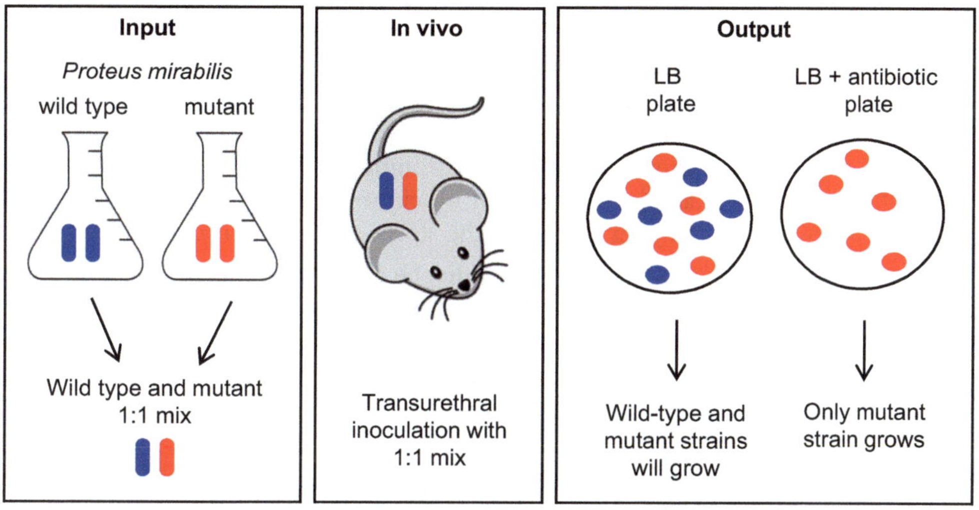

Fig. 1 Preparation, delivery, and output of cochallenge inoculum. *Input*: Preparation of inoculum with a 1:1 ratio of wild-type and mutant. Each culture needs to be diluted to an OD_{600} equal to 2×10^8 CFU/mL prior to mixing. Ensure OD_{600} of each culture is within 0.05 of the other for similar levels of CFU/mL input. In vivo Transurethrally inoculate mice with 1:1 mixture of wild-type and mutant bacteria. *Output*: Plate organ homogenate onto both LB plates and LB plates containing antibiotics. The wild-type and mutant strain will both grow on LB, and only the mutant strain will grow on the plate containing the antibiotic

a final volume of 10 mL. Ensure the OD_{600} of each culture is within 0.05 of the other for similar levels of CFU/mL input.

5. Combine a minimum of 3 mL of each inoculum to be challenged in a 1:1 ratio. Mix well. This is the input (Fig. 1).
6. Pipette 1 mL of the mixed culture into a sterile microcentrifuge tube to use for quantifying the input.
7. Dilute input(s) 1:10,000 (−4 dilution) and use this for plating. Plate duplicates on both LB and antibiotic plates (*see* **Note 4**).
8. After the plates dry, incubate them at 37 °C overnight.
9. Count the colonies and calculate the input CFU of both strains (e.g., wild type and mutant); *see* Subheading 3.4.

3.2 Infection Procedure

1. Cut a piece of absorbent underpad approximately 15″ × 20″ and place in the safety cabinet with the infusion pump positioned in the middle.
2. Place one sterile, empty petri dish on either side. In the petri dish on the left, place enough povidone iodine to cover at least half of the plate. Place an applicator inside so it can start to soak up the iodine. Leave the other petri dish empty.
3. Vortex or mix the diluted bacteria by inverting the tube. Pour the contents into a sterile basin.
4. Using a 10 mL syringe, load the contents of the basin into the syringe and then ensure any air bubbles are removed by flicking the tip.
5. Place one end of the extension set onto the end of the 10 mL syringe and place a 30 G needle onto the other end. Using forceps, insert a 14 mm sterilized catheter onto the needle. Push the inoculum through the extension set and catheter until a small amount of inoculum dribbles out (*see* **Note 8**). Place this set up inside the sterile, empty petri dish on the right.
6. Place the syringe into the infusion pump and set to appropriate volume (50 μL per mouse unless otherwise indicated). Set the infuse rate to 0.2 mL/min, ensuring a slow inoculum delivery. Run the apparatus until that volume of liquid consistently dribbles out, typically up to five times, prior to the transurethral inoculation.
7. Weigh each mouse and dose appropriately intraperitoneally (IP) with stock anesthetic solution. Wait until the mice are fully anesthetized prior to attempting to infect them (*see* **Note 9**).
8. Perform a pedal reflex test by gently grasping the feet and applying pressure. If the mouse does not pull her foot away, she has received an adequate dose of anesthetic (*see* **Note 10**).
9. Lay the mouse dorsally in front of the apparatus with her head pointing towards the left.
10. Gently blot the periurethral region with the iodine-soaked applicator, then carefully blot the area again with a tissue wipe to remove any excess iodine (*see* **Note 11**).
11. Using your nondominant hand, gently grasp the periurethral area with the blunt-ended forceps. Grasp close enough to the vulva so that it spreads apart and has a tent-like appearance. Gently insert the catheter in the middle of the tent. This is where the urethral opening lies. Sometimes, small redirection of the catheter is necessary either from side to side or top to bottom to find the orifice. Apply gentle pressure and the catheter should enter the urethra smoothly (*see* **Note 12**).

12. If you meet a large amount of resistance or the catheter cannot be inserted all the way to the hub, the catheter is not in the urethra. If this happens, gently remove the catheter and try again. Sometimes, there will be a very small amount of pressure needed to get from the urethra to the bladder, but initially, there should be little, if any, resistance (*see* **Notes 13–16**).
13. Keeping the catheter in place, push the run button on the apparatus and wait until it has completed.
14. After completion, wait an additional 5 s prior to slowly removing the urinary catheter. While removing it, pause a minimum of three times prior to full removal and let the catheter rest for 1 s at each pause. This gives the inoculum time to fully exit the catheter and stabilize itself in the bladder (*see* **Note 17**).
15. Place the mouse on a paper towel dorsally so the inoculum stays in place as long as possible.
16. Place a small amount (1 mm or less) of sterile ophthalmic ointment into each eye (*see* **Note 18**).
17. Use a heat source to maintain body temperature as the anesthesia wears off. Typically, the mice remain anesthetized for 20–30 min depending on the depth of onset anesthesia and variances between individual anesthetic reactions.

3.3 Necropsy

1. The day before the experimental endpoint (typically 4–7 dpi), label all agar plates with mouse number, organ, and dilution information.
2. Fill culture tubes with 3 mL of sterile PBS. One tube is needed for each organ (bladder, kidneys, or spleen). Store tubes at 4 °C overnight.
3. Assemble the homogenizer in a biosafety cabinet. Fasten the sterile rotor-stator probe to the homogenizer.
4. Turn on the infrared loop sterilizer in the biosafety cabinet 10 min prior to the necropsy to allow the temperature to reach 815 °C.
5. If collecting urine for CFUs, label a sterile microcentrifuge tube for each sample (*see* **Note 19**). Place a microcentrifuge tube rack onto your work surface. Place the tube on the lower portion and position the mouse on the upper so her periurethral region is hovering over the tube. The mouse can grasp the rack while you gently massage the lower abdominal region, encouraging her to release urine into the tube. Contamination from the periurethral region will occur if the urine catch is not clean, so ensure no fecal matter or hair is in the sample (*see* **Note 20**). Repeat for your cohort of mice and place the urine at 4 °C in between collections (*see* **Note 21**).
6. Place a 15″ × 20″ piece of absorbent underpad in the biological safety cabinet along with the sterilized dissecting instruments

and autoclaved weigh boats (*see* **Note 22**). The balance should also be ready for use. Place a small beaker with sterile water near the infrared loop sterilizer and have a tissue wipe on hand to clean the instruments in between uses.

7. Humanely euthanize mice by cage as directed in your approved animal use protocol. We use isoflurane and a drop jar (*see* **Note 23**), as this method allows the mice to become anesthetized until breathing ceases with minimal side effects.
8. Place the mouse horizontally on the absorbent underpad dorsally with her head pointing to the left. Using a squirt bottle filled with 100% ethanol, apply a small amount to the abdominal area and smooth the hair. Wetting it down not only keeps the hair out of the way but also reduces the potential for contamination.
9. Using the blunt-ended dissecting scissors, make a small tent in the diaphragm region of the mouse and make a superficial 5–10 mm incision. Gently grasp both sides of the incision and pull apart in a smooth, steady motion. The entire first layer of hair and flesh should easily be removed from the necropsy area, leaving the organs intact and the sublayer of fascia exposed (*see* **Note 24**).
10. Retrieve the urinary bladder first and place it gently into a sterile weigh boat. It is located in the distal abdominal cavity nestled in between the adipose tissue. Record the weight and repeat for all other organs. The spleen is located on the left side of the mouse nestled near the liver in close proximity to the left kidney (that is, it will be on the right side from the dissector's perspective). The kidneys are located on either side of the mouse in the lower abdominal wall close to the iliocostal muscle. Note any gross abnormalities such as bladder or kidney stones, which may appear as lighter areas within the tissue or external nodules on the tissue itself (*see* **Note 25**).
11. Place each organ in a sterile culture tube pre-filled with 3 mL of sterile PBS. Vortex or tap each tube gently so the organ falls to the bottom.
12. Using the maximum rpm setting, homogenize bladders for 60 s, kidneys for 30 s and spleens for 15 s, ensuring sterility is maintained throughout the process. Keep organs at 4 °C if you have a large number in between groups of mice.
13. After each organ is homogenized, clean the homogenizer by placing the tip into a plastic beaker of ethanol and running for 5–10 s. Then, repeat using a plastic beaker of sterile water (*see* **Note 26**).
14. Using a micropipette, measure each urine volume obtained and bring up to 150 μL with sterile PBS (e.g., if 20 μL of urine was

collected, add 130 μL of sterile PBS to the tube). Record both the amount of urine collected and the amount of PBS added for calculation of dilution factor later (*see* **Note 27**).

15. Transfer 1 mL of each organ homogenate into sterile 1.5 mL tubes. Use sterile PBS to make 1:100 (that is, 10^{-2}) dilutions of the bladder and kidney homogenates and urine samples.
16. Use the automated plater to spiral plate 0 and −2 dilutions for the bladders, kidneys, and urine, and 0 for the spleens. Plate duplicate dilutions on LB agar with and without antibiotic to quantify mutant bacteria and total bacteria, respectively (Fig. 1, output) (*see* **Note 6**).
17. Allow plates to dry, then incubate all plates at 37 °C overnight.

3.4 Data Formatting

1. The next day, count each plate using a colony counter, or count by hand if needed. Record colony numbers for each sample both with and without antibiotic selection, as you will need to know this for data calculations later (Table 2). Properly label the plate identities to note if antibiotics were used, as you will need to know this for data calculations later.
2. Save or export the data as a spreadsheet. The limit of detection for a typical spiral plated sample is 100 CFU, so any zeros should be graphed as 100 for statistical measurement purposes.
3. To calculate CFU/g of tissue, multiply the raw counts by 3 (the amount of PBS the organs were homogenized in, in mL). Divide that number by the weight of the organ (in grams) (*see* Table 3 and **Note 28**).

Table 2
Colony count example with proper labeling

Plate no	Plate ID	Dilution	Total count	Count used
1.1	1 Kidneys Plain	2.00	2.37E + 06	235
2.1	1 Kidneys Amp	2.00	1.39E + 05	232

This is an example readout from a QCount colony counter

Table 3
Example of data formatting for kidneys

		Raw data	Subtract	×3 PBS	Weight (g)	CFU/g
Mouse	plain	2.37E + 06	2.23E + 06	6.69E + 06	0.2345	2.85E + 07
1	amp	1.39E + 05	1.39E + 05	4.17E + 05	0.2345	1.78E + 06

Kidneys from Mouse 1 weighed 0.2345 g and were homogenized in 3 mL of PBS. Both raw data counts were plated from the same homogenate, on one plain LB plate and one plate containing ampicillin

4. To calculate CFU/mL of urine, divide **150 μL** by the amount of urine collected (in μL), then multiply by the raw count (*see* **Note 29**).
5. To determine the difference in bacterial load per organ or urine for wild-type and mutant bacteria, subtract the total CFU on the antibiotic plate (representing the mutant growth) from the total CFU on the plain plate (representing wild type) (*see* **Note 30**).
6. A competitive index (CI) is calculated by dividing the ratio of mutant to wild-type bacteria in the output by the ratio of mutant to wild-type bacteria in the input.
7. Graph the data. Graph raw CI on a log scale or log-transformed on a linear scale [11].
8. Calculate statistical significance using the Wilcoxon signed rank test (with a hypothetical value of 0) on log-transformed CI values.

4 Notes

1. Use an identification method to differentiate the individual mice (i.e., ear tagging, tail marking, or ear notching). We mark the tails with permanent marker and ensure the marks remain by touching up the preexisting marks every other day until completion of the experiment.
2. Higher concentrations of NaCl, common in many LB recipes, cause *P. mirabilis* isolates to swarm, making isolation of individual colonies impossible.
3. Typical mouse cohorts for urinary tract cochallenge infections are a minimum of 6–10 mice. This allows an adequate number of infected mice for statistical purposes as well as human and host error (i.e., unable to properly infect the mouse, the mouse has an anatomical abnormality, unexpected death, etc.).
4. This protocol relies on antibiotic resistance markers to distinguish wild type from mutant bacteria. Commonly used antibiotic resistance markers used in *P. mirabilis* include those listed in Table 1.
5. Ketamine–HCl is a controlled substance requiring careful documentation (e.g., in the USA, it is a Schedule III substance). In order to obtain ketamine, licensing may be required. Individual institutions may have varying policies regarding its distribution and storage.
6. CFU can be manually plated without the use of a spiral plater. In this case, individually spread-plate **100 μL** serial dilutions.

Alternatively, drip plating can be used for multiple dilutions on one plate [13].

7. On our spectrophotometer, $OD_{600} = 0.20$ is equivalent to 2×10^8 CFU/mL = 1×10^7 CFU/mouse. The OD_{600} that corresponds to 2×10^8 CFU/mL may vary depending on the machine used and should be determined independently by others following this protocol.
8. The priming volume needed for the 20″ extension set is 2.0 mL. When preparing inoculum, take this volume into account. We aim for a total volume of 10 mL to ensure there is enough, but the least amount needed for any transurethral inoculation is 4 mL.
9. It is extremely important that the mice are appropriately dosed with anesthesia. As they wake up, their bladders contract which can lead to premature expelling of inoculum if they are not under anesthesia for an appropriate amount of time. However, if given too much, they may have tremors which also cause bladder contraction.
10. If the mice are having trouble falling asleep, sometimes covering their eyes with a paper towel or cupping them with your gloved hand allows them to relax more quickly.
11. Only a small amount of povidone iodine is necessary to aseptically clean the periurethral region. Blot the excess; do not rub it off or wipe it, as you risk entry of the antiseptic into the urinary tract.
12. Depending on the size of the mouse that is being inoculated, the distance from the urethral opening to the bladder may be difficult to gauge. It is important to be as consistent as possible with where the catheter is being placed. You can use sterile scissors to cut the catheter if necessary. If the catheter is too long, it may be difficult to gently guide into the urethra. If the catheter is too short, there is the potential for rupture of the urethra and/or injury to the delicate tissues of the urogenital region.
13. Female mice sometimes have genital abnormalities which inhibit normal urethral catheterization. The most common abnormality I have encountered is a small lump of skin I call a "nub" which lies in the way of the urethra. There is no simple answer to infecting these mice successfully. Gently probe underneath the "nub" where the urethral opening should be. Do not force the catheter; simply move it in different directions until the opening is found. Always mark these mice specifically so if the infection is not successful they may be excluded from the data set.

14. The typical weight of 7–8-week-old CBA/J mice is 20 g. However, it is occasionally necessary to use older mice that are heavier for some experiments. Larger mice tend to be more difficult to infect due to the larger amount of abdominal adipose tissue on either side of and/or surrounding the urethra.
15. Due to the close proximity of the urethra and the vagina, it is common to mistake the vagina for the urethra. One way to ensure the catheter is in the urethra is by advancing the catheter gently all the way to the hub of the needle. It should easily glide in. If, however, the catheter is in the vagina instead, you will not be able to insert the catheter fully. Do not force it. Simply remove it and try again. If you believe the catheter is in the urethra and proceed with the inoculation, but the catheter is in the vagina instead, the inoculum will start to dribble out within a few seconds.
16. The tissue around the urogenital region is delicate. Even with careful manipulation, the area is easily irritated and becomes swollen, making inoculation difficult. If the catheter does not easily glide into the urethra, gently try to find the opening. If, with repeated manipulation, the region becomes increasingly inflamed, it will become impossible for a successful infection to take place.
17. Once the catheter has been inserted into the bladder, and the inoculum has been given, let the catheter remain in place for 5 s prior to removal. This allows the full volume to enter the bladder. When pulling the catheter out, do so slowly so the inoculum stays in place and does not follow the catheter back out.
18. Since ketamine is a dissociative anesthetic which does not allow the mice to close their eyes, a paper towel placed lightly over the mice while they are waking up, as well as protective lubrication, prevents corneal drying and damage. The paper towel also helps maintain body heat during the recovery process.
19. If a cage is soiled, change the bedding a minimum of 20 min before starting an experiment. This allows for less chance of secondary contamination while the mice are recovering from anesthesia. Conversely, once a urinary tract infection has been established, mice tend to urinate more frequently but when the cages are freshly changed, it excites them, causing them to release urine. If urine is needed for an experiment, allow for extra time so the mice calm down prior to collection. Typically, it takes a minimum of 20 min for urine to accumulate in the bladder again once it has been emptied, and multiple collection attempts may be needed.
20. Throughout the infection procedure and necropsy, it is common for gloves to accumulate mouse hair and debris. Instead of

using alcohol to clean them off, change gloves to minimize the risk of accidentally killing the bacteria or getting alcohol onto an organ.

21. When collecting urine, it is important to remain consistent in the collection process. Too much manual bladder manipulation may lead to irritation and inflammation. Unless required for experimental design (RNA studies, urine being collected at intermediate time points throughout the course of an experiment, etc.), the maximum number of urine collections we will perform is 3, spaced 30 min apart. Too much manual manipulation may cause excessive inflammation and discomfort. Even if enough urine has been collected from one mouse, manipulate the entire cohort the same number of times as the control.
22. Aluminum weighing dishes can be cleaned with water and autoclaved for multiple uses.
23. A drop jar consists of a plexiglass or glass container with isoflurane-soaked cotton balls in the bottom. The mice need to be prevented from coming into direct contact with the liquid inhalant with the use of a mesh grid covering the cotton.
24. Removal of the entire first layer of skin and hair from the mouse prior to organ removal leads to less chance of hair contamination.
25. During organ removal it is important that sterility is maintained. The intestinal tract will need to be gently moved out of the way to gain access to the kidneys. Ensure you do not accidentally puncture the intestines while removing the kidneys and avoid touching them excessively with your instruments during the procedure.
26. Use plastic beakers to avoid possible shatter hazards from the homogenizer. Change the alcohol and sterile water every 20 mice, or as necessary, if the liquids start retaining organ remnants or become turbid.
27. We use a spiral plater to dispense 50 μL per plate. Bringing the volume to 150 μL allows plating 50 μL in duplicate (plain LB and LB with antibiotic). If using another plating method, adjust the volume accordingly. Ensure consistency within the experiment and dilute every urine sample within a cohort the same to reduce mistakes during data analysis.
28. CFU/g of tissue calculation example: If the original output is 1.80×10^6 CFU, 3 mL of PBS is used, and the organ weight is 0.2356 g: $(1.80e6 \times 3)/0.2356 = 2.29 \times 10^7$ CFU/g.
29. CFU/mL of urine calculation example: If the CFU output is 1.88×10^4, the amount of urine collected is 20 μL, and 130 μL of PBS is added to bring the volume to 150 μL: $(150/20) \times 1.88e4 = 1.41 \times 10^5$ CFU/mL of urine.

30. Bacterial load calculation example: The plain plate has 6.02×10^8 CFU and the antibiotic plate has 4.89×10^6 CFU. The bacterial load for wild type is (6.02e8-- 4.89e6) = 5.97×10^8 CFU. The bacterial load for the mutant is the count from the antibiotic plate, or 4.89×10^6 CFU.

References

1. Hagberg L, Engberg I, Freter R, Lam J, Olling S, Svanborg Edén C (1983) Ascending, unobstructed urinary tract infection in mice caused by pyelonephritogenic *Escherichia coli* of human origin. Infect Immun 40 (1):273–283
2. Zunino P, Geymonat L, Allen AG, Legnani-Fajardo C, Maskell DJ (2000) Virulence of a *Proteus mirabilis* ATF isogenic mutant is not impaired in a mouse model of ascending urinary tract infection. FEMS Immunol Med Microbiol 29(2):137–143
3. Pearson MM, Mobley HLT (2007) The type III secretion system of *Proteus mirabilis* HI4320 does not contribute to virulence in the mouse model of ascending urinary tract infection. J Med Microbiol 56 (Pt 10):1277–1283. https://doi.org/10.1099/jmm.0.47314-0
4. Himpsl SD, Lockatell CV, Hebel JR, Johnson DE, Mobley HLT (2008) Identification of virulence determinants in uropathogenic *Proteus mirabilis* using signature-tagged mutagenesis. J Med Microbiol 57(Pt 9):1068–1078. https://doi.org/10.1099/jmm.0.2008/002071-0
5. Li X, Johnson DE, Mobley HLT (1999) Requirement of MrpH for mannose-resistant *Proteus*-like fimbria-mediated hemagglutination by *Proteus mirabilis*. Infect Immun 67 (6):2822–2833
6. Armbruster CE, Smith SN, Johnson AO, DeOrnellas V, Eaton KA, Yep A, Mody L, Wu W, Mobley HLT (2017) The pathogenic potential of *Proteus mirabilis* is enhanced by other uropathogens during polymicrobial urinary tract infection. Infect Immun 85(2): e00808–e00816. https://doi.org/10.1128/IAI.00808-16
7. Himpsl SD, Pearson MM, Arewång CJ, Nusca TD, Sherman DH, Mobley HLT (2010) Proteobactin and a yersiniabactin-related siderophore mediate iron acquisition in *Proteus mirabilis*. Mol Microbiol 78(1):138–157. https://doi.org/10.1111/j.1365-2958.2010.07317.x
8. Nielubowicz GR, Smith SN, Mobley HLT (2010) Zinc uptake contributes to motility and provides a competitive advantage to *Proteus mirabilis* during experimental urinary tract infection. Infect Immun 78(6):2823–2833. https://doi.org/10.1128/IAI.01220-09
9. Armbruster CE, Mobley HLT, Pearson MM (2018) Pathogenesis of *Proteus mirabilis* infection. EcoSal Plus 8(1). https://doi.org/10.1128/ecosalplus.ESP-0009-2017
10. Hagan EC, Mobley HLT (2009) Haem acquisition is facilitated by a novel receptor Hma and required by uropathogenic *Escherichia coli* for kidney infection. Mol Microbiol 71(1):79–91. https://doi.org/10.1111/j.1365-2958.2008.06509.x
11. Smith SN, Hagan EC, Lane MC, Mobley HLT (2010) Dissemination and systemic colonization of uropathogenic *Escherichia coli* in a murine model of bacteremia. mBio 1(5): e00262–e00210. https://doi.org/10.1128/mBio.00262-10
12. Obernier JA, Baldwin RL (2006) Establishing an appropriate period of acclimatization following transportation of laboratory animals. ILAR J 47(4):364–369
13. Thomas P, Sekhar AC, Upreti R, Mujawar MM, Pasha SS (2015) Optimization of single plate-serial dilution spotting (SP-SDS) with sample anchoring as an assured method for bacterial and yeast cfu enumeration and single colony isolation from diverse samples. Biotechnol Rep (Amst) 8:45–55. https://doi.org/10.1016/j.btre.2015.08.003

Chapter 17

Indwelling Urinary Catheter Model of *Proteus mirabilis* Infection

Sara N. Smith and Chelsie E. Armbruster

Abstract

Uncomplicated urinary tract infections, especially those caused by *Escherichia coli*, have historically been widely studied. However, complicated urinary tract infections are presenting ever increasing healthcare challenges, particularly with *Proteus mirabilis*. *P. mirabilis* is often found on indwelling urinary catheters causing monomicrobial and polymicrobial catheter-associated urinary tract infection (CAUTI). Widespread antibiotic resistance, combined with the ability of *P. mirabilis* to form urinary calculi during infection, warrants further investigation of this pathogen and its host interaction in an infection model that more closely mimics the presence of an indwelling urinary catheter. Here, we describe the methods necessary to establish a murine model of *P. mirabilis* CAUTI.

Key words *Proteus mirabilis*, Catheter-associated urinary tract infection, Indwelling catheter, Mouse infection model

1 Introduction

Urinary tract infections (UTIs) constitute a major healthcare burden, and in 2011 alone, there were approximately 400,000 hospitalizations for UTIs with an estimated cost of $2.8 billion [1]. Complicated infections are increasing at an alarming rate and *Proteus mirabilis* is the pathogen most noted for causing complicated UTIs, specifically catheter-associated urinary tract infections (CAUTIs). In fact, a recent study confirmed *P. mirabilis* as the number one causative agent of CAUTI in both single-species and polymicrobial infections from twelve nursing homes [2].

P. mirabilis CAUTIs present a unique challenge to treatment strategies for a variety of reasons, including the formation of urinary and bladder stones (uroliths) and the ability of *P. mirabilis* to form biofilms on catheters. Bacteria residing in biofilms and uroliths are more resistant to host defenses and antimicrobial agents, propelling the need for further investigation into the infection process and host interactions to identify better targets for therapeutic intervention.

Melanie M. Pearson (ed.), *Proteus mirabilis: Methods and Protocols*, Methods in Molecular Biology, vol. 2021, https://doi.org/10.1007/978-1-4939-9601-8_17,

The murine model of CAUTI enables the study of the progression of infection in a live host. A model that mimics the inflammation and infection severity seen with catheterized human patients and gives consistent, reliable results is imperative for teasing apart the various components of this complicated infection process. In fact, recent studies employing the use of a murine CAUTI model for *P. mirabilis* uncovered numerous differences in the pathology of CAUTI compared to UTI, and further identified differences in the genes required by *P. mirabilis* for fitness in each infection model [3, 4].

This optimized procedure for induction of CAUTI has proven successful in identifying contributing virulence factors for complicated infections with *P. mirabilis*. The ability to investigate the pathogenesis of CAUTI infections employing a live animal model enables investigation of novel antimicrobial strategies and targeted therapeutics to combat this rising healthcare concern.

2 Materials

2.1 Prior to Infection

For smooth experimental implementation, pour agar plates one week before the experiment and presterilize all supplies. Methods of sterilization vary depending on the item but include ethylene oxide for pre-cut urinary catheters, filter sterilization for anesthesia, and autoclave sterilization for silicone catheter segments, PBS, metal weigh boats, surgical instruments, beakers, and flasks. Prior to transurethral instillation of bacteria into mice, always perform in vitro growth curves to identify any in vitro growth defects that may confound experimental results.

1. Black or blue permanent marker (e.g., Sharpie) (*see* **Note 1**).
2. Spectrophotometer.
3. Spectrophotometer cuvettes, 1.5 mL semi-micro.
4. 37 °C incubator plus 37 °C shaker incubator.
5. Sterile phosphate-buffered saline (PBS), pH 7.4.
6. Low salt Luria broth (LB): 5 g of yeast extract, 10 g of trypticase peptone, 0.5 g of NaCl; bring up to 1 L in deionized water. Mix well and sterilize by autoclaving (*see* **Note 2**).
7. Low salt LB agar: LB plus 15 g of agar/L; autoclave to sterilize, cool to 55 °C, and pour into Petri dishes, approximately 20–25 mL/plate (*see* **Note 2**).
8. Sterile culture tubes, one per bacterial strain to be tested.
9. Female CBA/J mice, age 5–6 weeks, should be ordered 2 weeks prior to the experimental date. This allows them proper time to both arrive at the facility and acclimate for 7 days to a new environment. This period of adjustment is

necessary due to the impact widespread stress has on physiological systems during, and directly after, transportation [5].

10. Bacterial strains to be tested (*see* **Note 3**).

2.2 Day of Infection

1. Sterile white petrolatum and mineral oil eye ointment.
2. Anesthetic stock solution (per 5 mL): 1 mL of 100 mg/mL ketamine, 0.05 mL of 100 mg/mL xylazine, 3.95 mL of 0.9% sterile NaCl. Mix all ingredients together and filter-sterilize with a 0.22 μm PEM membrane filter unit. Every 100 μL of diluted anesthesia is equal to 0.02 mL of ketamine. This stock solution has a shelf life of 7 days (*see* **Note 4**).
3. Sterile 1 mL disposable Luer-lock tip syringes with clear barrel.
4. 26 G × ½ (0.45 mm × 13 mm) needles.
5. 0.22 μm PEM membrane filter unit.
6. Harvard Apparatus Pump 11 Elite infuse-only single syringe pump.
7. Absorbent underpad (polyethylene, 750 mL/m^2 absorbency).
8. 20″ Sterile extension set (approx. priming volume 2.0 mL, latex-free, non-DEHP, kink resistant). This is the tubing that connects the infusion pump to the catheter.
9. Sterile 25 mL single use pipet basin (gamma-irradiated).
10. Sterile dissecting dressing forceps (autoclavable, latex free, stainless steel, straight, serrated tip).
11. Sterile 10 mL disposable Luer-lock tip syringes with clear barrel.
12. Sterile polyethylene tubing: I.D. 0.28 mm (0.011″), O.D. 0.61 mm (0.024″), 30.5 m (100′). Cut tubing into 9 mm catheter segments and sterilize using ethylene oxide gas.
13. Silicone rubber tubing 0.025″ OD × 0.012″ ID (e.g., RenaSil tubing from BrainTree, supplied as a 25 ft coil); cut into 4 mm segments and sterilize by autoclaving (*see* **Note 5**).
14. 10% povidone iodine solution.
15. Sterile polystyrene 100 × 15 mm Petri dishes.
16. Tissue wipes (e.g., Kimwipes).
17. 1.5 mL microcentrifuge tubes (autoclavable, polypropylene).
18. Polyester-tipped applicators (6 in., sterile, plastic shaft, single use).
19. 30 G × ½ (0.3 × 13 mm) needles.

2.3 Necropsy

1. CO_2 or isoflurane with drop jar.
2. Dissecting blunt-pointed forceps (autoclavable, latex free, chrome-plated steel, straight, serrated, blunt point tip).

3. Sharp-pointed dissecting scissors (autoclavable, surgical grade, stainless steel, sharp/sharp tip).
4. Blunt-end dissection scissors (autoclavable, surgical grade, stainless steel, straight, blunt/blunt tip).
5. Infrared loop sterilizer.
6. 100% ethanol in a squirt bottle.
7. Aluminum weighing dishes (20 mL capacity, 52 mm diameter, fluted sides).
8. Fine analytical balance.
9. Sterile 5 mL clear polystyrene culture tubes.
10. Rotor-stator mechanical laboratory homogenizer (5000–28,000 rpm, capacity 0.03 mL–10 L).
11. Sawtooth rotor-stator homogenizer probe (stainless steel, 7 mm). Sterilize by enclosing in a sealable sterilization pouch and steam autoclave.
12. Sterile 500 mL plastic beakers.
13. Sterile ddH_2O.
14. Automated spiral plater (*see* **Note 6**).
15. Bacterial colony counter.
16. Sealable sterilization pouches.

3 Methods

All biohazardous work should be completed in a Class II biological safety cabinet, and all animal procedures should be institutionally reviewed and approved. Sterile technique should be used for all protocols using live bacteria.

3.1 Inoculum Preparation

1. Inoculate cultures from a fresh plate or frozen stock of *P. mirabilis* and the mutant strain into 5 mL of LB, and incubate overnight with aeration (225 rpm) at 37 °C. Add antibiotics to the culture(s) if needed.
2. Centrifuge 1.8 mL of each culture in 2 mL microfuge tubes (3 min at 8,000–10,000 × *g*).
3. Resuspend each culture in 1 mL of sterile PBS.
4. Dilute 500 μL of each culture into 4.5 mL of PBS.
5. Take the OD_{600} reading using the spectrometer and adjust to a culture density of 2×10^8 CFU/mL (approximately $OD_{600} = 0.2$) (*see* **Note 7**).
6. Ensure each culture has the same approximate OD_{600} (*see* **Note 8**). Then, dilute 1:100 for a culture density of approximately 2×10^6 CFU/mL.

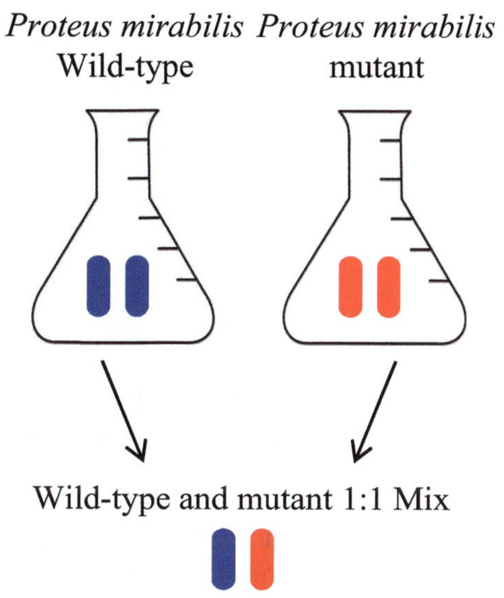

Fig. 1 Preparation of inoculum with a 1:1 ratio of wild-type and mutant. Each culture needs to be diluted to 2×10^6 CFU/mL prior to mixing. Ensure the OD_{600} of each undiluted culture is within 0.05 of the other for similar levels of CFU/mL input

7. For an individual challenge, use each diluted inoculum as is. For a cochallenge experiment, to determine if a mutant strain is outcompeted by the wild-type strain, combine a minimum of 3 mL of each inoculum to be challenged, in a 1:1 ratio (Fig. 1). Vortex well.
8. Pipette 1 mL of the inoculum culture into a sterile microcentrifuge tube to use for quantifying the input.
9. Dilute input(s) 1:100 (−2 dilution) and spiral plate to determine the input CFU. For cochallenge experiments, plate duplicates on both LB and antibiotic plates.
10. After the plates dry, incubate them overnight at 37 °C.

3.2 Infection Procedure

1. Cut a piece of absorbent underpad approximately 15″ × 20″ and place in the safety cabinet with the infusion pump positioned in the middle.
2. Place one sterile, empty petri dish on either side. In the petri dish on the left, place enough povidone iodine to cover at least half of the plate. Place an applicator inside so it can start to soak up the iodine. Leave the other petri dish empty.
3. Vortex or mix the diluted bacteria by inverting the tube. Pour the contents into a sterile basin.
4. Using the 10 mL syringe, fill the contents of the basin into the syringe and then ensure any air bubbles are removed by flicking the tip.

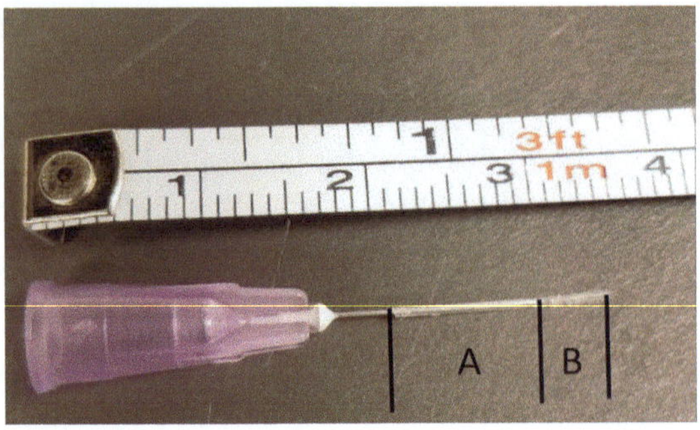

Fig. 2 Catheter setup for CAUTI model. (**a**) The polypropylene segment of catheter. (**b**) The silicone segment of catheter

5. Place one end of the extension set onto the end of the 10 mL syringe and place a 30 G needle onto the other end. Insert a 9 mm pre-cut, gas sterilized segment of polyethylene tubing onto the needle.
6. In front of the 9 mm tubing, insert a 4 mm pre-cut, autoclaved silicone catheter onto the needle (Fig. 2).
7. Push the inoculum through the extension set and catheter until a small amount of inoculum dribbles out (*see* **Note 9**). Place this setup inside the sterile, empty petri dish on the right.
8. Place the syringe into the infusion pump and set to appropriate volume (50 μL per mouse unless otherwise indicated). Set the infuse rate to 0.2 mL/min, ensuring a slow inoculum delivery. Run the pump until that volume of liquid consistently dribbles out, typically up to five times prior to the transurethral inoculation.
9. Weigh each mouse and dose appropriately intraperitoneally (IP) with stock anesthetic solution. Wait until the mice are fully anesthetized prior to attempting to infect them (*see* **Note 10**).
10. Perform a pedal reflex test by gently grasping the feet and applying pressure. If the mouse does not pull her foot away, she has received an adequate dose of anesthetic (*see* **Note 11**).
11. Lay the mouse dorsally in front of the apparatus with her head pointing toward the left.
12. Gently blot the periurethral region with the iodine-soaked applicator, then carefully blot the area again with a tissue wipe to remove any excess iodine (*see* **Note 12**).
13. Using your nondominant hand, gently grasp the periurethral area with the blunt-ended forceps. Grasp close enough to the vulva so that it spreads apart and has a tent-like appearance. Gently insert the catheter in the middle of the tent. This is

where the urethral opening lies. Sometimes, small redirection of the catheter is necessary either from side to side or top to bottom to find the orifice. Apply gentle pressure and the catheter should enter the urethra smoothly (*see* **Note 13**).

14. If you meet a large amount of resistance or the catheter cannot be inserted all the way to the hub, the catheter is not in the urethra. If this happens, gently remove and try again. Sometimes, there will be a very small amount of pressure needed to get from the urethra to the bladder, but initially, there should be little, if any resistance (*see* **Notes 13–17**).
15. Keeping the catheter in place, push the run button on the apparatus and wait until it has completed. After completion, wait an additional 5 s, then gently push the 9 mm catheter, with 4 mm silicone segment attached, into the bladder an additional 9 mm. This pushes the silicone segment into the bladder, where it remains for the duration of the experiment. Then, gently remove the 9 mm segment, pausing a minimum of three times prior to full removal and let the catheter rest for 1 s at each pause. This gives the inoculum time to fully exit the catheter and stabilize itself in the bladder (*see* **Note 18**). If done correctly, the silicone segment should remain in the bladder.
16. Place the mouse on a paper towel dorsally so the inoculum stays in place as long as possible.
17. Place a small amount (1 mm or less) of sterile ophthalmic ointment into each eye (*see* **Note 19**). Use a heat source to maintain body temperature as the anesthesia wears off. Typically, the mice remain anesthetized for 20–30 min depending on the depth of onset anesthesia and variances between individual anesthetic reactions.

3.3 Necropsy

1. The day before the experimental endpoint (typically at 72 h post-inoculation for a 96-h study, *see* **Note 8**), label all agar plates with mouse number, organ, and dilution information.
2. Fill culture tubes with 3 mL of sterile PBS. One tube is needed for each organ (bladder, kidneys, or spleen). Store tubes at 4 °C overnight.
3. Turn on the infrared loop sterilizer in the biosafety cabinet 10 min prior to the necropsy to allow temperature to reach 815 °C.
4. If collecting urine for CFUs, label a sterile microcentrifuge tube (*see* **Note 20**). Place a microcentrifuge tube rack onto your work surface. Place the tube on the lower portion and position the mouse on the upper so her periurethral region is hovering over the tube. The mouse can grasp the rack while you gently massage the lower abdominal region, encouraging

her to release urine into the tube. A sterile swab can also be used for gentle massage of the lower abdominal region to promote urination. Some contamination from the periurethral region will be noted if the urine catch is not clean so ensure no fecal matter or hair is in the sample. Repeat for your cohort of mice and place the urine at 4 °C in between collections (*see* **Note 21**).

5. Place a 15″ × 20″ piece of absorbent underpad in the biological safety cabinet along with the sterilized dissecting instruments and autoclaved weigh boats (*see* **Note 22**). The balance should also be ready for use. Place a small beaker with sterile water near the infrared loop sterilizer and have a tissue wipe on hand to clean the instruments in between uses.
6. Humanely euthanize mice by cage as directed in your approved animal use protocol. We use isoflurane and a drop jar (*see* **Note 23**), as exposure to this method allows the mice to become anesthetized until breathing ceases with minimal side effects.
7. Place the mouse horizontally on the absorbent underpad dorsally with her head pointing to the left. Using a squirt bottle with 100% ethanol, apply a small amount to the abdominal area and smooth the hair. Wetting it down not only keeps the hair out of the way, but also reduces the potential for contamination.
8. Using the blunt-ended dissecting scissors, make a small tent in the diaphragm region of the mouse and make a superficial 5–10 mm incision. Gently grasp both sides of the incision and pull apart in a smooth, steady motion. The entire first layer of hair and flesh should easily be removed from the necropsy area, leaving the organs intact and the sublayer of fascia exposed (*see* **Note 24**).
9. Retrieve the urinary bladder first and place it gently into a sterile weigh boat. It is located in the distal abdominal cavity nestled in between the adipose tissue. Record the weight and repeat for all other organs. The spleen is located on the left side of the mouse nestled near the liver in close proximity to the left kidney (that is, it will be on the right side from the dissector's perspective). The kidneys are located on either side of the mouse in the lower abdominal wall close to the iliocostal muscle. Note any gross abnormalities such as bladder or kidney stones, which may appear as lighter areas within the tissue or external nodules on the tissue itself.
10. Place each organ in a sterile culture tube prefilled with 3 mL of sterile PBS. Vortex or tap each tube gently so the organ falls into the bottom.

11. Using the maximum rpm, homogenize bladders for 60 s, kidneys for 30 s and spleens for 15 s, ensuring sterility is maintained throughout the process (*see* **Note 25**). Keep organs at 4 °C if you have a large number in between groups of mice. Change the alcohol and sterile water every 20 mice, or as necessary, if the liquids start retaining organ remnants or become turbid.
12. Measure each urine volume obtained and bring up to 150 μL with sterile PBS (e.g., if 20 μL of urine was collected, add 130 μL of sterile PBS to the tube). Record both the amount of urine collected and the amount of PBS added for calculation of dilution factor later (*see* **Note 26**).
13. Transfer 1 mL of each homogenate to sterile 1.5 mL tubes. Use sterile PBS to make 1:100 (that is, 10^{-2}) dilutions of the bladder and kidney homogenates and urine samples.
14. Use the automated plater to spiral plate 0 and −2 dilutions for the bladders, kidneys, and urine, and 0 for the spleens. For cochallenge experiments, plate duplicate dilutions on LB agar with and without antibiotic to quantify mutant bacteria and total bacteria, respectively.
15. Allow plates to dry, then incubate all plates at 37 °C overnight.

3.4 Data Formatting

1. The next day, count each plate using a colony counter, or count by hand if needed. Save or export the data as a spreadsheet. Properly label the Plate ID if antibiotics were used, as you will need to know this for data calculations later. The limit of detection is 100 CFU, so any zeros should be graphed as 100 for statistical measurement purposes.
2. To calculate CFU/g of tissue, multiply the raw colony counts by 3 (the volume of PBS the organs were homogenized in, in mL). Divide that number the weight of the organ (in grams). To calculate CFU/mL of urine, divide 150 μL by the amount of urine collected (in μL), then multiply by the raw count (*see* **Notes 27** and **28**).
3. For an independent challenge experiment, graph the data. Use the two-tailed Mann–Whitney test for statistical analysis. Otherwise, for a cochallenge experiment, proceed to the next step.
4. To determine the difference in bacterial load per organ or urine from wild type and mutant in a cochallenge experiment, subtract the CFU on the antibiotic plate (representing the mutant growth) from the CFU on the plain plate (representing total CFUs) to obtain CFUs of the wild type (*see* **Note 29**).
5. For a cochallenge experiment, calculate a competitive index (CI) by dividing the ratio of mutant to wild-type bacteria in the output organ by the ratio of mutant to wild-type bacteria

from the input inoculum. The raw CI can be graphed on a log scale, or a $\log_{10}$-transformation can be used and the resulting data can be graphed on a linear scale [6].

6. To determine if a mutant has a significant fitness defect during a cochallenge experiment, use the Wilcoxon signed rank test (with a hypothetical value of 1 for raw CIs or a hypothetical value of 0 for log-transformed CI values).

4 Notes

1. Use an identification method to differentiate the individual mice (i.e., ear tagging, tail marking, or ear notching). We mark the tails with permanent marker and ensure the marks remain by touching up the preexisting marks every other day until completion of the experiment.
2. Higher concentrations of NaCl, common in many LB recipes, cause *P. mirabilis* isolates to swarm, making isolation of individual colonies impossible.
3. This protocol relies on antibiotic resistance markers to distinguish wild-type from mutant bacteria. Commonly used antibiotic resistance markers in *P. mirabilis* include those listed in Table 1.
4. Ketamine–HCl is a controlled substance requiring careful documentation (e.g., in the USA, it is a Schedule III substance). In order to obtain ketamine, licensing may be required. Individual institutions may have varying policies regarding its distribution and storage.
5. Silicone catheter segments are pliable, which makes grasping them during an infection challenging. To ensure sterility, if a segment is accidentally dropped while trying to place it onto the end of the needle, discard it. Thus, always cut a minimum of five extra segments, in the case of loss.

Table 1
Antibiotic use with *P. mirabilis*

Antibiotic	Concentration of stock solution (mg/mL)	Final concentration (μg/mL)
Ampicillin	100	100
Kanamycin	50	25
Chloramphenicol	20	20

Prepare antibiotic stock solutions at 1000–2000× concentration, filter-sterilize, portion into 1 mL aliquots, and store at −20 °C. Prepare ampicillin and kanamycin in distilled water, and chloramphenicol in ethanol

6. CFU can be manually plated without the use of a spiral plater. Serial dilutions can be individually spread-plated using 100 μL. Alternatively, drip plating can be used for multiple dilutions on one plate [7].
7. On our spectrophotometer, $OD_{600} = 0.20$ is equivalent to 2×10^8 CFU/mL $= 1 \times 10^7$ CFU/mouse. The correct OD_{600} to obtain this number of CFU may vary depending on the machine used and the bacterial strain, and should be determined independently by others following this protocol.
8. The inoculum used for *P. mirabilis* experimental CAUTI is 100-fold lower than that used in the ascending UTI model because the presence of the catheter within the bladder promotes increased colonization and resistance to host defenses. Thus, mice inoculated with a higher dose of *P. mirabilis* experience significant morbidity and mortality in this model, as early as 48 h, while reducing the inoculum 100-fold allows for survival of the majority of the mice to a 96 h timepoint. Further reduction of the inoculating dose may be required for experiments of longer duration (such as 7 days).
9. The priming volume needed for the 20″ extension set is 2.0 mL. When preparing the inoculum, take this volume into account. We aim for a total volume of 10 mL to ensure there is enough, but the lowest amount needed for any transurethral inoculation is 4 mL.
10. It is extremely important that the mice are appropriately dosed with anesthesia. As they wake up, their bladders contract, which can lead to premature expelling of inoculum if they are not under anesthesia for an appropriate amount of time. However, if given too much, they may have tremors which also cause bladder contraction.
11. If the mice are having trouble falling asleep, sometimes covering their eyes with a paper towel or cupping them with your gloved hand allows them to relax more quickly.
12. Only a small amount of povidone iodine is necessary to aseptically clean the periurethral region. Blot the excess; do not rub it off or wipe it, as you risk entry of the antiseptic into the urinary tract.
13. Depending on the size of the mouse that is being inoculated, the distance from the urethral opening to the bladder may be difficult to gauge. It is important to be as consistent as possible with where the catheter is being placed. Use sterile scissors to cut the catheter if necessary. If the catheter is too long, it may be difficult to gently guide into the urethra. If the catheter is too short, there is the potential for rupture of the urethra and/or injury to the delicate tissues of the urogenital region. The pliable nature of the silicone catheter segment makes entry

into the urethra more challenging. Gentle pressure and redirection is the best method to ensure successful entry into the bladder. Sometimes, the silicone catheter segment falls off. If this occurs, discard it and use a fresh one to ensure sterility.

14. Female mice sometimes have genital abnormalities which inhibit normal urethral catheterization. The most common abnormality we have encountered is a small lump of skin we call a "nub," which lies in the way of the urethra. There is no simple answer to infecting these mice successfully. Gently probe underneath the nub where the urethral opening should be. Do not force the catheter; simply move it in different directions until the opening is found. Always mark these mice specifically so if the infection is not successful they may be excluded from the data set.
15. The typical weight of 7–8 week old mice is 20 g. However, it is occasionally necessary to use older mice that are heavier for some experiments. Larger mice tend to be more difficult to infect due to the larger amount of abdominal adipose tissue on either side of and/or surrounding the urethra.
16. Due to the close proximity of the urethra and the vagina, it is common to mistake the vagina for the urethra. One way to ensure the catheter is in the urethra is by advancing the catheter gently all the way to the hub of the needle. It should easily glide in. If, however, the catheter is in the vagina instead, you will not be able to insert the catheter fully. Do not force it. Simply remove it and try again. If you believe the catheter is in the urethra and proceed with the inoculation but if it is in the vagina instead, the inoculum will start to dribble out within a few seconds.
17. The tissue around the urogenital region is delicate. Even with careful manipulation, the area is easily irritated and becomes swollen, making inoculation difficult. If the catheter does not easily glide into the urethra, gently try to find the opening. If, with repeated manipulation, the region becomes increasingly inflamed, it will become impossible for a successful infection to take place.
18. Once the catheter has been inserted into the bladder, and the inoculum has been given, let the inoculation apparatus remain in place for 5 s prior to removal. This allows the full volume to enter the bladder. Then, using sterile forceps, slowly and carefully advance the tubing closest to the base of the needle to push the silicone catheter fragment fully into the bladder. Remove the inoculation apparatus slowly so the inoculum stays in place and does not follow the tubing back out.
19. Since ketamine is a dissociative anesthetic which does not allow the mice to close their eyes, a paper towel placed lightly over the mice while they are waking up, as well as protective

lubrication, prevents corneal drying and damage. The paper towel also helps maintain body heat during the recovery process.

20. If urine is needed for an experiment, allow for extra time so the mice calm down prior to collection. Typically, it takes a minimum of 20 min for urine to accumulate in the bladder again once it has been emptied, and multiple collection attempts may be needed.
21. When collecting urine, it is important to remain consistent in the collection process. Too much manual bladder manipulation may lead to irritation and inflammation. Unless required for experimental design (RNA studies, urine being collected at intermediate time points throughout the course of an experiment, etc.), the maximum number of urine collections we will perform is 3, spaced 30 min apart. Too much manual manipulation may cause excessive inflammation and discomfort. Even if enough urine has been collected from one mouse, manipulate the entire cohort the same number of times as the control.
22. Aluminum weighing dishes can be cleaned with water and autoclaved for multiple uses.
23. A drop jar consists of a plexiglass or glass container with isoflurane-soaked cotton balls in the bottom. The mice need to be prevented from coming into direct contact with the liquid inhalant, generally by the use of a mesh grid covering the cotton.
24. Removal of the entire first layer of skin and hair from the mouse prior to organ removal leads to less chance of hair contamination.
25. During organ removal, it is important that sterility is maintained. The intestinal tract will need to be gently moved out of the way to gain access to the kidneys. Ensure you do not accidentally puncture it while removing the kidneys and avoid touching it excessively with your instruments during the procedure. While homogenizing, in between each organ, run the homogenizer tip first in a plastic beaker of ethanol, then in a beaker of sterile water. Use plastic beakers to avoid possible shatter hazards from the homogenizer.
26. Urine volume is typically brought up to 150 μL using PBS. This is to ensure an adequate amount of urine is present for scenarios where multiple plates are used for one sample. If more than 2 plates are being used with one urine sample, you may need to bring up all the samples more than 150 μL. Ensure consistency within the experiment and dilute every urine sample within a cohort to the same total volume to reduce the possibility of mistakes during data analysis.

27. CFU/g of tissue calculation example: If original output is 1.80×10^6 CFU and the organ weight is 0.2356 g: $(1.80e6 \times 3)/0.2356 = 2.29 \times 10^7$ CFU/g.
28. CFU/mL of urine calculation example: If CFU output is 1.88×10^4, the amount of urine collected is 20 μL, and 130 μL of PBS is added to bring the volume to 150 μL: $(150/20) \times 1.88e4 = 1.41 \times 10^5$ CFU/mL of urine.
29. Example bacterial load calculation from a cochallenge experiment: The plain plate has 6.02×10^8 CFU and the antibiotic plate has 4.89×10^6 CFU. The raw bacterial load for wild type is (6.02e8–4.89e6) = 5.97e8 CFU. The raw bacterial load for the mutant is the count from the antibiotic plate, or 4.89e6 CFU.

Acknowledgments

The authors would like to thank Harry Mobley and members of the Mobley laboratory for their helpful comments and critiques during the initial development of this protocol. This work was supported by the National Institute of Diabetes Digestive and Kidney Disorders (R00 DK105205 to C.E.A.).

References

1. Simmering JE, Tang F, Cavanaugh JE, Polgreen LA, Polgreen PM (2017) The increase in hospitalizations for urinary tract infections and the associated costs in the United States, 1998–2011. Open Forum Infect Dis 4(1): ofw281. https://doi.org/10.1093/ofid/ofw281
2. Armbruster CE, Prenovost K, Mobley HLT, Mody L (2017) How often do clinically diagnosed catheter-associated urinary tract infections in nursing homes meet standardized criteria? J Am Geriatr Soc 65(2):395–401. https://doi.org/10.1111/jgs.14533
3. Armbruster CE, Forsyth-DeOrnellas V, Johnson AO, Smith SN, Zhao L, Wu W, Mobley HLT (2017) Genome-wide transposon mutagenesis of *Proteus mirabilis*: essential genes, fitness factors for catheter-associated urinary tract infection, and the impact of polymicrobial infection on fitness requirements. PLoS Pathog 13(6): e1006434. https://doi.org/10.1371/journal.ppat.1006434
4. Armbruster CE, Smith SN, Johnson AO, DeOrnellas V, Eaton KA, Yep A, Mody L, Wu W, Mobley HLT (2017) The pathogenic potential of *Proteus mirabilis* is enhanced by other uropathogens during polymicrobial urinary tract infection. Infect Immun 85(2): e00808–e00816. https://doi.org/10.1128/IAI.00808-16
5. Obernier JA, Baldwin RL (2006) Establishing an appropriate period of acclimatization following transportation of laboratory animals. ILAR J 47(4):364–369
6. Smith SN, Hagan EC, Lane MC, Mobley HLT (2010) Dissemination and systemic colonization of uropathogenic *Escherichia coli* in a murine model of bacteremia. mBio 1(5): e00262–e00210. https://doi.org/10.1128/mBio.00262-10
7. Thomas P, Sekhar AC, Upreti R, Mujawar MM, Pasha SS (2015) Optimization of single plate-serial dilution spotting (SP-SDS) with sample anchoring as an assured method for bacterial and yeast cfu enumeration and single colony isolation from diverse samples. Biotechnol Rep (Amst) 8:45–55. https://doi.org/10.1016/j.btre.2015.08.003

Chapter 18

Vaccination to Protect Against *Proteus mirabilis* Challenge Utilizing the Ascending Model of Urinary Tract Infection

Sara N. Smith, Stephanie D. Himpsl, and Harry L. T. Mobley

Abstract

Proteus mirabilis is a major cause of complicated urinary tract infections (UTIs). *P. mirabilis*' urease activity hydrolyzes urea and raises urine pH levels, which can catalyze bladder and kidney stone formation. This urolithiasis leads to harder-to-treat infections, possible urinary blockage, and subsequent septicemia. Development of a mucosal vaccine against *P. mirabilis* urinary tract infections is critical to protect against this potentially deadly infection process. Here, we describe the methodology necessary to produce a vaccine candidate conjugated to cholera toxin, administer the vaccine via the intranasal route, and test efficacy in a murine transurethral *P. mirabilis* infection model.

Key words *Proteus mirabilis*, Mucosal vaccine, Transurethral infection, Intranasal immunization, UTI vaccine

1 Introduction

Urinary tract infections caused by *Proteus mirabilis* complicate an already complex infection process through a variety of sometimes intersecting and overlapping factors [1]. *Proteus*' ability to attach and swarm on indwelling catheters and develop biofilms is mediated by fimbriae, encoded in some cases by fimbrial operons which have the ability to switch "on" or "off" dependent on cues that can control swarming versus swimming phenotypes [2]. *P. mirabilis* increases urine pH through urease activity, which hydrolyzes urea and heightens the likelihood of urolithiasis within the bladder and kidneys, causing chronic, hard-to-treat infections. These infections, left untreated or without effective treatment options available, may progress to bacteremia and sepsis, which combined with an ever-increasing ability to evade current drug therapies, make *P. mirabilis* an attractive target for vaccine studies.

Currently, there are no approved vaccines to prevent urinary tract infections that are licensed in the United States. An available UTI vaccine in Europe called Solco-Urovac uses heat-killed

Melanie M. Pearson (ed.), *Proteus mirabilis: Methods and Protocols*, Methods in Molecular Biology, vol. 2021, https://doi.org/10.1007/978-1-4939-9601-8_18,

bacteria including *Proteus* and demonstrates limited protection against recurrent infections [3]. Other studies have identified potential vaccine targets such as outer membrane protein antigens [4], the MR/P fimbrial tip adhesin MrpH [5], as well as Proteus toxin agglutinin (Pta) [6]. Recently, a multipeptide subunit vaccine administered subcutaneously showed protection in a murine bladder challenge model [7]. Here, we describe methods employed to develop and test an intranasal *P. mirabilis* vaccine using a murine ascending transurethral infection model.

2 Materials

The materials listed here are required for the expression and purification of an insoluble protein under denaturing conditions, with subsequent wash and elution steps for refolding of the protein. Optimization of the expression and purification of any other protein of interest is required.

Female CBA/J mice, age 5–6 weeks, should be ordered 2 weeks prior to the experimental start date. This allows them proper time to both arrive at the facility and acclimate for 7 days to a new environment. This period of adjustment is necessary due to the impact widespread stress has on physiological systems during and directly after transportation [8].

2.1 Protein Expression and Purification

1. Phosphate-buffered saline (PBS) (per liter), pH 7.4: 8 g of NaCl, 0.2 g of KCl, 1.44 g of Na_2HPO_4, 0.24 g of KH_2PO_4. Autoclave or filter-sterilize.
2. *Escherichia coli* transformed with vector encoding C-terminally 6×His-tagged protein.
3. Lysogeny broth (LB) (per liter): 10 g of tryptone, 5 g of yeast extract, 0.5 g of NaCl. Autoclave to sterilize.
4. Inducer to control protein expression (*see* **Note 1**).
5. 37 °C shaking incubator.
6. 2 L flask.
7. Nickel-nitriloacetic acid agarose (Ni-NTA agarose) (*see* **Note 2**).
8. Polypropylene column, 5 mL.
9. Lysis Buffer B: 100 mM NaH_2PO_4, 10 mM Tris–HCl, 8 M urea. Adjust to pH 8.0 using NaOH (*see* **Note 3**).
10. Wash Buffer C: 100 mM NaH_2PO_4, 10 mM Tris–HCl, 8 M urea. Adjust to pH 6.3 using HCl (*see* **Note 3**).
11. Elution Buffer D: 100 mM NaH_2PO_4, 10 mM Tris–HCl, 8 M urea. Adjust to pH 5.9 using HCl (*see* **Note 3**).

12. Elution Buffer E: 100 mM NaH_2PO_4, 10 mM Tris–HCl, 8 M urea. Adjust to pH 4.5 using HCl (*see* **Note 3**).
13. Dialysis Buffer 1: 100 mM NaH_2PO_4, 5 mM Tris–HCl, 6 M urea, 50 mM NaCl, pH 5.5 (*see* **Note 4**).
14. Dialysis Buffer 2: 100 mM NaH_2PO_4, 2.5 mM Tris–HCl, 4 M urea, 100 mM NaCl, 0.005% zwittergent, pH 6.5 (*see* **Note 4**).
15. Dialysis Buffer 3: 100 mM NaH_2PO_4, 2 M urea, 150 mM NaCl, 0.01% zwittergent, pH 7.4 (*see* **Note 4**).
16. Dialysis Buffer 4: 1× PBS, 0.05% zwittergent, pH 7.4 (*see* **Note 4**).
17. 1 L beaker with magnetic stir bar.
18. 4-(2-hydroxyethyl)-1-piperazineethanesulfonic acid (HEPES), 10 mM, pH 7.0.
19. Benzonase or other nuclease.
20. French press and French pressure cell (*see* **Note 5**).
21. Microcentrifuge, superspeed centrifuge, and ultracentrifuge.
22. Zwittergent 3-14 detergent, 10% (Calbiochem).
23. Vortex.
24. Rotisserie shaker.
25. Bicinchoninic acid (BCA) protein assay kit (e.g., Pierce).
26. 1.5 mL microcentrifuge tubes, 15 and 50 mL conical tubes.
27. Slide-A-lyzer dialysis cassette with float buoy (*see* **Note 6**).
28. Sterile 10 mL disposable Luer-lock tip syringes with clear barrel.
29. 18G × 1 (1.2mm × 40mm) needle.
30. SDS-PAGE gel; western supplies and reagents.
31. Centrifugal filter units.

2.2 Cross-Linking of Antigen to Cholera Toxin

1. PBS–EDTA: 1 mM EDTA in PBS.
2. Cholera holotoxin (CT) (Sigma).
3. *N*-succinimidyl-3-(2-pyridyldithio)propionate (SPDP).
4. Dimethylformamide (DMF).
5. Dithiothreitol (DTT).
6. Centrifugal filter units.
7. Microcentrifuge.

2.3 Intranasal Immunization and Boost

1. Anesthetic solution (per 5 mL): 1 mL of 100 mg/mL ketamine, 0.05 mL of 100 mg/mL xylazine, 3.95 mL of 0.9% sterile NaCl. Mix all ingredients together and filter sterilize with a 0.22 μm PEM membrane filter unit (*see* **Note 7**).

2. Female CBA/J mice, age 5–6 weeks. 5–10 mice per control group and, minimally, ten per experimental group.
3. CT-linked antigen.
4. 10–20 mL volume pipette and sterile tips.
5. Sterile 1 mL disposable Luer-lock tip syringes with clear barrel.
6. 26G × ½ (0.45 mm × 13 mm) needles.
7. Fine analytical balance with a range to 0.1 mg.
8. Sterile white petrolatum and mineral oil eye ointment.
9. 1.1–1.2 mm micro hematocrit capillary tubes.
10. Microtainer serum separator tubes.
11. Paper towels (*see* **Note 8**).
12. Heat source for mouse recovery from anesthesia (*see* **Note 9**).

2.4 Transurethral Inoculation

1. LB agar (per liter): 5 g of yeast extract, 10 g of tryptone, 0.5 g of NaCl, 15 g of agar. Autoclave to sterilize. Cool to 55 °C and pour approximately 20–25 mL into 100 mm petri dishes (*see* **Note 10**).
2. Sterile PBS, pH 7.4.
3. Sterile white petrolatum and mineral oil eye ointment.
4. 1.1–1.2 mm micro hematocrit capillary tubes.
5. Microtainer serum separator tubes.
6. Anesthetic solution (per 5 mL): 1 mL of 100 mg/mL ketamine, 0.05 mL of 100 mg/mL xylazine, 3.95 mL of 0.9% sterile NaCl. Mix all ingredients together and filter sterilize with a 0.22 μm PEM membrane filter unit (*see* **Note** 7).
7. Sterile 1 mL disposable Luer-lock tip syringes with clear barrel.
8. 26G × ½ (0.45 mm × 13 mm) needles.
9. Pump 11 Elite infuse-only single syringe pump (Harvard Apparatus).
10. Absorbent underpad (polyethylene, 750 mL/m^2 absorbency).
11. 20″ Sterile Extension Set (approx. priming volume 2.0 mL, latex-free, Non-DEHP, kink resistant). This is the tubing that connects the infusion pump to the inoculation catheter.
12. Sterile 25 mL single use pipet basin (sterilized by gamma irradiation).
13. Sterile dissecting dressing forceps (autoclavable, latex free, stainless steel, straight, serrated tip).
14. Sterile 10 mL disposable Luer-lock tip syringes with clear barrel.
15. Polyethylene tubing: I.D. 0.28 mm (0.011″), O.D. 0.61 mm (0.024″), 30.5 m (100′).

16. Povidone iodine 10% solution.
17. Sterile polystyrene 100 × 15 mm petri dishes.
18. Tissue wipes (e.g., Kimwipes).
19. 1.5 mL microcentrifuge tubes (autoclavable, polypropylene).
20. Polyester-tipped applicators (6 in., sterile, plastic shaft, single use).
21. 30G × ½ (0.3 mm × 13 mm) needles.
22. *P. mirabilis* culture prepared for transurethral inoculum. Use LB to dilute an overnight culture to 2×10^8 CFU/mL, which corresponds approximately to $OD_{600} = 0.2$. The exact OD_{600} reading should be determined empirically for a given spectrophotometer and strain of bacteria.
23. Fine analytical balance with a range to 0.1 mg.
24. Paper towels (*see* **Note 8**).
25. Heat source for mouse recovery from anesthesia (*see* **Note 9**).

2.5 Necropsy

1. CO_2 or isoflurane with drop jar.
2. Blunt-pointed dissecting forceps (autoclavable, latex free, chrome-plated steel, straight, serrated, blunt point tip).
3. Sharp-pointed dissecting scissors (autoclavable, surgical grade, stainless steel, sharp/sharp tip).
4. Blunt-end dissection scissors (autoclavable, surgical grade, stainless steel, straight, blunt/blunt tip).
5. Infrared micro-sterilizer.
6. 100% ethanol in a squirt bottle.
7. Aluminum weighing dishes (20 mL capacity, 52 mm diameter, fluted sides).
8. Tissue wipes (e.g., Kimwipes).
9. Fine analytical balance with a range to 0.1 mg.
10. Sterile 5 mL clear polystyrene culture tubes.
11. General laboratory homogenizer (5000–28,000 rpm, capacity 0.03 mL–10 L).
12. Rotor-stator probe (stainless steel, 7 mm, saw tooth). Autoclave in sterilization pouch to sterilize.
13. Two sterile, plastic 500 mL beakers to serve as cleaning reservoirs during homogenization.
14. 1 L of sterile ddH_2O.
15. 100% ethanol.
16. Autoplate spiral plating system (*see* **Note 11**).
17. Bacterial colony counter (*see* **Note 11**).

18. 1.5 mL microcentrifuge tubes (autoclavable, polypropylene).
19. LB agar plates.

3 Methods

Expression and purification of any protein of interest will need to be empirically determined, as no single method is appropriate for all proteins. Native purification is best for soluble proteins that are C-terminally tagged, while insoluble proteins will require purification under denaturing conditions which necessitate additional steps for refolding of the protein. Proper temperature for expression, amount of inducer, method of cell lysate preparation, incubation time and the amount of resin, binding and elution buffer components and concentrations, etc., are all variables that need to be empirically tested for each protein.

All biohazardous work should be completed in a Class II biological safety cabinet and all animal procedures should be institutionally reviewed and approved. Aseptic technique should be used for all protocols using live bacteria.

3.1 Expression and Preparation of a Lysate

1. Subculture an overnight culture of *E. coli* carrying the 6×His-tagged protein into 1 L of LB (1:100 dilution) containing antibiotic and shake at 37 °C until mid-logarithmic phase is reached (OD_{600} = 0.5–1.0; approximately 2.5 h). Collect 1 mL of sample and label as the uninduced control for SDS-PAGE analysis (*see* **Note 12**).
2. Add required inducer for expression of protein and grow for 3 h at 37 °C with shaking (*see* **Note 1**).
3. Remove 1 mL of induced sample for SDS-PAGE analysis. Using a microcentrifuge, spin down uninduced control and induced samples (4 min, 9,600 × *g*, 4 °C) and store pellets at −20 °C (*see* **Note 12**).
4. Spin down remaining induced LB culture (15 min, 10,820 × *g*, 4 °C) and store this pellet at −80 °C.
5. Thaw the cell pellet on ice for 15 min and resuspend in approximately 20 mL of 10 mM HEPES, pH 7.0.
6. Add 40 μL of Benzonase nuclease to digest DNA and RNA.
7. Pass twice through a French pressure cell at 20,000 psi to lyse cells, or until lysate clears (*see* **Notes 5, 13**).
8. Spin (30 min, 4300 × *g*, 4 °C) to remove cell debris. Place supernatant into a new tube and spin for a second time.
9. Ultracentrifuge the supernatant (1 h, 120,000 × *g*, 4 °C).
10. Remove supernatant (soluble fraction) from the pellet and retain 10 μL for SDS-PAGE analysis (*see* **Note 12**).

11. Add 5 mL of Buffer B (*see* **Note 3**) to cover the pellet (insoluble fraction). Add 10 μL of 10% zwittergent and freeze at −20 °C to help dissolve the membrane pellet (*see* **Note 14**).

3.2 Purification of 6× His-Tagged Protein Under Denaturing Conditions

1. Thaw the membrane pellet on ice and resuspend in 10 mL of additional Buffer B (*see* **Note 3**); collect 20 μL as an insoluble fraction sample for SDS-PAGE analysis (*see* **Note 12**).
2. Gently mix or vortex until dissolved (*see* **Note 15**).
3. Add 2 mL of the 50% Ni-NTA slurry (*see* **Note 16**) to the membrane suspension and mix gently by shaking (e.g., 8 rpm on a rotisserie shaker) for 30 min at room temperature.
4. Load the lysate–Ni-NTA mixture from **step 3** onto a polypropylene column with the bottom cap still attached. Allow the lysate-resin to settle by gravity, remove the bottom cap of the column, and collect flow-through into a conical tube. Retain 20 μL of the flow-through sample for SDS-PAGE analysis.
5. Wash the lysate-resin four times with 4 mL of Wash Buffer C and save 50 μL of each wash step for SDS-PAGE analysis.
6. Elute the recombinant protein four times with 1 mL of Elution Buffer D and collect fractions for SDS-PAGE analysis.
7. Elute eight times with 1 mL of Elution Buffer E and collect fractions for SDS-PAGE analysis.
8. Analyze collected fractions by SDS-PAGE (Fig. 1) and combine eluted fractions containing visibly pure protein.
9. Hydrate a dialysis cassette (*see* **Note 6**) with Dialysis Buffer 1 and carefully remove air using a syringe with an 18-gauge needle (*see* **Note 17**). Add the combined protein fractions into the dialysis cassette using the syringe and needle. Attach a float buoy to cassette, place into a 1 L beaker with Dialysis Buffer 1, and slowly stir at 4 °C for 24 h.
10. Perform dialysis with the remaining Dialysis Buffers 2–4 at 4 °C consecutively for 24 h each.
11. Following dialysis with Dialysis Buffer 4, carefully extract the protein from the dialysis cassette using a 10 mL syringe with an 18-gauge needle (*see* **Note 17**).
12. Concentrate the protein to 1–2 mL using a centrifugal filter unit.
13. Quantify the concentration of protein using a BCA assay to determine the yield (*see* **Note 18**).

3.3 Preparation of Cross-Linking Reagents

1. Resuspend cholera toxin (CT) in PBS–EDTA to 2 mg/mL (*see* **Note 19**).
2. Prepare 20 mM SPDP stock in DMF (*see* **Note 20**).
3. Prepare 100 mM DTT in PBS–EDTA (*see* **Note 20**).

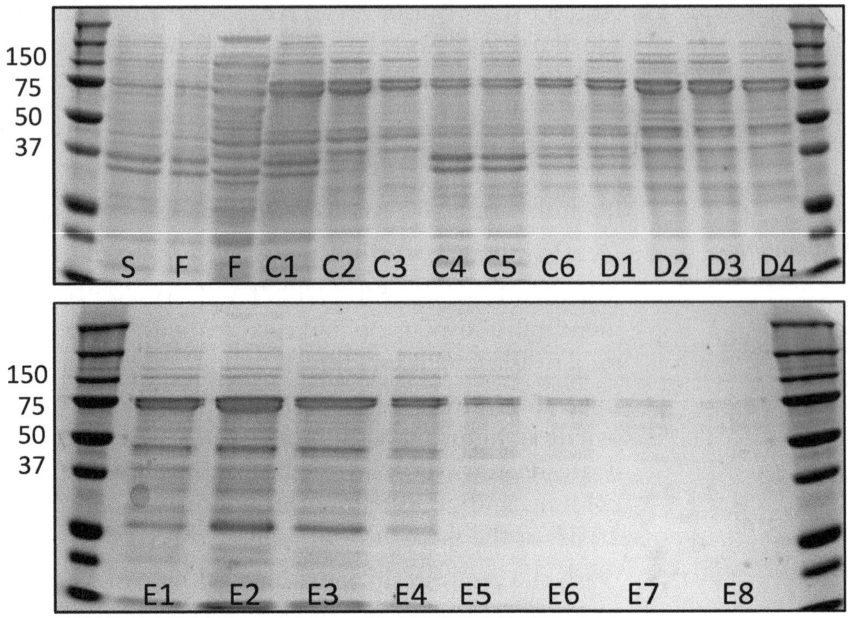

Fig. 1 Analysis of purified 6×His recombinant protein. A 12% SDS-PAGE gel was loaded with 1/100 of the soluble fraction (S), Ni-NTA flow-through (F), fractions eluted with Wash Buffer C (C), fractions eluted with Wash Buffer D (D), and subsequent elutions with Elution Buffer E (E). Elutions 5–8 containing purified protein were combined, dialyzed to allow protein refolding, and quantified. Molecular mass standards are indicated in kDa on the left

3.4 Cholera Toxin Activation

1. Add 25 μL of 20 mM SPDP to 1 mL of 2–5 mg CT in PBS–EDTA (*see* **Note 19**).
2. Incubate 30–60 min at room temperature.
3. Remove nonreacted SPDP and reaction byproducts by splitting the solution evenly into centrifugal filter units and spinning in a microcentrifuge (30 min, 10,000 × *g*, 23 °C) to concentrate.
4. Recover CT by adding PBS–EDTA to each filter, wait 1–2 min, invert filter into a new tube and spin (3 min, 1000 × *g*, 23 °C).
5. Add 500 μL of 100 mM DTT to approximately 1 mL of modified CT and incubate for 30 min at room temperature.
6. Desalt to completely remove DTT on centrifugal filter units and recover reduced modified CT with PBS–EDTA to reach desired concentration (*see* **Note 21**).

3.5 Antigen Activation

Purified protein is covalently coupled to CT at a ratio of 10:1 (antigen: CT). A concentration of 25–200 μg of cross-linked antigen to CT can be administered per mouse. For example, we use 100 μg of antigen for primary immunization and 25 μg of antigen for boosts. Determine the total protein required by multiplying the desired dosage (in μg per mouse) by the total number of mice (n) + 1 in the experimental group. The CT concentration will be

1/10th of the total of antigen. Determine the volume of CT needed by dividing the required CT concentration by the stock concentration of activated CT. Determine the total volume needed to immunize the group of mice and add PBS–EDTA to achieve the required volume. The desired protein concentrations are different for the primary immunization and subsequent boosts. For the primary immunization of a group of ten mice via the intranasal route, 100 μg in 20 μL (5 mg/mL) per mouse is required and a total of 25 μg in 20 μL (1.25 mg/mL) per mouse for subsequent boost is required.

1. Add 2.5 μL of 20 mM SPDP for each 200 μg of protein and incubate 30–60 min at room temperature.
2. Split into centrifugal filter units and desalt using a microcentrifuge (30 min, 10,000 × *g*, 23 °C).
3. Recover antigen by inverting the column into a new tube and spin (3 min, 1000 × *g*, 23 °C) (*see* **Note 22**).
4. Check the final volume and add PBS–EDTA if needed to achieve the desired concentration.

3.6 Cross-Linking of Antigen to CT

1. Add the required volume of CT prepared in Subheading 3.4 (1/10th the concentration of antigen) to antigen and mix well (*see* description in Subheading 3.5).
2. Incubate for 1 h at room temperature.
3. Store at 4 °C until use.

3.7 Immunization and Boosting

A control group of 5–10 mice should be mock-immunized with adjuvant alone and challenged in every experiment. A minimum of 10 mice should be used in every experimental group. Permanent identification of mice by ear punch or ear tag may be necessary to follow individual immune system responses over the duration of the experiment.

1. On the day of immunization (day 0) and day of each boost (day 7 and 14) (Fig. 2), perform the steps in Subheadings 3.5 and 3.6.
2. Prepare the work surface in a biosafety cabinet with a piece of absorbent underpad approximately 15″ × 20″.
3. Properly label cages to identify the control and experimental groups of mice. Collect urine from each individual mouse. Urine may be pooled by groups depending on your experimental requirements.
4. Prior to intranasal administration of CT-antigen or CT–PBS–EDTA (adjuvant alone), weigh each mouse and dose appropriately via the intraperitoneal (IP) route with stock anesthetic solution. Wait until the mice are fully anesthetized prior to attempting any procedures.

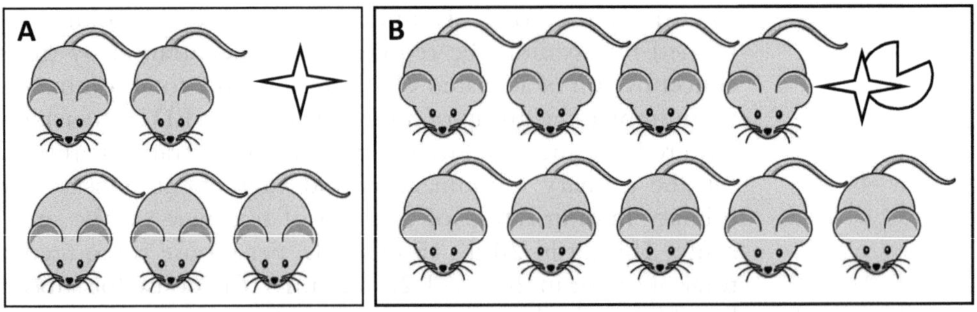

C	Day 0	Day 7	Day 14	Day 21	Day 28
Immunize (100 μg)	X				
Boost (25 μg)		X	X		
Urine	X	*	*	X	X
Blood	X	*	*	X	
Challenge				X	
Harvest Organs					X

Fig. 2 Experimental design indicating times of intranasal immunization, subsequent boosts, sample collection, challenge, and organ harvest for naïve and experimental mice. (**a**) Naïve mice receive adjuvant only (star shape) administered via the intranasal route. This control cohort is needed with every experiment. (**b**) Experimental mice receive the adjuvant-linked antigen (star-pie shape) administered via the intranasal route. (**c**) Both groups of mice receive the same procedures on each day (0, 7, 14, 21, 28) as indicated by an X. Urine and blood are collected prior to immunization (Day 0) and prior to challenge (Day 21). Urine and blood may be collected (∗) prior to each boost (Day 7 and 14). Urine must be collected prior to harvest of organs (Day 28)

5. Collection of preimmune serum by infraorbital ocular bleed (*see* **Note 23**) as well as method of choice to permanently identify individual mice should be performed prior to the initial immunization.
6. On day 0, administer the primary immunization via the intranasal route (Fig. 2). Using a pipette, carefully place 10 μL of CT-antigen or CT–PBS–EDTA into each nostril (total of 20 μL). The administration should be dripped in slowly, allowing the mice to respire in between drops until the full 10 μL is given (*see* **Note 24**).
7. On days 7 and 14, repeat **steps 1–4** and **6** for the two boost inoculations via the intranasal route (Fig. 2). On boost days, weigh mice and anesthetize with half of the initial dose to reach "twilight sleep" (*see* **Note 25**).
8. After administration, place each mouse under a heat source (*see* **Note 9**) and on a paper towel ventrally to ensure ease of respiration while anesthesia wears off (*see* **Note 8**).

3.8 Challenge with *P. mirabilis*

A broad outline of the experimental UTI model is included below. A more thorough protocol for performing experimental UTI using the ascending murine model is included in Chapter 15 of this volume.

1. On day 21, prepare the bacterial inoculum for infection (Subheading 2.4, **item 22**).
2. Pipette 1 mL of the inoculum into a sterile microcentrifuge tube to plate for input CFU.
3. Dilute the input 1:10,000 (10^{-4} dilution) and plate onto LB agar plates in triplicate using an automated spiral plater (*see* **Note 11**) to verify inoculum dose. Incubate at 37 °C overnight, and count the colonies to verify the inoculum.
4. Prior to transurethral infection, collect urine from each mouse into sterile, labeled microcentrifuge tubes and store at −20 °C (*see* **Note 26**).
5. Weigh mice and anesthetize with a full dose of anesthesia (*see* Subheading 3.7, **step 4**).
6. Collect serum by infraorbital ocular bleed and place in a microtainer serum separator tube. Place serum from each mouse into sterile, labeled microcentrifuge tubes and store at −20 °C (*see* **Note 23**).
7. Challenge vaccinated mice and naive or mock-immunized mice (Fig. 2) via the transurethral route with 50 μL of prepared bacterial suspension (e.g., 1×10^7 CFU per mouse).
8. Following challenge, place each mouse dorsally onto a paper towel to help aid in inoculum dissemination within the urinary tract and reduce discharge at the urethra. Place a small amount (1 mm column from the tube or less) of sterile ophthalmic ointment into each eye. Use a heat source to maintain body temperature while anesthesia wears off (*see* **Note 9**).
9. On day 28, collect urine from mice. Sacrifice mice with an approved method. Harvest and homogenize organs of interest (e.g., bladder, kidneys, spleen) (Fig. 2).
10. Dilute samples, plate on plain LB agar (*see* **Note 11**), and incubate overnight at 37 °C.

3.9 Data Formatting

1. Count the colonies on each plate using a bacterial colony counter, or by hand if needed (*see* **Note 11**). The limit of detection from homogenized organs is 100 CFU; a count of zero should be plotted as 100 for statistical measurement purposes.
2. To calculate CFU/g of tissue, multiply the raw CFU count by the volume of PBS (in mL) used to homogenize the organ and

divide the resulting number by the weight (in g) of the organ (*see* **Note 27**).

3. To calculate CFU/mL of urine, divide the total volume of urine (adjusted with PBS to 150 μL if necessary) by the amount of urine collected and multiply the resulting number with the raw CFU count (*see* **Note 28**).
4. Analyze data with your preferred analysis and graphing software. Use a two-tailed Mann–Whitney test for statistical analysis.

4 Notes

1. The expression of recombinant proteins requires the correct inducer to be added. For example, expression systems containing the *araBAD* promoter require arabinose for induction, while the *lac* promoter requires IPTG.
2. Nickel-NTA agarose is designed for the purification of 6×His-tagged recombinant proteins under native or denaturing conditions, dependent on the solubility of the protein of interest.
3. These buffers are required for denaturing conditions for the purification of insoluble membrane proteins. Due to the dissociation of urea, the pH of Buffers B, C, D, and E should be adjusted immediately prior to use. In our experience, the membrane protein refolds following step-wise removal of urea and increased pH and salt.
4. Dialysis buffers should be made the day of dialysis and stored at 4 °C. Perform all dialysis steps at 4 °C.
5. Sonication can be used to lyse cells if a French press is not available.
6. The molecular weight cutoff and volume size of the dialysis cassette is dependent on the size of the protein of interest and the volume of purified protein to be dialyzed.
7. Ketamine–HCl is a schedule III controlled substance (USA) requiring careful documentation and proper storage. Observe local regulations, such as state or federal licensing. Once prepared, the ketamine–xylazine anesthesia solution has a shelf life of 7 days. Refer to your institution's policies for proper discarding of unused anesthesia solution.
8. Placing the mice onto paper towels serves multiple purposes; it limits the possibility of the mice inhaling bedding dust as they wake up from anesthesia as well as providing a source to limit heat loss during recovery.
9. Mice require supportive heating to maintain body temperature while under anesthesia. During recovery, a lamp with an

incandescent light bulb may be used as a heat source. Maintain compliance with your institution's animal care and use policies.

10. Higher concentrations of NaCl, common in many LB recipes, cause *P. mirabilis* isolates to swarm, making counting of individual colonies difficult.
11. Samples can be manually plated and CFUs counted without the use of an automated spiral plater and bacterial colony counter; however, because automated platers can plate dilutions on a single plate, more plates will be required for manual plating. Tenfold serial dilutions of 100 μL volume can be individually spread-plated onto plain agar. Drip plating can be used for multiple dilutions on one agar plate [9].
12. Collecting uninduced control, induced control, and soluble and insoluble fraction samples for SDS-PAGE analysis is an optional step to verify expression and location of the tagged protein.
13. Lysis is complete when the solution becomes translucent. If overexpressing proteins in expression systems other than pET in *E. coli* BL21 (DE3) pLysS cells, such as pBAD, additional lysozyme (20 mg/mL) may be required for complete cell lysis.
14. The pellet can be kept at −20 °C for an indefinite period of time until ready to proceed with the next steps. This freeze and subsequent thaw step helps in dissolving the membrane pellet.
15. If necessary, use an 18-gauge needle to break up small chunks.
16. Ni-NTA agarose should be resuspended in its bottle by inverting and gently tapping the bottle repeatedly.
17. Tips for dialysis: wet the dialysis cassette with Dialysis Buffer 1 prior to use. Insert a needle attached to an empty 10 mL syringe through the cassette gasket via one of the corner ports and remove approximately 7 mL of air (cassette volume 3–12 mL); the membrane of the cassette will barely touch in the very middle of the cassette. Be careful to keep the needle straight as to not pierce the membrane of the cassette and gently remove the needle from the cassette port. Add the protein sample via the needle attached to the 10 mL syringe and insert it into second corner port of the cassette (it is important not to enter a port previously used). Gently dispense the sample into the cassette and remove excess air by tilting the cassette to shift sample towards one corner. Leave a small amount of air in the cassette and remove the needle. To remove the sample following dialysis, insert the syringe with needle into the third corner port of the cassette.
18. For primary immunization of 10 mice via the intranasal route, 1 mg of protein is required. For subsequent boost of ten mice, 250 μg of protein is required.

19. In our experiments, 2 mg of cholera toxin (CT) was resuspended in 1 mL of PBS–EDTA for a final concentration of 2 mg/mL. Resuspension and activation of CT can be performed days ahead of the experiment and stored at −20 °C.
20. For each experiment, freshly prepare SPDP in DMF and 100 mM DTT in PBS–EDTA. Do not store long term.
21. It is important to remove most of the DTT solution from CT before bringing it up in PBS–EDTA (aim for approximately 50 μL volume remaining). The concentration of CT is dependent on the final volume recovered from desalting columns; add PBS–EDTA to reach the desired concentration. In our experiments, a final volume of 600 μL yielded 3.3 mg/mL CT.
22. Add PBS–EDTA only if the filter reservoir is dry.
23. Collection of serum both before vaccination and prior to transurethral challenge (day 21) allows for the systemic humoral immune response to be evaluated. Levels of serum antigen-specific antibodies can be measured and compared between the vaccinated versus mock-immunized mice.
24. To ensure proper inhalation, we anesthetize all mice in one cage at the same time. Then, we instill the vaccine in the left nostril of the entire group, lay them ventrally and give them a few minutes to completely absorb the first 10 μL. Next, we instill the vaccine into the right nostril of each mouse. Typically, the first side is inhaled more quickly and with more ease than the second. The second instillation requires a longer time in between drops to instill the full 10 μL. Giving the vaccine too quickly can cause the mice to suffocate.
25. Give half the dosage required by weight to anesthetize for transurethral infection. Utilizing "twilight sleep" for the vaccine boost gives the mice enough anesthesia to be immobilized but does not suppress respiration as intensely, which enables easier inhalation of the vaccine antigen(s).
26. Following vaccination and prior to transurethral infection, collection of urine on day 21 allows for the evaluation of the humoral immune response at the primary site of infection. Levels of IgA in the urine can be measured by ELISA and compared between the vaccinated versus mock-immunized cohorts of mice.
27. CFU/g of tissue calculation example: If the output is 1.80×10^6 CFU, the organ weight is 0.2356 g and the organ is homogenized in 3 mL of PBS, then $(1.80e6 \times 3)/0.2356 = 2.29 \times 10^7$ CFU/g of tissue.
28. The total volume of urine can be adjusted with PBS for your experimental specifications. For this protocol, a total volume of 150 μL of urine is desired for plating of bacterial outputs with

an automated spiral plater. CFU/mL urine calculation example: If the output is 1.88×10^4 CFU, the amount of urine collected is 20 μL and 130 μL of PBS is added, then $(150/20) \times 1.88e4 = 1.41 \times 10^5$ CFU/mL of urine.

Acknowledgments

The authors would like to thank Christopher Alteri for helpful comments and critiques. This work was supported Public Service Grants AI116791 and AI059722 from the National Institutes of Health to H.L.T.M.

References

1. Armbruster CE, Mobley HLT, Pearson MM (2018) Pathogenesis of *Proteus mirabilis* infection. EcoSal Plus 8(1). https://doi.org/10.1128/ecosalplus.ESP-0009-2017
2. Kuan L, Schaffer JN, Zouzias CD, Pearson MM (2014) Characterization of 17 chaperone-usher fimbriae encoded by *Proteus mirabilis* reveals strong conservation. J Med Microbiol 63 (Pt 7):911–922. https://doi.org/10.1099/jmm.0.069971-0
3. Kochiashvili D, Khuskivadze A, Kochiashvili G, Koberidze G, Kvakhajelidze V (2014) Role of the bacterial vaccine Solco-Urovac(R) in treatment and prevention of recurrent urinary tract infections of bacterial origin. Georgian Med News 231:11–16
4. Nielubowicz GR, Smith SN, Mobley HLT (2008) Outer membrane antigens of the uropathogen *Proteus mirabilis* recognized by the humoral response during experimental murine urinary tract infection. Infect Immun 76 (9):4222–4231. https://doi.org/10.1128/IAI.00533-08
5. Li X, Lockatell CV, Johnson DE, Lane MC, Warren JW, Mobley HLT (2004) Development of an intranasal vaccine to prevent urinary tract infection by *Proteus mirabilis*. Infect Immun 72 (1):66–75
6. Alamuri P, Eaton KA, Himpsl SD, Smith SN, Mobley HLT (2009) Vaccination with Proteus toxic agglutinin, a hemolysin-independent cytotoxin in vivo, protects against *Proteus mirabilis* urinary tract infection. Infect Immun 77 (2):632–641. https://doi.org/10.1128/IAI.01050-08
7. Choubini E, Habibi M, Khorshidi A, Ghasemi A, Asadi Karam MR, Bouzari S (2018) A novel multi-peptide subunit vaccine admixed with AddaVax adjuvant produces significant immunogenicity and protection against *Proteus mirabilis* urinary tract infection in mice model. Mol Immunol 96:88–97. https://doi.org/10.1016/j.molimm.2018.03.001
8. Obernier JA, Baldwin RL (2006) Establishing an appropriate period of acclimatization following transportation of laboratory animals. ILAR J 47(4):364–369
9. Thomas P, Sekhar AC, Upreti R, Mujawar MM, Pasha SS (2015) Optimization of single plate-serial dilution spotting (SP-SDS) with sample anchoring as an assured method for bacterial and yeast CFU enumeration and single colony isolation from diverse samples. Biotechnol Rep (Amst) 8:45–55. https://doi.org/10.1016/j.btre.2015.08.003

Chapter 19

Characterization of *Proteus mirabilis* Lipopolysaccharide Samples by Infrared Spectroscopy and Serological Methods

Katarzyna Durlik, Grzegorz Czerwonka, Paulina Żarnowiec, and Wiesław Kaca

Abstract

Methods of lipopolysaccharide extraction, purification, and sample validation are presented. Based on serological reaction in ELISA, immunoblotting, and infrared spectra, identities of two LPS preparations from smooth *P. mirabilis* (O18) PrK 34/57 are presented.

Key words *Proteus mirabilis*, Lipopolysaccharide, Infrared spectroscopy, Serology, ELISA

1 Introduction

Lipopolysaccharide (LPS, endotoxin) is a major component of the *Proteus mirabilis* outer membrane and is part of cell wall structures. Amphipathic LPS molecules are composed from hydrophobic lipid A and hydrophilic polysaccharide parts. The classical method of extraction is based on the Westphal procedure and relies on hydrophilic, smooth form LPS collected from the water phase of a phenol–water mixture [1]. Due to the complex procedures of extraction and purification of LPS samples, methods are needed to verify reproducibility of the obtained samples. Here we present a scheme of experiments for cultivation, extraction, purification, and validation of LPS samples. SDS-PAGE electrophoresis, immunoblotting, ELISA and infrared spectra are used as validation methods. *P. mirabilis* (O18) strain PrK 34/57 is used as a model representing *P. mirabilis* strains with LPSs carrying noncarbohydrate substitutes [2]. The O-polysaccharide part of LPS of this tested strain is characterized by the presence of noncarbohydrate components, phosphocholine residues [3]. The presented protocols are suitable for LPS analysis from Gram-negative bacterial cells with different LPS structures.

Melanie M. Pearson (ed.), *Proteus mirabilis: Methods and Protocols*, Methods in Molecular Biology, vol. 2021, https://doi.org/10.1007/978-1-4939-9601-8_19,

2 Materials

1. *Proteus mirabilis* (O18) strain PrK 34/57 [Czech National Collection of Type Cultures (CNCTC), Czech Republic].
2. Control *Salmonella enterica* ser. Abortus-equi LPS (Sigma-Aldrich).
3. LB medium (per liter): 10 g of tryptone, 5 g of yeast extract, 5 g of sodium chloride, pH at 25 °C: 7.0 ± 0.2. Suspend 20 g of medium in distilled water, mix well and dissolve by heating with frequent agitation. Sterilize in autoclave at 121 °C for 15 min.
4. Anti-LPS antiserum. In this example, rabbit O-antiserum against *P. mirabilis* O18 is obtained by immunization of rabbits with heat-inactivated bacteria (100 °C, 2.5 h). To prepare the immunogen, lyophilized bacteria are suspended in saline (aq 0.85% NaCl) at a concentration of 1 mg/mL. Animals are injected intravenously with 50, 100, 100, 200, and 500 μg of immunogen in 500 μL of saline on days 1, 4, 7, 11, and 16, respectively. Seven days after the last immunization, blood is taken by cardiac puncture. The obtained serum is stored at −20 °C. Methods described in [4] (*see* **Note 1**).
5. 90% phenol (*see* **Note 2**).
6. Methanol with $MgCl_2$ (per 30 mL): Make a 20% solution of $MgCl_2$ in ethanol. Add 0.3 mL of the $MgCl_2$ solution to 30 mL of methanol. Chill to 4 °C before use.
7. DNase and RNase.
8. Sample buffer: 100 mM Tris, 20 mM EDTA, 20% glycerol, 8% sodium dodecyl sulfate (SDS), and 0.005% bromophenol blue. Mix buffer components in a few mL of deionized water. After mixing, adjust pH to pH 6.8 using HCl. Bring up to 100 mL with deionized water.
9. 30% acrylamide solution (per 100 mL): 30 g of acrylamide and 0.8 g of bisacrylamide in deionized water.
10. Tris–HCl, pH 8.8 (resolving gel buffer): dissolve 13.627 g of Tris in 80 mL of deionized water, adjust pH to 8.8 with 1 N HCl, then dissolve 300 mg of SDS and make up the final volume to 100 mL with deionized water.
11. Tris–HCl, pH 6.8 (stacking gel buffer): dissolve 3.028 g of Tris in 80 mL of deionized water. Adjust pH to 6.8 with 1 N HCl, then dissolve 200 mg of SDS and make up the final volume to 100 mL with deionized water.
12. Tris–glycine electrophoresis running buffer (per liter): dissolve 12 g of Tris, 57.6 g of glycine, and 4 g of SDS in a small amount of deionized water and top up to 1000 mL deionized

water. Before using, mix with deionized water in proportion 1:3 by vol.

13. 10% ammonium persulfate (APS).
14. TEMED (*N,N,N′,N′*-tetramethylethane-1,2-diamine).
15. Fixation solution: 40 mL of 100% methanol, 10 mL of 100% acetic acid, and 50 mL of deionized water.
16. Sensitizer solution: dissolve 70 mg of sodium periodate in 100 mL of water (*see* **Note 3**).
17. Staining solution: add 1 g of $AgNO_3$ to 5 mL of deionized water. Add this $AgNO_3$ drop by drop to a mixture of 1.4 mL of 2 N NaOH, 2 mL of ammonia water (35.20%), and 141.6 mL of deionized water.
18. Developer solution: dissolve 50 mg of citric acid and 0.5 mL of formaldehyde (37%) in 900 mL of deionized water, then top up to 1000 mL.
19. PVDF (polyvinylidene difluoride) membrane.
20. Filter paper.
21. 2× transfer buffer: mix 3.0275 g of Tris, 28.1513 g of glycine, and 500 mL of methanol with a small amount of deionized water, and top up to 1000 mL with deionized water. Before using, mix with deionized water in proportion 1:1 by vol to make 1× transfer buffer.
22. 3× Tris-buffered saline (TBS) buffer: 20 mM Tris, 50 mM NaCl. Dissolve in a small amount of deionized water. After complete mixing, adjust pH using hydrochloric acid to pH 7.5 and bring solution to 1000 mL. Before using, mix with deionized water in proportion 1:2 by vol.
23. Blocking solution: dissolve 1 g of bovine serum albumin (BSA) per 100 mL of 1× TBS (for western blot) or PBS (for ELISA). Prepare fresh for each experiment.
24. TBS with 0.1% Tween 20.
25. Secondary antibody (e.g., peroxidase-conjugated goat anti-rabbit IgG (Sigma)).
26. 4-chloro-1-naphthol solution: dissolve 3 mg in 10 mL of methanol. Mix with TBS in proportion 1:5 by vol. Add 0.5 μL of 30% hydrogen peroxide per mL of solution immediately before use.
27. Phosphate-buffered saline (PBS), pH 7.2.
28. 5% BSA in PBS.
29. *O*-phenylenediamine dihydrochloride (OPD) in urea hydrogen peroxide/buffer: 0.4 mg/mL OPD, 0.4 mg/mL urea hydrogen peroxide, and 0.05 M phosphate-citrate, pH 5.0 (e.g., SIGMA*FAST* OPD, Sigma).

30. 0.5 M H_2SO_4.
31. Sterile 10 L bottle.
32. Centrifuge.
33. Evaporator (e.g., Heidolph).
34. Bacterial incubator.
35. Adjustable water bath.
36. Dialysis bags.
37. Electrophoresis apparatus with gel holders and cassettes for electrophoretic transfer (e.g., Bio-Rad).
38. Microtiter plates.
39. Multimode microplate reader (e.g., Infinite M200 PRO (Tecan)).
40. Freeze dryer (e.g., VirTis BenchTop K).
41. Infrared spectrometer, Spotlight 400 FTIR Imaging System (Perkin Elmer, Waltham, MA, USA).

3 Methods

3.1 Bacterial Cultivation

1. Inoculate 5 mL of LB broth with the bacterial strain.
2. Incubate for 18 h at 37 °C with shaking (160 rpm).
3. Add the inoculum to 6 L of fresh LB in a 10 L bottle and incubate for a further 18 h at 37 °C with shaking (80 rpm).
4. Kill bacteria by adding 18 mL of 90% phenol to the entire volume of the bacterial culture (*see* **Note 2**).
5. Centrifuge bacteria (3000 × *g*, 20 min, 4 °C).
6. Wash the obtained biomass four times with distilled water (3000 × *g*, 20 min, 4 °C).
7. Lyophilize the bacterial mass using a freeze dryer.

3.2 LPS Isolation

This lipopolysaccharide extraction method from a dry bacterial mass is modified from the phenol-water method [1].

1. Add 30 mL of distilled water to 1 g of the obtained dry bacterial mass.
2. Immerse in a water bath at 70 °C.
3. When the suspension becomes homogeneous, add 30 mL of 90% phenol (*see* **Note 2**).
4. Stir until the mixture is heated to 70 °C, and continue stirring for an additional 10 min.
5. Cool the mixture at 4 °C for 30 min.
6. Centrifuge (3000 × *g*, 30 min, 4 °C).

7. Separate the supernatant into a fresh vessel and set aside.
8. To the residue, again add 30 mL of fresh 90% phenol, heat to 70 °C and repeat **steps 4**–7.
9. Repeat extraction (**steps 4**–7) for a third time, but using 30 mL of distilled water instead of phenol.
10. Collect all water phases together, transfer to dialysis bags with a 0.5–1.0 kD MWCO pore size, and dialyze at 4 °C against distilled water for 3 days. Change distilled water twice a day.

3.3 LPS Purification

This method is according to Rezania [5] with modifications.

1. After dialysis, use an evaporator to concentrate the contents of the bags to 10 mL.
2. Add 30 mL of cold methanol containing $MgCl_2$ to the concentrated solution while stirring in a refrigerator at 4 °C. A precipitate will quickly become visible.
3. Continue mixing at 4 °C for 1 h.
4. Centrifuge at 3000 × *g* for 30 min at 4 °C and discard the supernatant.
5. Dissolve the precipitate in 10 mL of distilled water and add to 20 mL of cold methanol (without $MgCl_2$).
6. Incubate the solution for 60 min at 4 °C, centrifuge at 3000 × *g* for 30 min at 4 °C, and discard the supernatant.
7. Repeat **steps 5** and **6**.
8. Resuspend the pellet in 10 mL of distilled water and concentrate in an evaporator.
9. Repeat **step 8** to completely remove methanol from the mixture.
10. Dilute the obtained solution to 100 mL with distilled water.
11. Digest with DNase and RNase overnight (18 h) at 37 °C in nuclease digestion buffer. Add 1 mg of DNase and 1 mg of RNase per 100 mL of LPS solution. This step removes nucleic acids from the LPS preparation.
12. Dialyze the lipopolysaccharide solution against distilled water for 2 days at 4 °C, replacing deionized water twice a day.
13. Lyophilize the LPS solution. It is now ready for analysis using SDS-PAGE (Subheading 3.4) or ELISA (Subheading 3.8).

3.4 SDS-PAGE

This method is according to Laemmli [6], with modifications.

3.4.1 Sample Preparation

1. Dissolve 1 mg of dry LPS sample in 1 mL of deionized water. This is used as an LPS stock solution.
2. Mix 1 mg/mL LPS stock, deionized water and sample buffer in a proportion of 2:3:5 by vol., and incubate at 96 °C for 5 min.

Table 1
Components of SDS-PAGE

	30% Acrylamide (mL)	Tris–HCl, pH 8.8 (mL)/Tris–HCl, pH 6.8 (mL)	Water (mL)	10% APS (μL)	TEMED (μL)
Resolving gel	7.5	4.95	2.465	75	7.5
Stacking gel	1	3.3	0.645	50	5

3.4.2 Polyacrylamide Gel Polymerization

1. Clean the glass plates and spacers of the gel casting unit with deionized water and ethanol.
2. Assemble the plates with the spacers on a stable, even surface.
3. Prepare resolving gel solution using the reagents shown in Table 1.
4. Pour the resolving gel solution into the plates assembled with spacers. To maintain an even and horizontal resolving gel surface, layer the surface with deionized water.
5. Allow the gel to set up for 30–40 min at room temperature.
6. Prepare stacking gel solution using the reagents shown in Table 1.
7. Discard the water from the resolving gel.
8. Add the stacking gel solution until it overflows. Insert the comb immediately, ensuring no air bubbles are trapped in the gel or near the wells.
9. Allow the gel to polymerize for 30–40 min at room temperature.

3.4.3 SDS-PAGE

1. Mount the gel in the electrophoresis apparatus.
2. Add Tris–glycine electrophoresis running buffer to the top and bottom reservoirs.
3. Remove any bubbles.
4. Load 10 μL of each sample in a predetermined order into the bottom of each well.
5. Attach the electrophoresis apparatus to an electric power supply.
6. Run at 40 mA for 10–15 min, then run at 60 mA for 30–45 min (until dye exits the gel).
7. Remove the glass plates from the electrophoresis apparatus and place them on a paper towel. Using a spatula, pry the plates apart.

8. Mark the orientation of the gel by cutting a corner from the bottom of the gel.
9. Remove the gel from the glass plates and incubate it in fixation solution (for silver staining, Subheading 3.5) for a minimum of 2 h, or incubate in western transfer buffer for approximately 10 min (for electrophoretic transfer, Subheading 3.6).

3.5 Silver Staining

This is a method for staining LPS in a polyacrylamide gel according to Tsai [7] with modifications.

1. Immerse gel in fixation solution and incubate overnight or for a minimum of 2 h.
2. Replace fixation solution with sensitizer solution (*see* **Note 3**).
3. Wash gel with deionized water three times for 10 min.
4. Incubate gel with staining solution for 10 min.
5. Wash gel with deionized water three times for 10 min.
6. Add developer solution with gentle mixing until LPS bands are visible.
7. Monitor the progress of the reaction carefully. When the bands are of the desired intensity, immediately add water to the gel and gently mix for 10 min to stop the reaction.
8. Repeat the water wash three more times.

Results of SDS-PAGE and silver staining are presented in Fig. 1. A typical smooth form of LPS, with a ladder-like pattern of migration, is clearly visible only for control *Salmonella* ser. Abortus-equi LPS. In contrast, only faint O-polysaccharide is visible for two separate preparations of *P. mirabilis* (O18) PrK 34/57 LPS (*see* **Note 4**). Two preparations of *P. mirabilis* (O18) PrK 34/57 LPSs slightly differ in migration patterns. Sample I-Rep differs from sample II-Rep by an additional band that may represent a fraction of core oligosaccharide substituted by one repeating unit of O-specific oligosaccharide [8]. Slight differences in Tsai-- Frasch staining of two O18 LPSs may have resulted from LPS preparation, electrophoresis, or in differences in storage temperature; the LPS in lane 2 was stored at 4 °C while the LPS in lane 3 was stored at room temperature. The results of a follow-up immunoblot indicate that both O18 LPSs are identical (Fig. 2).

3.6 Electrophoretic Transfer

This method for transfer of LPS to a PVDF membrane is according to Towbin [9], with modifications.

1. Fill an ice cooling unit with water and store it in −20 °C until ready to use. After use, return the cooling unit to the freezer for storage.

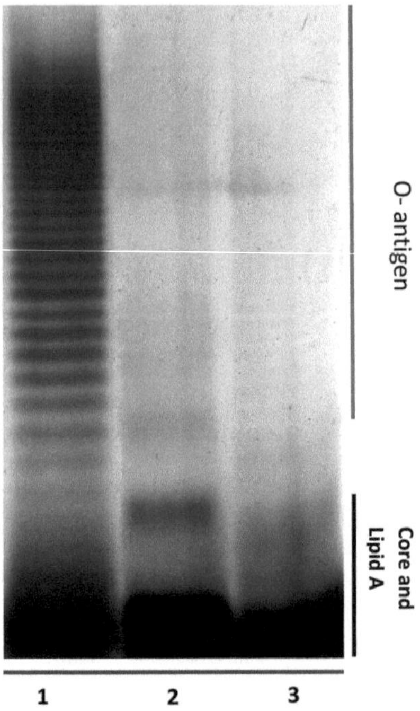

Fig. 1 LPS profiles obtained by SDS-PAGE and silver staining. Lane 1, *Salmonella* ser. Abortus-equi LPS. Lanes 2 and 3, *P. mirabilis* (O18) Prk 34/57 LPSs from two sample preparations, I-rep. and II-rep., respectively

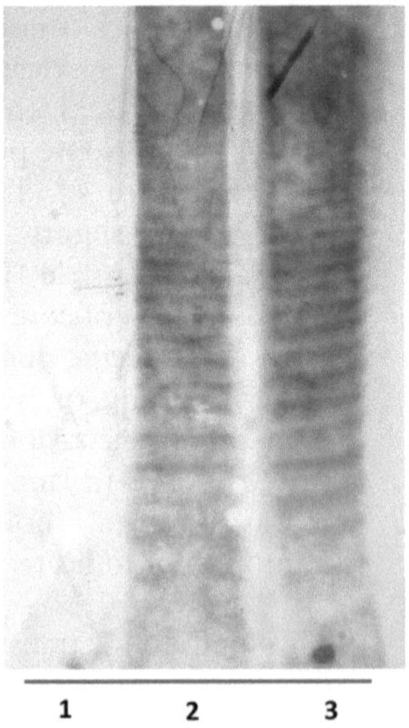

Fig. 2 Western immunoblotting of LPS with anti-LPS serum against *P. mirabilis* (O18) Prk 34/57. Lane 1, *Salmonella* ser. Abortus-equi; lane 2, *P. mirabilis* Prk (O18) 34/57 (I-rep.); lane 3, *P. mirabilis* (O18) Prk 34/57 (II-rep)

2. Immerse the gel (**step 9** in Subheading 3.4.3) in transfer buffer for 20 min.
3. Cut a PVDF membrane and filter paper to the dimensions of the gel.
4. Soak filter papers in transfer buffer for at least 30 s.
5. Prepare the PVDF membrane: place the membrane in methanol for 15 s. The membrane should uniformly change from opaque to semitransparent. Carefully place the membrane in deionized water and soak for 15 min, then carefully place the membrane in transfer buffer and let equilibrate for at least 5 min.
6. Prepare the gel sandwich:
   (a) Open the cassette holder.
   (b) Place a fiber pad on the black side of the cassette.
   (c) Place three sheets of filter paper on the fiber pad.
   (d) Place the equilibrated gel on the filter paper.
   (e) Place the membrane on the gel.
   (f) Complete the sandwich by placing three more pieces of filter paper on the membrane. Remove any air bubbles which may have formed; this is very important to ensure an even transfer. Use a glass tube, pipette or stirring rod to gently roll air bubbles out. Do not apply excessive pressure, to prevent damaging the membrane and gel.
   (g) Add a second fiber pad.
   (h) Close the cassette firmly, being careful not to move the gel and filter paper sandwich. Lock the cassette closed. Place the cassette in its holder.
7. Place the cassette holder in the transfer tank so that the gel side of the cassette holder is facing the black cathode lead (−) and the membrane side is facing the red anode lead (+).
8. Add the frozen cooling unit and adequate transfer buffer into the tank to cover the cassette holder.
9. Insert the black cathode lead (−) into the cathode jack and the red anode lead (+) into the anode jack on the transfer unit. Connect the anode lead and cathode lead to their corresponding power outputs.
10. Run at 100 mA for 80 min.
11. After the transfer, unclamp the blot sandwich and remove the sheets of blotting paper, exposing the blot membrane. Mark the side of the membrane that was facing the gel.
12. Mark the position of the prestained markers, since they may fade away during detection.

3.7 Immunodetection

This method for immunodetection of LPS is according to Burnette [10] with modifications.

1. Place the membrane in blocking solution at 37 °C for 1 h.
2. Wash in TBS with 0.1% Tween 20 for 5 min. Repeat two more times.
3. Place the membrane in rabbit anti-*P. mirabilis* O18 serum 1:500 diluted in TBS with 1% BSA for 1 h at 37 °C.
4. Wash in TBS with 0.1% Tween 20 for 5 min. Repeat two more times.
5. Place the membrane in the secondary antibody 1:1000 diluted in TBS with 1% BSA for 1 h at 37 °C.
6. Wash in TBS with 0.1% Tween 20 for 5 min. Repeat two more times.
7. Place the membrane in 4-chloro-1-naphthol solution.
8. Monitor the progress of the reaction carefully. When the bands are of the desired intensity (2–5 min), wash with deionized water.

Immunoblotting results are presented in Fig. 2. As expected with control *Salmonella* ser. Abortus-equi LPS, no reaction was detected. Two *P. mirabilis* Prk (O18) 34/57 LPSs obtained in two independent experiments were very similar, with more than 20 repeating units of O-oligosaccharides (Fig. 2, compare lanes 2 and 3) (*see* **Note 5**).

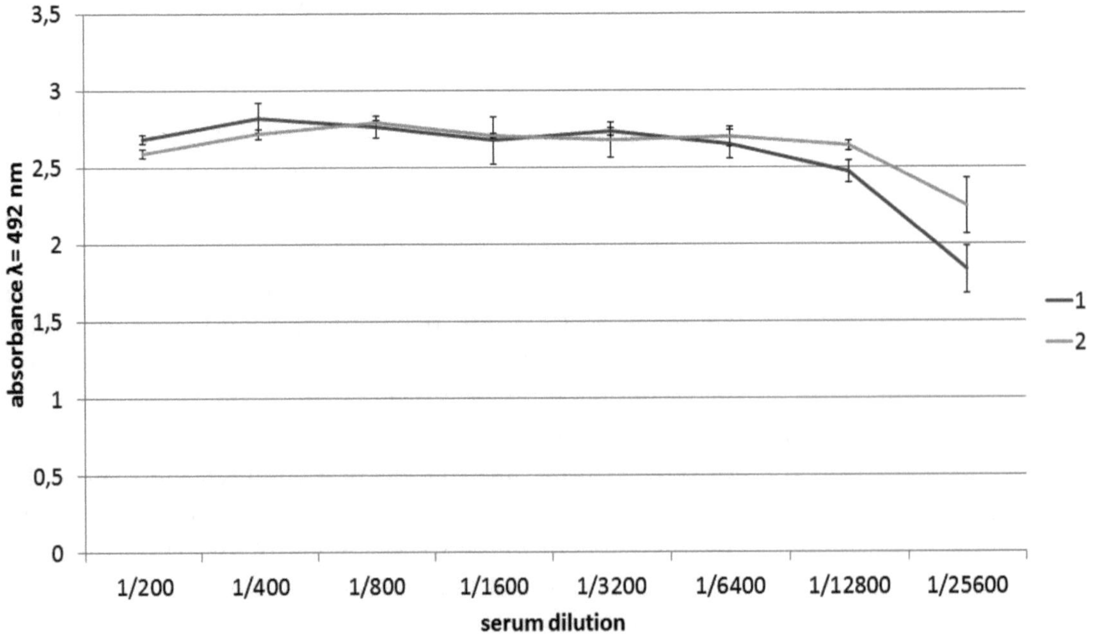

Fig. 3 Reactivity of anti-LPS serum against *P. mirabilis* Prk 34/57 measured by ELISA assay. 1, *P. mirabilis* (018) Prk 34/57 (I-rep.); 2, *P. mirabilis* (018) Prk 34/57 (II-rep)

3.8 ELISA

To confirm the identity of LPS, ELISA may be used. This ELISA method is according to Limet [11] with modifications. An example of two *P. mirabilis* (O18) Prk 34/57 LPS ELISAs is presented here.

1. Add 0.5 μg of LPS in 50 μL of PBS to individual microtiter plate wells, in triplicate, and incubate at 4 °C overnight. Add PBS (without LPS antigen) to another set of wells, as a control (*see* **Note 6**).
2. Remove the coating solution and wash the plate three times with PBS (200 μL/well). Remove the washes by flicking the plate over a sink. Remove the remaining drops by patting the plate on a paper towel.
3. Add 100 μL/well of 5% BSA in PBS and incubate at 37 °C for 2 h, to block the remaining LPS-binding sites.
4. Wash the plate three times with PBS (200 μL/well).
5. Add 100 μL/well of rabbit anti-*P. mirabilis* O18 serum (diluted from 1:200 to 1:125600 in PBS with 1% BSA). Incubate the plate at 37 °C for 1 h with gentle shaking.
6. Wash the plate five times with PBS (200 μL/well).
7. Add 50 μL/well of secondary antibody, diluted 1:1000 in 1% BSA in PBS and incubate at 37 °C for 1 h with gentle shaking.
8. Wash the plate five times with PBS (200 μL/well).
9. Add 50 μL/well of freshly prepared OPD buffer. Incubate the plate at 37 °C in the dark for 20 min.
10. After incubation add 50 μL/ well of 0.5 M H_2SO_4 to stop the reaction.
11. Read the absorbance of each well at 492 nm with a plate reader.

The ELISA results should confirm the immunoblotting pattern (Fig. 3). In this example, both LPS samples reacted equally with rabbit anti-*P. mirabilis* (O18) Prk 34/57 serum, to the titer below 1/25600.

3.9 ATR-FTIR Spectra Acquisition

This is a method for LPS spectra acquisition using attenuated total reflection (ATR) Fourier-transform infrared spectroscopy (FTIR) according to Żarnowiec [12].

1. Clean the ATR crystal with 70% ethanol or isopropanol and ensure the crystal is dry before a background acquisition.
2. Record a background spectrum.
3. Place the lyophilized sample on the crystal of the ATR device and press with a pressure applicator (*see* **Note 7**).
4. Acquire a sample spectrum with a resolution of 4 cm^{-1} and 32 co-added scans (*see* **Note 8**).

IR spectra of two preparations of *P. mirabilis* (O18) PrK 34/57 LPSs were almost identical across the whole range of spectra, with only differences observed at wavenumbers of 964, 913, and 796 cm^{-1} in the polysaccharide and fingerprinting window (Fig. 4).

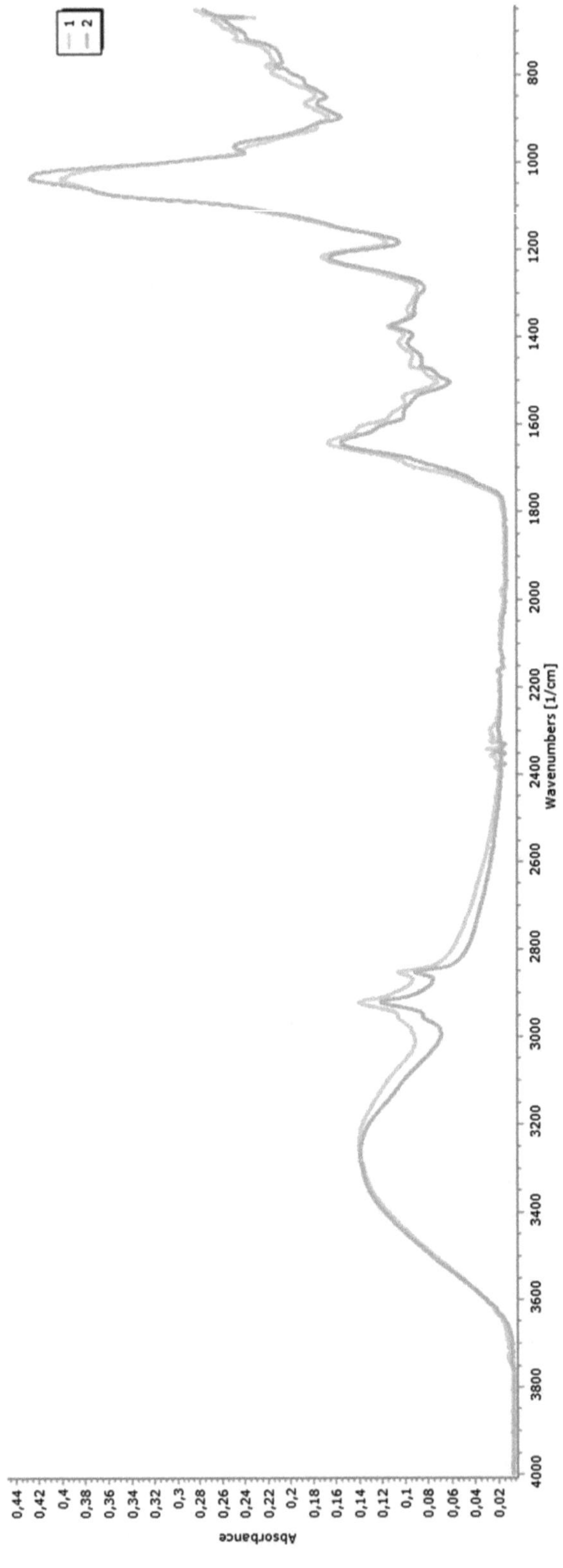

Fig. 4 IR absorbance spectra of LPS extracted from *P. mirabilis* (O18) Prk 34/57 (I-rep.); 2, *P. mirabilis* (O18) Prk 34/57 (II-rep)

4 Notes

1. Obtain appropriate ethical approval for experiments involving animals. Rabbit anti *O*-polysaccharide specific serum is polyclonal and animal-to-animal differences might be observed. It is recommended to immunize more than one rabbit. Antibody cross-reactivity with heterologous antigens, based on structural epitope similarities, might be observed.
2. Phenol water solution is very toxic; protective gloves and goggles are mandatory.
3. Sensitizer solution of sodium periodate must be prepared freshly, just before using.
4. Depending on the LPS preparation and structures, *O*-specific parts might not be visible as resolved bands or only faint Tsai–Frasch staining might be visible. The principle of LPS silver staining is the oxidation of carbohydrates with sodium periodate and staining by metallic silver. LPS structures may influence the Tsai–Frasch staining. Compare two S form LPSs in Fig. 2. Despite high sensitivity, silver staining does not always give a full picture of the stripe pattern of the tested LPS and results should be extended with other tests such as western blot with the use of homologous serum.
5. As with silver staining, the immunoblotting pattern of the *O*-polysaccharide part of *P. mirabilis* LPS might not be visible as a ladder-like pattern. Instead, broad, nonresolved bands may be visible. That might be due to differences in LPS sample preparation, electrophoresis conditions or O-polysaccharide LPS structures.
6. Serum can react with the plate nonspecifically, so the obtained result is calculated by the subtraction of the background from the reaction with antigen.
7. Ensure that the same pressure is applied to all the samples.
8. After each sample spectral acquisition, wash the crystal with isopropanol or 70% ethanol and ensure the crystal is dry.

Acknowledgments

The presented work was supported by a grant from the Ministry of High Education, Poland (BS2018 for WK).

References

1. Westphal O, Jann K (1965) Bacterial lipopolysaccharides. Extraction with phenol-water and further applications of the procedures. Meth Carbohydr Chem 5:83–91
2. Knirel YA, Perepelov AV, Kondakova AN, Senchenkova SN, Sidorczyk Z, Rozalski A, Kaca W (2011) Structure and serology of

O-antigens as the basis for classification of *Proteus* strains. Innate Immun 17:70–96

3. Fudala F, Kondakova AN, Bednarska B, Senchenkova SN, Shashkov AS, Knirel YA, Zaehringer U, Kaca W (2003) Structure and serological characterization of the O-antigen of *Proteus mirabilis* O18 with a phosphocholine-containing oligosaccharide phosphate repeating unit. Carbohydr Res 338:1835–1842
4. Kondakova AN, Fudala R, Senchenkova SN, Shashkov AS, Knirel YA, Kaca W (2003) Structural and serological studies of the O-antigen of *Proteus mirabilis* O-9. Carbohydr Res 338:1191–1196
5. Rezania S, Amirmozaffari N, Tabarraei B et al (2011) Extraction, purification and characterization of lipopolysaccharide from *Escherichia coli* and *Salmonella typhi*. Avicenna J Med Biotechnol 3:1–7
6. Laemmli UK (1970) Cleavage of structural proteins during the assembly of the head of bacteriophage T4. Nature 227:680–685
7. Tsai CM, Frasch CE (1982) A sensitive silver stain for detecting lipopolysaccharides in polyacrylamide gels. Anal Biochem 119:115–119
8. Holzer SU, Schlumberger MC, Joeckel D, Hensel M (2009) Effect of the O-antigen length of lipopolysaccharide on the functions of type III secretion systems in *Salmonella enterica*. Infect Immun 77(12):5458–5470
9. Towbin H, Staehelin T, Gordon J (1979) Electrophoretic transfer of proteins from polyacrylamide gels to nitrocellulose sheets: procedure and some applications. Proc Natl Acad Sci 76 (9):4350–4354
10. Burnette S (1982) "Western blotting": electrophoresis of protein from sodium dodecyl sulphate polyacrylamide gels to unmodified nitrocellulose and radiographic detection with antibody and radioiodinated protein A. Anal Biochem 112(2):195–203
11. Limet JN, Bosseray N, Garin-Bastuji B, Dubray G, Plommet M (1989) Humoral immunity in mice mediated by monoclonal antibodies against the A and M antigens of *Brucella*. J Med Microbiol 30(1):37–43
12. Żarnowiec P, Czerwonka G, Kaca W (2017) Fourier transform infrared spectroscopy as a tool in analysis of *Proteus mirabilis* endotoxins. In: Microbial toxins: methods and protocols, Methods in Molecular Biology. Springer Science + Business Media, New York, NY, pp 113–124

Chapter 20

Isolation and Purification of *Proteus mirabilis* Bacteriophage

Agnieszka Maszewska and Antoni Różalski

Abstract

Bacteriophages specifically targeting different strains of bacteria can be isolated from urban sewage using properly modified enrichment techniques. This chapter provides a detailed protocol for isolation of *Proteus mirabilis*-specific bacteriophages. Briefly, prefiltered sewage is mixed with double-concentrated tryptic soy broth containing the target strain and incubated. Subsequently, the suspension is spread on phage nutrient agar, and if needed, supplemented with swarming motility inhibitor, for the induction of bacterial growth and phage multiplication. Phages infecting bacteria are identified by plaques (patches of dead bacteria) in the confluent bacterial lawn. A pure phage preparation is obtained by cutting out a single plaque from a double-layer agar plate and subsequent virus propagation five times on a given *P. mirabilis* strain.

Key words *Proteus mirabilis*, Bacteriophage, Bacteriophage isolation, Sewage, Enrichment technique, Double-layer agar method, Plaque

1 Introduction

Bacteriophages exist in various environmental niches (e.g., soil, lakes, seas, sewage, or the human body), and their number on earth is estimated at 10^{31} [1]. The standard technique of bacteriophage isolation from environmental samples is enrichment culture [2–4]. This method is based on incubation of a target bacterial culture with a liquid sample devoid of endogenous flora, which is expected to contain phages. An appropriate modification of the method allows the isolation of various types of bacteriophages, depending on their intended use [5–7].

One of the plentiful sources of phages, especially those infecting intestinal bacteria, is urban sewage [8, 9]. The volume of used sewage depends on whether the sought-after phages occur in large numbers or not; for example, for *P. mirabilis* phage isolation, 10–100 mL of wastewater is usually used [7, 10–12]. The contaminating bacteria and other cells are removed from liquid source material by passage through a 0.45 or 0.2 μm membrane filter, or

Melanie M. Pearson (ed.), *Proteus mirabilis: Methods and Protocols*, Methods in Molecular Biology, vol. 2021, https://doi.org/10.1007/978-1-4939-9601-8_20, © Springer Science+Business Media, LLC, part of Springer Nature 2019 by treatment with chloroform. However, some phages exhibit sensitivity to this chemical [13]. The time, temperature of incubation, and type of growth medium used is adjusted to the given bacteria. After incubation, culture fluid is examined for the presence of phages infecting the used host. If phages are present in the sample, they are detected as a plaque by standard plating techniques. For this purpose, the suspension is poured directly onto a suitable agar medium, or dilutions of sample (devoid of bacteria) are spotted onto a lawn of potential hosts, or the double-layer agar method is used [10–12]. The last technique is also used to obtain a pure phage preparation. A single plaque is picked and the phages are propagated in broth cultures of the host strain. This procedure is repeated five times.

Plaque selection is a very important step and depends on the purpose for which the phages are intended. The plaque morphology may indicate some features of the phages, and it should be taken into account during plaque selection [5, 14]. Phages can form small ($\leq$1 mm) to large (a few mm) clear plaques typical for lytic phages (Fig. 1). Certain plaques are surrounded by a semitransparent ring, called a halo zone, which may be the result of phage enzymes degrading bacterial capsular polysaccharide [15, 16]. Sometimes, viruses form turbid plaques, which suggests that they may be temperate phages, although lytic phages may also form this type of plaques, for example, when part of the bacterial population exhibits resistance to them. Some plaques display a "bull's-eye" morphology, with rings of growth in the middle of clear regions of complete lysis, which is the result of decreasing lytic efficiency caused by aging of the bacterial lawn or the lysis inhibition phenomenon, observed among T-even phages [5, 17].

In this chapter, the isolation of *P. mirabilis*-specific phages from prefiltered municipal sewage mixed with double-concentrated tryptic soy broth and a target *P. mirabilis* strain is described. The suspension is spread on a nutrient agar plate. It is important to supplement the medium with a swarming motility inhibitor (e.g., phenol or glycerol), if the used strain exhibits this feature. To obtain a pure line of bacteriophages, a double-layer agar method is used. A single well-defined plaque is selected and phages are propagated on the target host strain. This step is repeated five times.

As a result of the described procedure of *P. mirabilis* phage isolation from 5 independent wastewater samples, we collected 48 phages specific for 31 out of the 50 *P. mirabilis* strains used [10]. For 6, 2, and 1 strains of *P. mirabilis*, there were 2, 3, and 4 phages isolated, respectively. Interestingly, for one of the tested *P. mirabilis* strains, phages were obtained in each of the five isolation cycles.

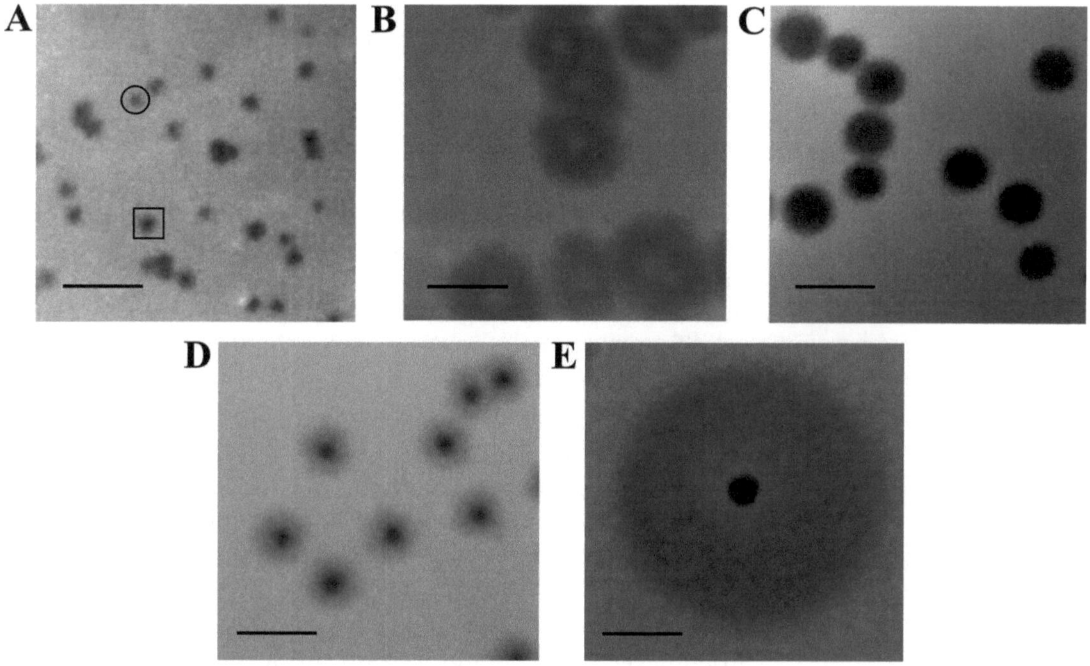

Fig. 1 Examples of plaque morphology. Small plaques: (○) turbid and (□) clear with halo (**a**), plaques with turbid center surrounded by a clear zone (**b**), large clear plaques (**c**), clear plaques with halo (**d** and **e**). Bar corresponds to 5 mm

2 Materials

2.1 Sewage

1. Collect municipal wastewater from the sewage treatment plant after pretreatment for a few days (*see* **Note 1**). Store it at 4 °C.
2. Mix wastewater samples and centrifuge at 9000 × *g*, 4 °C for at least 30 min.
3. Filter the supernatant through a paper filter and subsequently through a sterile filter with a pore diameter of 0.2 μm (*see* **Note 2**) into a sterile bottle. Use the purified wastewater as soon as possible. A preparation received in this way should be regarded as one sample.

2.2 Media, Buffers and Chemicals

1. Tryptic soy broth (TSB), pH 7.3 ± 0.2 (per L): 17.0 g of pancreatic digest of casein, 3.0 g of soy peptone, 5.0 g of NaCl, 2.5 g of glucose, 2.5 g of K_2HPO_4. Prepare medium according to the manufacturer's instructions.
2. 2× TSB: double concentrated TSB.
3. Phage nutrient agar (PNA), pH 7.6–7.8 [18]: Weigh 5 g of casein hydrolysate, 1.5 g of meat extract, 2.5 g of NaCl, 1.5 g of Na_2HPO_4 and transfer to a 1 L Erlenmeyer flask. Add approx 450 mL of distilled water, heat up and mix to dissolve the

ingredients. Next, adjust the pH with 10% NaOH. Add 6 g of agar and bring up to 0.5 L with distilled water. If the *P. mirabilis* strain used exhibits swarming motility (characteristic for most *P. mirabilis* strains), 0.5 mL of phenol or 2.5 mL of glycerol should be added to the medium before sterilization (*see* **Note 3**). Sterilize the medium by autoclaving (115 °C, 20 min, ¾ ATM). After sterilization, pour approx 20 mL of the liquid medium into sterile 9 cm petri dishes. Plate dimensions may differ among manufacturers. Adjust the volume of the medium so that the agar thickness is about 3–4 mm. After solidification of the medium, store it at 4 °C, for up to 3 weeks (*see* **Note 4**).

4. Nutrient broth (NB) supplemented with Ca^{2+} and Mg^{2+}, pH 7.1 (*see* **Note 5**): Prepare medium according to the manufacturer's instructions. Adjust the pH to 7.1 with HCl. After sterilization by autoclaving (115 °C, 20 min, ¾ ATM), add 1 mL of filter-sterilized 1 M $CaCl_2$ and 1 M $MgSO_4$ per 100 mL of medium. Store at 4 °C.
5. Soft agar 0.7%, pH 7.4: Put 2.1 g of nutrient broth in a 1 L Erlenmeyer flask and add 450 mL of distilled water. Mix and adjust the pH with 10% NaOH. Add 1.4 g of agar and bring up to 0.5 L with distilled water. Autoclave for 20 min, 115 °C, ¾ ATM. Pour 3.2 mL of liquid medium into sterile heat-resistant tubes. Store at 4 °C, for up to 3 weeks (*see* **Note 4**).
6. SM buffer, pH 7.5: 50 mM Tris–HCl, pH 7.5, 100 mM NaCl, 10 mM $MgSO_4$, 0.01% gelatin. Dissolve 6.06 g of Tris in 900 mL of pure water (*see* **Note 6**). Mix and adjust pH with 37% HCl. Next, add 5.84 g of NaCl and make up to 990 mL with pure water. Split equally into four 250-mL bottles and carefully pour 25 mg of gelatin into each of them. Autoclave for 20 min, 115 °C, ¾ ATM. After cooling the buffer to room temperature, add 2.5 mL of filter-sterilized 1 M $MgSO_4$ to each bottle. Store at room temperature.
7. 5% ammonium molybdate, pH 6.0 adjusted with 2 M NaOH.
8. 2% uranyl acetate.

2.3 Equipment

1. Water bath or Koch's steam sterilizer.
2. Incubator.
3. Transmission electron microscope.
4. Formvar-coated copper grid stabilized with carbon, 200 square mesh, for transmission electron microscopy.

3 Methods

Carry out all procedures at room temperature, unless specified otherwise.

3.1 Isolation of Phages

This method is from Maszewska et al. [10].

1. Mix 10 mL of purified wastewater with an equal volume of 2× TSB in a 50 mL tube, and add 1 mL of an 18 h *P. mirabilis* culture in TSB.
2. Incubate the tube containing prepared suspensions for 3 h at 37 °C, with shaking (150 rpm).
3. Apply about 1 mL of the suspension to PNA or, if the *P. mirabilis* strain exhibits swarming motility, PNA containing 0.1% phenol or 0.5% glycerol (*see* **Note** 7). Spread the suspension on the plate and remove the excess.
4. Dry the plates at room temperature (*see* **Note 8**) and incubate for 24 h at 37 °C (*see* **Note 9**).
5. Store the tube with the used suspension at 4 °C until the next day (*see* **Note 10**).
6. Observe individual plaques on the plates. Select plaques with different morphologies (Fig. 1), cut out a single plaque using a Pasteur pipette, and transfer it into a tube containing 8 mL of NB.
7. Mix gently using a vortex every 10 min for half an hour to release the phage from the agar.
8. Add 50 μL of an 18 h culture of the same bacterial strain as used in **step 1** on nutrient broth pH 7.1.
9. In order to bind the phage to the bacterial surface, mix the sample gently using a vortex every 10 min for half an hour.
10. Incubate the cultures overnight at 37 °C (*see* **Note 11**).
11. Centrifuge the culture at 4000 × *g*, 4 °C for 30 min.
12. Filter the supernatant through a sterile syringe filter with a pore diameter of 0.2 μm (*see* **Note 12**).
13. Store the phage lysate at 4 °C.

3.2 Purification of Phages

1. To obtain pure phage lysate, the single plaque isolation has to be repeated five times (*see* **Note 13**). For this purpose, a double-layer agar method can be used [19]. Place soft agar tubes into a hot-water bath (100 °C) or Koch's steam sterilizer to melt the agar.
2. Transfer the tubes into 50 °C water; this allows soft agar to remain in liquid form until it is used.

3. Prepare serial tenfold dilutions of phage lysate in SM buffer in the range from 10^1 to 10^7 (*see* **Note 14**).
4. Aseptically add 100 μL of an 18 h *P. mirabilis* culture and 200 μL of the first lysate dilution to a soft agar tube taken from the water bath (50 °C) and mix gently.
5. Pour the mixture into a PNA plate with or without 0.1% phenol or 0.5% glycerol (*see* **Note 7**). This will create a thin layer of agar containing bacteria and viruses.
6. Repeat this procedure for the rest of the serial dilutions.
7. Leave the plates for a few minutes to allow the soft agar to solidify.
8. Carefully invert the plates to make sure the soft agar will not fall off the PNA surface. Incubate plates for 24 h at 37 °C (*see* **Note 9**).
9. After incubation, the plate should seem turbid (due to bacterial growth) with clear spots (plaques).
10. Repeat Subheading 3.1, **steps 6–13** and Subheading 3.2, **steps 1–9** four times.
11. Store the final pure phage lysate at 4 °C (*see* **Note 15**).
12. Plates containing between 30 and 300 plaques can be used for viral titer determination [19], which is expressed as plaque forming units (pfu) per mL. Each plaque represents one virus particle. Use the following formula to determine the phage titer: pfu/mL = number of plaques × dilution × 5 (conversion per 1 mL).

3.3 Observation of Phages

1. Perform observation of phages using a transmission electron microscope to confirm virion homogeneity (*see* **Note 16**) [20, 21]. Centrifuge the phage lysate (titer of 10^8 or higher) at 24,500 × *g* for 3 h or longer at 4 °C (*see* **Note 17**).
2. Wash the phages twice with 5% ammonium molybdate, pH 6.0 using the same spin conditions.
3. Suspend the phage pellet in 5% ammonium molybdate, pH 6.0 to obtain a titer of 10^{10}–10^{11} pfu per mL.
4. Put 2–3 μL of the suspension onto a formvar and carbon-coated copper grid.
5. Leave the grid on filter paper for 5 min to absorb the sample.
6. Stain the phages with 2% uranyl acetate for 1 min.
7. Examine under a TEM at 80 kV.

4 Notes

1. Samples of wastewater for testing should be collected for several days and then mixed. The use of mixed samples of sewage increases the probability of isolating bacteriophages specific for more bacterial strains and, depending on the expectations, with a narrow or broad host range.
2. If a sewage sample is very contaminated or thick, a sterile filter with a pore diameter of 0.45 μm can be used for filtration followed by sterilization with a 0.2 μm filter.
3. If a strain demonstrates swarming motility, characteristic of *P. mirabilis*, phenol or glycerol should be added to the medium. It was observed that plaques on PNA supplemented with phenol were smaller than plaques on PNA with glycerol. Glycerol might contribute to increasing plaque size in solid media [22]. In some cases, hazy and blurred plaques on PNA were observed. Unfortunately, it depends on the bacterial strain used. A suitable increase of glycerol concentration may solve this problem [22]. Isolation of phages from strains showing strong swarming motility should be carried out on a medium with phenol. This chemical should be used with special caution. Read the manufacturer's precautions before use. It should also be remembered that phenol may have a bactericidal effect on certain *P. mirabilis* strains. *See* also Chapter 2 of this volume for more suggestions on swarming inhibition.
4. The use of old or overdried medium may result in a lack of plaques.
5. Nutrient broth is supplemented with $CaCl_2$ and $MgSO_4$, because some phages require divalent ions either for adsorption or during phage multiplication [23, 24].
6. Pure water means double-distilled water or deionized water.
7. The surface of PNA should be dried before use, but do not use overdried plates. If necessary, dry plates before use in a laminar cabinet for about 5 min.
8. To speed up the drying process of plates, they can be placed in a laminar cabinet for about 5 min. Be careful not to overdry the plates.
9. Strains which do not require a swarming inhibitor can be incubated for 5 h at 37 °C and then left overnight at 4 °C. The plaques will be more visible.
10. The tubes with suspension should be stored at 4 °C until the next day. In some cases, complete lysis of bacteria can be seen on the PNA plate. This indicates a very large number of phages in the test sample, and the isolation should be repeated.

Centrifuge the suspension at 4000 × *g*, 4 °C for 30 min and filter-sterilize the supernatant. Prepare serial tenfold dilutions of the phage lysate in SM buffer in the range from 10^1 to 10^7 and perform a plaque assay (Subheading 3.2, **steps 3–9**).

11. During reisolation, plaque samples can be incubated for 4 h at 37 °C with shaking (110 rpm) and subsequently overnight at 4 °C. This procedure allows obtainment of a higher titer phage lysate (10^8–10^{10} pfu per mL). Remember this when preparing dilutions used for the double-layer agar method.
12. The filter-sterilization step can be replaced by 2% chloroform treatment. This method is generally used to kill any bacteria present in the phage lysate, but chloroform may inactivate some phages [13], and should be used with caution and proper controls.
13. This procedure allows elimination of the influence of bactericidal activity of some chemicals present in wastewater, and isolation of a single phage strain.
14. Most of the propagated phages (at this stage) reach a titer between 10^7 and 10^8 pfu per mL (range from 10^3 to 10^9 pfu per mL); therefore, dilutions between 10^3 and 10^7 can be used for the double-layer agar test. The dilution series of phage lysates can be stored at 4 °C until the following day. If no plaques are observed the next day, perform the test using lower dilutions and the initial sample.
15. Due to the high variation of phage sensitivity to various storage conditions, there is no universal phage storage method. Keeping crude or purified phage lysates at 4 °C is a standard bacteriophage preservation technique. Freezing and storage at −80 °C, or in liquid nitrogen, and storage of dried or lyophilized phages have also been used [25, 26]. Golec et al. [27] described a method where phages were frozen in LB supplemented with 10 mM $MgSO_4$, 10 mM $CaCl_2$, and 15% glycerol inside the infected bacterial cells at −80 °C without a major loss of phage and host viability. This method seems to be most appropriate for long-term storage of different types of phages. Here, *P. mirabilis*-specific phages were frozen in NB medium with 15% glycerol inside the infected bacteria and parallel lysates were stored at 4 °C.
16. The choice of the appropriate stain, staining time, etc. depends on the type of tested phage and often needs to be determined experimentally. Many technical variants of the basic negative staining method are available. Negative staining of purified phages with phosphotungstate or uranyl acetate is fully described in [20].
17. The spin time depends on the phages. Tested phages, with a genome size of approx 160 kb, were centrifuged for 3 h.

References

1. Suttle CA (2005) Viruses in the sea. Nature 437:356–361. https://doi.org/10.1038/nature04160
2. Adams MH (1959) Bacteriophages. Interscience Publishers, New York
3. Ackermann HW, Nguyen TM (1983) Sewage coliphages studied by electron microscopy. Appl Environ Microbiol 45:1049–1059
4. Seeley ND, Primrose SB (1982) The isolation of bacteriophages from the environment. J Appl Bacteriol 53:1–17
5. Jurczak-Kurek A, Gąsior T, Nejman-Faleńczyk B, Bloch S, Dydecka A, Topka G, Necel A, Jakubowska-Deredas M, Narajczyk M, Richert M, Mieszkowska A, Wróbel B, Węgrzyn G, Węgrzyn A (2016) Biodiversity of bacteriophages: morphological and biological properties of a large group of phages isolated from urban sewage. Sci Rep 6:34338. https://doi.org/10.1038/srep34338
6. Yu P, Mathieu J, Li M, Dai Z, Alvarez PJJ (2016) Isolation of polyvalent bacteriophages by sequential multiple-host approaches. Appl Environ Microbiol 82:808–815. https://doi.org/10.1128/AEM.02382-15
7. Lehman SM, Donlan RM (2015) Bacteriophage-mediated control of a two-species biofilm formed by microorganisms causing catheter-associated urinary tract infections in an in vitro urinary catheter model. Antimicrob Agents Chemother 59 (2):1127–1137. https://doi.org/10.1128/AAC.03786-14
8. Weber-Dąbrowska B, Jończyk-Matysiak E, Żaczek M, Łobocka M, Łusiak-Szelachowska M, Górski A (2016) Bacteriophage procurement for therapeutic purposes. Front Microbiol 7:1177. https://doi.org/10.3389/fmicb.2016.01177
9. Muniesa M, Imamovic L, Jofre J (2011) Bacteriophages and genetic mobilization in sewage and faecally polluted environments. Microb Biotechnol 4(6):725–734. https://doi.org/10.1111/j.1751-7915.2011.00264.x
10. Maszewska A, Wójcik E, Ciurzyńska A, Wojtasik A, Piątkowska I, Dastych J, Różalski A (2016) Differentiation of polyvalent bacteriophages specific to uropathogenic *Proteus mirabilis* strains based on the host range pattern and RFLP. Acta Biochim Pol 63(2):303–310. https://doi.org/10.18388/abp.2015_1114
11. Schmidt WC, Jeffries CD (1974) Bacteriophage typing of *Proteus mirabilis, Proteus vulgaris*, and *Proteus morganii*. Appl Microbiol 27 (1):47–53
12. Nzakizwanayo J, Hanin A, Alves DR, McCutcheon B, Dedi C, Salvage J, Knox K, Stewart B, Metcalfe A, Clark J, Gilmore BF, Gahan CGM, Jenkins ATA, Jones BV (2016) Bacteriophage can prevent encrustation and blockage of urinary catheters by *Proteus mirabilis*. Antimicrob Agents Chemother 60 (3):1530–1536. https://doi.org/10.1128/AAC.02685-15
13. Ackermann HW (2006) Classification of bacteriophages. In: Calendar R (ed) The bacteriophages, 2nd edn. Oxford University Press, New York, pp 8–16
14. Gallet R, Kannoly S, Wang IN (2011) Effects of bacteriophage traits on plaque formation. BMC Microbiol 11:181. https://doi.org/10.1186/1471-2180-11-181
15. Latka A, Maciejewska B, Majkowska-Skrobek-G, Briers Y, Drulis-Kawa Z (2017) Bacteriophage-encoded virion-associated enzymes to overcome the carbohydrate barriers during the infection process. Appl Microbiol Biotechnol 101(8):3103–3119. https://doi.org/10.1007/s00253-017-8224-6
16. Pires DP, Oliveira H, Melo LD, Sillankorva S, Azeredo J (2016) Bacteriophage-encoded depolymerases: their diversity and biotechnological applications. Appl Microbiol Biotechnol 100(5):2141–2151. https://doi.org/10.1007/s00253-015-7247-0
17. Abedon ST (2009) Phage evolution and ecology. Adv Appl Microbiol 67:1–45
18. Ślopek S, Durlakowa I, Kucharewicz-Krukowska A, Krzywy T, Ślopek A, Weber B (1972) Phage typing of *Shigella flexneri*. Arch Immunol Ther Exp 20:1–60
19. Kropinski AM, Mazzocco A, Waddell TE, Lingohr E, Johnson RP (2009) Enumeration of bacteriophages by double agar overlay plaque assay. In: Clokie MR, Kropinski AM (eds) Bacteriophages, Methods in molecular biology, vol 501. Humana Press, New York, NY, pp 69–76
20. Ackermann HW (2009) Basic phage electron microscopy. In: Clokie MR, Kropinski AM (eds) Bacteriophages, Methods in molecular biology, vol 501. Humana Press, New York, NY, pp 501:113–501:126
21. Kwiatek M, Parasion S, Rutyna P, Mizak L, Gryko R, Niemcewicz M, Olender A, Łobocka M (2017) Isolation of bacteriophages and their application to control *Pseudomonas aeruginosa* in planktonic and biofilm models. Res Microbiol 168(3):194–207. https://doi.org/10.1016/j.resmic.2016.10.009

22. Santos SB, Carvalho CM, Sillankorva S, Nicolau A, Ferreira EC, Azeredo J (2009) The use of antibiotics to improve phage detection and enumeration by the double-layer agar technique. BMC Microbiol 9:148. https://doi.org/10.1186/1471-2180-9-148
23. Garen A, Puck TT (1951) The first two steps of the invasion of host cells by bacterial viruses. II. J Exp Med 94:177–189
24. Paranchych W (1966) Stages in phage R 17 infection: the role of divalent cations. Virology 28:90–99
25. Ackermann HW, Tremblay D, Moineau S (2004) Long-term bacteriophage preservation. World Fed Cult Coll Newslett 38:35–40
26. Fortier LC, Moineau S (2009) Phage production and maintenance of stocks, including the expected stock lifetime. In: Clokie MR, Kropinski AM (eds) Bacteriophages, Methods in molecular biology, vol 501. Humana Press, New York, NY, pp 203–219
27. Golec P, Dąbrowski K, Hejnowicz MS, Gozdek A, Łoś JM, Węgrzyn G, Łobacka MB, Łoś M (2011) A reliable method for storage of tailed phages. J Microbiol Methods 84:486–489. https://doi.org/10.1016/j.mimet.2011.01.007

Chapter 21

Detection of Neutrophil Extracellular Traps in Urine

Yanbao Yu, Keehwan Kwon, and Rembert Pieper

Abstract

Neutrophils are important mediators of the antimicrobial defense during urinary tract infections (UTIs). When activated at the site of infection, these innate immune cells phagocytose and neutralize an invading pathogen. Another neutrophil defense strategy is the release of effectors, such as antimicrobial peptides and proteins stored in neutrophil granules and reactive oxygen species. Their release can be facilitated by cellular signals that trigger chromatic decondensation and the disruption of nuclear membranes, followed by granule and plasma membrane disintegration, DNA release into the extracellular milieu, and neutrophil cell death. Neutrophil extracellular traps (NETs) form. If microbial pathogens are the cause of neutrophil infiltration, they are entrapped in the network of DNA fibers that characterize NETs and are exposed to antimicrobial granule effectors and histones that bind to the extracellular DNA fibers. Here, we describe nonmicroscopic methods applied to clinical (urine sediment) samples to identify and characterize NETs associated with UTI. A stepwise extraction procedure using PBS, deoxyribonuclease I digestion and SDS-based solubilization is described. This is followed by native gel analysis to visualize protein–DNA macromolecular assemblies and proteomic analysis to identify signature proteins and their quantities in NETs. Microbes observed to be entrapped in NETs in the process of the innate immune response to the infection are *Escherichia coli*, *Klebsiella pneumoniae*, *Proteus mirabilis*, *Staphylococcus aureus*, and *Enterococcus faecalis*.

Key words Neutrophil extracellular traps, Neutrophils, Urinary tract infection, Urine, Inflammation, Proteomics, *Proteus mirabilis*

1 Introduction

Neutrophil extracellular traps (NETs) were discovered by pioneering studies of Zychlinsky et al. in 2004 [1]. In a neutrophil-rich inflamed environment, NET formation may be triggered by Gram-positive and Gram-negative bacteria as well as fungal pathogens. In neutrophils stimulated by the presence of these organisms or specific molecules such as bacterial lipopolysaccharide (LPS) and phorbol myristate acetate, the nuclear chromatin decondenses, nuclear membranes are disrupted, and the release of histones and DNA into the cytoplasm occurs. These events finally result in the disintegration of granule and plasma membranes and cell death.

Melanie M. Pearson (ed.), *Proteus mirabilis: Methods and Protocols*, Methods in Molecular Biology, vol. 2021,
https://doi.org/10.1007/978-1-4939-9601-8_21,

Chromosomal DNA fibers and other cellular matter are released into the extracellular milieu. Immunostaining experiments have revealed that granular effector proteins such as neutrophil elastase (NE), myeloperoxidase (MPO), and histones bind to the extracellular DNA fibers and that histones are degraded in that process [1, 2]. The pathogen's cells are entrapped in a network of disintegrating cells, DNA fibers and positively charged protein effectors. NE and MPO are regulators of NET formation [2]. This microbial growth-inhibiting process is different from phagocytosis, degranulation of neutrophils without the loss of cellular structure, necrosis, and apoptosis [3, 4]. NETs also form in vitro, suggesting that their formation does not require a multicellular inflamed tissue environment. By enzymatic degradation with deoxyribonuclease (DNase) I, DNA fibers were shown to form the critical scaffold of NETs, thus allowing the release of well-characterized basic granular effector proteins, cytoplasmic proteins (e.g., calprotectin S100-A8 and S100-A9), and nuclear proteins (e.g., histones). Colocalization studies confirmed binding of these proteins to the extracellular DNA [5].

Pathogens have developed strategies to escape from NETs, as reported for Group A *Streptococcus* via secretion of a DNase and for *Staphylococcus aureus* via activities of a secreted nuclease and an adenosine synthase [6]. The mechanism resulting in NET formation, also termed NETosis, includes the key role for reactive oxygen species (ROS) to dissociate a granular membrane complex termed the azurosome [7]. NETs not only play a role in the host defense against pathogens but also in autoimmune diseases such as gout [8]. Among the infections where NETs were discovered are urinary tract infections (UTIs). Especially, the analysis of viscous human urine sediments revealed evidence of NETs associated with a uropathogen, including *Proteus mirabilis* [9].

NETs can be characterized using various methods. Immunofluorescence microscopy (IF) is frequently used to detect cellular evidence of NETs in the context of human and murine studies [1, 3, 5, 9, 10]. There are additional methods for the molecular investigation of NETs associated with UTI, and these methods are the focus of this contribution. We describe differential protein extraction methods for samples containing NETs, gel-based analysis, and metaproteomic methods to characterize NETs and the pathogens trapped in NETs simultaneously.

2 Materials

All solutions and buffers pertaining to proteomics experiments should be made using LCMS grade water. Low protein-binding tips and tubes are recommended for these experiments as well. Urea-containing solutions should be prepared freshly before use. Chemicals for all experiments should be ACS reagent grade.

2.1 Materials for Extraction and Gel-Based Experiments

1. SDS-PAGE precast polyacrylamide gels (e.g., 4–12% Bis-Tris polyacrylamide gels).
2. Native-PAGE precast polyacrylamide gels.
3. Polyvinylidene difluoride (PVDF) membranes.
4. SDS-PAGE sample buffer: 25% glycerol (v/v), 10% β-mercaptoethanol (v/v), 3% SDS (w/v), 100 mM Tris–HCl (pH 8.1).
5. Gel electrophoresis and protein transfer system for SDS-PAGE and immunoblotting.
6. SDS running buffer.
7. Western blot transfer buffer: 14.4 g of glycine and 3 g of Tris dissolved in water, 10% methanol (V/V).
8. Western blot membrane wash buffer: Tris-buffered saline (TBS), 0.05% Tween.
9. Dry milk.
10. Extraction solutions for urine sediment samples:
    (a) ES1: PBS buffer (137 mM NaCl, 2.7 mM KCl, 8 mM Na_2HPO_4, and 2 mM KH_2PO_4).
    (b) ES2: PBS, 50 mM dithiothreitol (DTT).
    (c) ES3: PBS, 2 μL of bovine deoxyribonuclease I/100 μL buffer (~10 Kunitz units per μL) and 5 mM $MgCl_2$.
    (d) ES4: 1% SDS (w/v), 10 mM Na-EDTA and 25 mM DTT.
11. EDTA–CHAPS extraction solution: PBS, 5 mM Na-EDTA and 0.4% CHAPS (w/v).
12. Molecular weight markers for SDS-PAGE: a protein mixture in the M_r range of 10 kDa to 200 kDa (e.g., Bio-Rad Precision Plus Dual Color Protein Standards).
13. Molecular weight markers for Native PAGE: a protein mixture in the M_r range of 10 kDa to 1000 kDa (e.g., Thermo-Fisher Native PAGE Unstained Protein Standard).
14. Centrifugal membrane concentrator device with a 10 kDa M_r cutoff (e.g., Sartorius Vivacon-10).
15. Horseradish peroxidase (HRP) chemiluminescent substrate.
16. Agarose gel (0.75% agarose gel electrophoresis).
17. Ethidium bromide (DNA staining).
18. Gel documentation/scanning system.
19. TBE buffer: 89 mM Tris, 89 mM boric acid, 2 mM EDTA.
20. 10× Blue Juice gel loading buffer: 10 mM Tris–HCl (pH 7.6), 60 mM EDTA, 0.03% bromophenol blue, 60% glycerol.
21. DNA molecular weight markers for Native PAGE: a DNA fragment mixture in size of 100 bp to 2.1 kbp (e.g., Thermo-Fisher 100 bp DNA Ladder).

2.2 Biologicals Used for Extraction, SDS-PAGE, and Western Blotting

1. Bovine deoxyribonuclease I (DNase) (Sigma-Aldrich).
2. Bovine serum albumin (BSA).
3. Anti-myeloperoxidase (MPO) heavy chain C-16 antibody (Santa Cruz Biotechnology).
4. Anti-lactotransferrin (LTF) antibody H-65 (Santa Cruz Biotechnology).
5. Anti-neutrophil elastase (NE) H-57 (Santa Cruz Biotechnology).
6. Anti-histone H4 H-97 (Santa Cruz Biotechnology).
7. Horseradish peroxidase (HRP) conjugate of goat anti-rabbit IgG (e.g., goat anti-rabbit IgG-HRP, product number sc-2004; Santa Cruz Biotechnology).

2.3 Materials for Proteomic Analyses

1. Ultrasonic cell disruptor (conditions and settings used here pertain to the Misonix Sonicator 3000 with sonication in water–ice bath) (*see* **Note 1**).
2. Trypsin solution: 10 ng/μL of sequencing-grade trypsin in 25 mM ammonium bicarbonate (ABC).
3. Filter aided sample preparation (FASP) filters (30 kDa cutoff; Millipore or Sartorius).
4. ABC buffer: 50 mM ammonium bicarbonate buffer in H_2O.
5. Urea buffer: 8 M urea in 100 mM Tris–HCl, pH 8.0.
6. Empore C18 disks for StageTip desalting (3 M).
7. Liquid chromatography–mass spectrometry (LC-MS)/MS workstation, such as Ultimate 3000 nLC solvent delivery system, and high-resolution mass spectrometer, such as Q-Exactive mass spectrometer (Thermo Scientific).
8. Acetonitrile.
9. 10 mM DTT.
10. 55 mM iodoacetamide (IAA) solution.
11. Extraction buffer: 50% acetonitrile and 0.1% trifluoroacetic acid in water.
12. HPLC solvent A: 0.1% formic acid in water.
13. HPLC solvent B: 0.1% formic acid in acetonitrile.

3 Methods

Several strategies have been used to detect and characterize NETs derived from a variety of conditions and environments [11]. NETs have been induced in vitro from primary neutrophils derived from blood samples upon use of different stimuli [12, 13]. In vivo detection of NETs in paraffin-embedded tissue by use of immunostaining methods has also been reported [10, 14]. Our emphasis here is the description of methods to identify and analyze NETs present in human urine sediment specimens.

3.1 Urine Sediment Sample Analysis by SDS-PAGE

1. NETs are fragile structures. It is best to use freshly collected urine specimens from individuals (patients) with clinical evidence of a UTI. After collection in the clinic, urine specimens should be maintained at 4 °C and processed quickly. Freezing, thawing, and agitation may degrade NETs and alter the experimental outcomes (*see* **Note 2**).
2. After gentle transfer of the specimen into a conical 15 mL tube (typically 2–10 mL), collect the urine sediment via centrifugation at approximately 1500 × *g* for 8 min at 4 °C. Retain the centrifugal urinary pellet (UP).
3. Add a tenfold volume of PBS (4 °C) and observe UP consistency. A viscous and aggregated appearance may indicate NET contents (Fig. 1).
4. Use gentle tube agitation or a spatula to allow the UP sample to float and be exposed to PBS. Pipetting may perturb the integrity of NET structures.
5. Gently turn the tube sideways or upside-down for a few minutes to observe if the aggregate maintains its consistency.

Fig. 1 Images of urinary pellets suspended in PBS with and without strong aggregation of solid matter. The top two samples (upper panel) show unaggregated urine pellets, indicative of the absence or high fragility of NETs. The bottom three samples (lower panel) show clear urine pellet aggregates, more likely to contain NETs

6. Recentrifuge the sample at 1500 × *g* for 8 min at 4 °C. If sample consistency has changed (e.g., a decreased pellet aggregation or homogenization), the supernatant of this spin should be collected. This "PBS wash" fraction can be stored on ice and also prepared for analysis by SDS-PAGE (*see* **Note 3**).
7. Move an aliquot of remaining UP (for NETs, typically an aggregated/clumpy pellet after the PBS wash step) gently to a new microcentrifuge tube and add a fivefold volume of solution ES4. Resuspend by vortexing or pipetting.
8. Incubate at 20 °C for 15 min.
9. Sonicate the UP suspension at amplitude 6.5 for ten 30 s on–off cycles in a Misonex 3000 (or equivalent) water–ice bath (*see* **Note 1**).
10. Move the UP lysate to a heat block and heat at 95 °C for 3 min.
11. Cool down and let the UP lysate sit for 5 min.
12. Vortex for 30s and let sit for another 5 min.
13. Centrifuge the lysate at maximum speed (16,000 × *g*) at 20 °C for 15 min. Discard the pellet.
14. To the lysate supernatant (only approximately 15–30 μL is needed), add an equal volume of SDS-PAGE sample buffer and heat to 90–98 °C in a boiling water bath/heat block for 2–5 min. Use the same volume for the PBS wash fraction described in **step 6**.
15. Load samples into wells of a precast gel and use standard procedures for SDS-PAGE.
16. Stain gel with Coomassie Brilliant Blue G-250, then destain to visualize protein.

This protocol and the resulting SDS-PAGE analysis serve to obtain preliminary data on urine sediment (UP) samples associated with suspected urinary tract infections and potential NET formation. A PBS-washed UP from a healthy patient normally has a small volume. A major band in the M_r range of ~80 kDa (gel-forming uromodulin) will be visible. Other less abundant bands may also be visualized. A NET-containing sample will have a streaky band pattern with the highest staining intensity at the bottom of the gel. Pellets associated with UTI, but no NETs, may show more even distribution of bands across the M_r spectrum and will likely also have a strong uromodulin band. Due to variability in protein/salt precipitation behaviors in urine, SDS-PAGE band patterns for samples not associated with NETs are variable. Figure 2 shows an SDS-PAGE gel with 4 solubilized UP samples; 3 of them have a streaky band pattern with more intense staining in the low M_r range. This pattern is indicative of degradation of proteins derived from the high proteolytic activities of activated neutrophils. Data from a

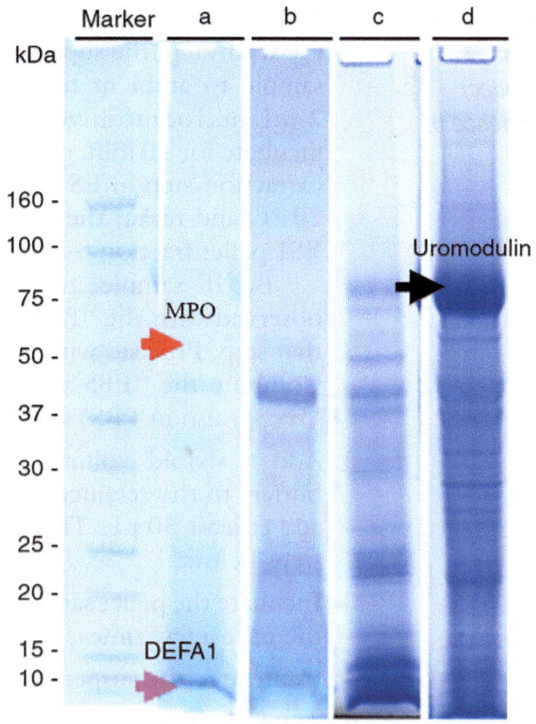

Fig. 2 Protein patterns of urinary pellet (UP) lysates visualized by SDS-PAGE. The samples **a**, **c**, and **d** were identified as cases of UTI caused by *P. mirabilis*. Samples **c** and **d** were derived from catheterized patients. Detection of the pathogen suggests catheter-associated UTI (CAUTI). Sample **b** was associated with a UTI caused by *Staphylococcus aureus* and *Candida albicans*. Three UP samples (**a**, **b**, and **c**) contained a viscous, aggregated urine sediment. Higher abundance of proteins in the low molecular weight (M_r) range suggests protein degradation, which is often observed in neutrophil-rich sediments. The data do not provide direct information on the mode of leukocyte (e.g., neutrophil) cell death (e.g., necrosis or NET formation). Sample **d** shows staining across a wider M_r range and high uromodulin abundance (black arrow; please *see* methods in Subheading 3.7). The red arrow points to a faint band for a protein fragment of myeloperoxidase (MPO; sample **b**). The pink arrow indicates a band for neutrophil α-defensin-1 (DEFA1). The bands were identified via gel-based LC-MS/MS (the majority of PSMs matched the abovementioned proteins). The SDS-PAGE gel had a 4–12% acrylamide gradient and was stained with Coomassie Brilliant Blue G-250. The left lane contains M_r standards

recent publication suggest that these proteolytic activities are extensive in NETs associated with UTI [9]. *See* Subheading 3.7 for identification of individual bands from Coomassie-stained gels.

Proceed with **steps 1–5** as described in Subheading 3.1. As described in **step 6** of Subheading 3.1, the "PBS wash" step (with a tenfold volume of UP sample) may result in a fraction that can be discarded or kept.

3.2 Fractionation of Urine Sediment Samples to Collect Molecular Evidence of NETs

1. **A.** If no sample turbidity and consistency change is observed in the wash step, the supernatant can be discarded. Adjust the UP sample to ambient temperature (~20 °C), and transfer to a 2-mL microcentrifuge tube. Add ES1 in a sixfold volume and incubate for 10 min, gently inverting the tube a few times (PBS extraction step in ES1). Centrifuge at 4000 × *g* for 8 min at 20 °C and retain the supernatant on ice (UP_{sol1}). Retain the ES1 pellet fraction for the extraction **step 2**.

   B. If sample turbidity and consistency changes are observed with the "PBS wash" step, retain the supernatant of that step. Proceed with the ES1 extraction step (as under 1A). Combine the "PBS wash" step supernatant and supernatant UP_{sol1} (also in PBS) and keep on ice.
2. Add a sixfold volume of extraction buffer ES2 (reducing buffer) to the retained pellet. If the pellet has less than 10 μL, add at least 50 μL. This minimum volume also pertains to the steps below.
3. Incubate the pellet sample for 20 min at 20 °C, gently inverting the tube a few times.
4. Centrifuge the suspension at 4000 × *g* for 8 min at 20 °C.
5. Transfer the supernatant to a new 2-mL tube (with the identifier UP_{sol2}) and keep on ice. Retain the ES2 pellet fraction for the next step.
6. Add a sixfold volume of extraction buffer ES3 to the retained pellet.
7. Gently invert the tube a few times and incubate for 30 min at 20 °C. Observe sample consistency changes (e.g., partial or complete homogenization).
8. Move the sample into a microtube shaker set at 880 rpm for 30 min at 37 °C.
9. Centrifuge the suspension at 4000 × *g* for 8 min at 20 °C.
10. Transfer the supernatant to a new 2-mL tube (with the identifier UP_{sol3}) and keep on ice. Retain the ES3 pellet fraction for the next step.
11. Add a sixfold volume of extraction solution ES4 to the retained pellet. Exposure of biological samples to ES4 denatures and solubilizes proteins.
12. Briefly vortex the suspension and incubate in a microtube shaker set at 880 rpm for 30 min at 37 °C.
13. Vortex the suspension again, and leave at 20 °C for 5 min.
14. Sonicate at amplitude 6.5 for ten 30 s on–off cycles in a Misonex 3000 water–ice bath.
15. Heat-denature the sonicated sample at 95 °C for 3 min, then vortex.

16. Centrifuge the sample at 16,000 × *g* for 15 min.
17. Transfer the supernatant to a new 2-mL tube (with the identifier UP_{sol4}). The ES4 pellet fraction may be retained (*see* **Note 4**).
18. Keep the UP_{sol1}, UP_{sol2}, and UP_{sol3} fractions on ice until ready for electrophoresis. All UP_{sol} fractions can be frozen long term prior to SDS-PAGE analysis.
19. Separate samples by agarose gel electrophoresis (to visualize DNA) or native PAGE (to visualize proteins). Loading quantity is dependent on gel well volumes (in mini-gels typically between 2 and 15 μL).
20. Stain PAGE-separated proteins with a Coomassie Brilliant Blue G-250 solution, and stain agarose-separated DNA fragments with ethidium bromide.

Figure 3 shows an SDS-PAGE gel with samples corresponding to 4 extraction steps for a NET-containing urine sediment sample.

3.3 Native Gel Electrophoresis

1. Insert a precast Native-PAGE gel into an electrophoresis workstation and submerge in TBE buffer.
2. Load 7.5 μL of a Native-PAGE protein standard (1048 kDa –20 kDa), and 5 μL of a 100 bp DNA ladder. Add 1 μL of 10× Blue Juice gel loading buffer to 9 μL of UP_{sol1}, UP_{sol2} and UP_{sol3} fractions, and load fractions on the gel.
3. Run Native-PAGE for about 1.5–2 h at a constant voltage of 150 V.
4. Remove the gel from the cassette and stain with a SYBR dye to visualize DNA (2 μL of SYBR-Safe in 50 mL of water for 30 min). Take a picture using an appropriate gel imaging system.
5. Stain the gel with Coomassie Brilliant Blue G-250 (CBB) to visualize proteins. Destain the gel and take an image using an appropriate imaging system.

Figure 4 shows the results of DNA and protein gel analysis.

3.4 Agarose Gel Electrophoresis

1. Prepare a 0.75% agarose gel in TBE buffer.
2. Load the gel with 2 μL of the DNA ladder (12 kbp to 100 bp) and UP_{sol1}, UP_{sol2}, and UP_{sol3} fractions in 2–10 μL quantities with an equal volume of Blue Juice gel loading buffer.
3. Run the agarose gel at a constant voltage of 80 V until the blue dye front reaches the gel end.
4. Remove the gel from the cassette and stain with 0.01% ethidium bromide to visualize DNA fragments. Extracellular DNA present in NETs shows a broad size range in agarose gels, with

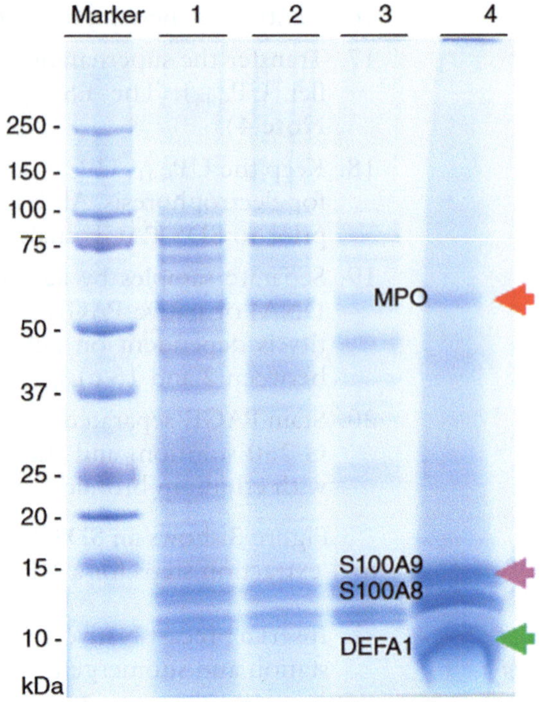

Fig. 3 A representative protein pattern of a stepwise-extracted UP sample in an SDS-PAGE gel. The extraction steps of $UP_{sol}1$, $UP_{sol}2$, $UP_{sol}3$, and $UP_{sol}4$ fractions are shown in lanes 1, 2, 3, and 4, respectively. The red arrow indicates the position of the large fragment of MPO. The green arrow indicates the position of DEFA1. The pink arrow indicates the position of the two calprotectin subunits S100-A8 and S100-A9. The data on stepwise extractions show moderate enrichment of NET-associated proteins in the DNase I digestion step ($UP_{sol}3$). In stable NETs, higher amounts of antibacterial effectors (e.g., MPO, S100A8, and S100A9) are observed in $UP_{sol}3$. In fragile NETs, proteins are solubilized to a significant extent in the first two extraction steps (as shown here). Proteins not solubilized by DNase I incubation remain in fraction $UP_{sol}4$. Many less soluble and aggregated neutrophil proteins remain in the $UP_{sol}4$ fraction. The SDS-PAGE gel had a 4–12% acrylamide gradient and was stained with Coomassie Brilliant Blue G-250

decreasing size ranges when NETs degrade in situ or degradation is initiated by digestion with DNase I.

3.5 Western Blotting

Detection of the proteins from Subheading 3.1 via western blotting can confirm the presence of effector proteins in UP_{sol} fractions and contribute to the identification of NETs [9].

1. The M_r weight range of proteins of interest determines the choice of the polyacrylamide concentration for the SDS-PAGE gel. A wide gradient gel (4–20%T) may allow western blots for a wide M_r range. For example, histones and defensins (and

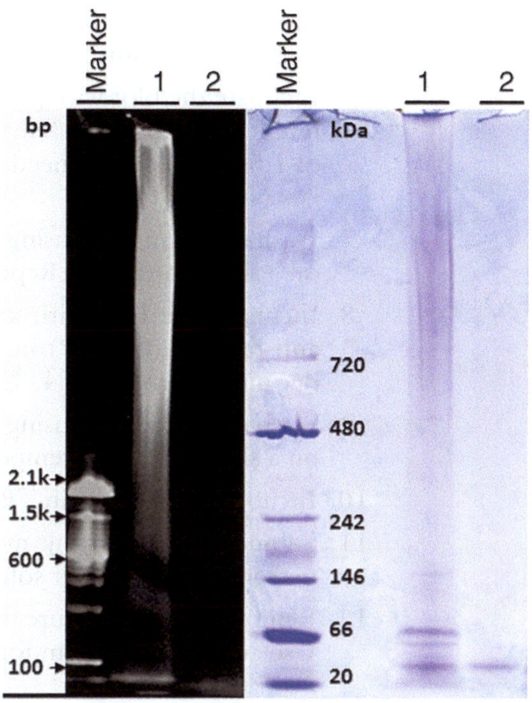

Fig. 4 DNA and protein staining patterns for DNA–protein complexes/aggregates present in NETs. A Native-PAGE gel was stained with SYBR-Safe (left panel) followed by staining with Coomassie Brilliant Blue G-250 (right panel). Lanes 1 and 2 are fractions $UP_{sol}1$ and $UP_{sol}3$, equivalent for both images. The marker lanes (DNA and native protein markers) with their respective sizes are included. The data support the presence of DNA–protein aggregates/assemblies in $UP_{sol}1$ that are degraded when incubated with DNase I ($UP_{sol}3$). DNA–protein aggregates are released as long streaks over a broad size range. The position of intensely stained DNA fragments (left panel) matches that of protein streaks in the high M_r area (>720 kDa). Further support for NETs was obtained by in-gel digestion (please *see* Subheading 3.7); among the most abundant proteins in the >720 kDa size range were histones, lactotransferrin, and myeloperoxidase

their maturation and degradation products) have low M_r values, requiring a high polyacrylamide percentage. Lactotransferrin, with a mass of 75 kDa, can be analyzed in 4–12%T gels.

2. Place PVDF blotting membrane on gel and sandwich between filter papers and blotting pads.
3. Use standard electroblotting conditions (semidry or wet) to transfer protein to the membrane.
4. Use a 1% acetic acid solution with a spatula tip of Ponceau S to stain the proteins on the blotting membrane. Use a pencil to mark M_r marker bands and sample lanes to allow associating visualized blot bands with M_r and sample following blot development.

5. Block PVDF membrane with 5% dry milk in western blot wash buffer for 90 min.
6. Incubate the blot with primary polyclonal antibodies (raised in rabbit) in TBS with 1% dry milk overnight at 4 °C in dilutions of 1:500 to 1:2500 (needs to be optimized for each antibody) (*see* **Note 5**).
7. Wash the membrane using western blot wash buffer for 10 min on a shaker at 20 °C. Repeat two more times.
8. Incubate the blot with secondary antibody conjugate (goat anti-rabbit IgG-HRP) in a 1:5000 dilution in TBS with 1% dry milk overnight at 4 °C or for 60 min on a shaker at 20 °C.
9. Wash the membrane using western blot wash buffer for 10 min on a shaker at room temperature. Repeat two more times.
10. Incubate membrane in TBS for 5 min.
11. Submerge the blotting membrane for 5 min in a chemiluminescence detection kit solution for 5 min at 20 °C.
12. Remove excess moisture from the blot and develop the image. Use an imaging system for chemiluminescent band detection.

3.6 FASP Digestion of UP Fractions

This method follows previously described procedures [15].

1. Follow Subheading 3.1, **steps 1–12** to process a urine sample and obtain a UP lysate.
2. Precipitate urine samples by centrifuging at 3000 × *g* for 15 min at 4 °C; discard the supernatant fraction and recover the sediment (*see* **Note 6**).
3. Aliquot a volume equivalent to 10~20 μg of total protein quantity (*see* **Note** 7) into a FASP filter and mix with 200 μL of urea buffer.
4. Spin at 14,000 × *g* for 15–20 min until the volume is reduced to less than 20 μL.
5. Add 200 μL of urea buffer and repeat **step 4** one more time.
6. Add 100 μL of IAA solution into the filter, and incubate in the dark for 20 min.
7. Centrifuge at 10,000 × *g* for around 10 min to remove IAA solution.
8. Wash filter with 200 μL of urea buffer.
9. Wash filter with 200 μL of ABC buffer.
10. Transfer the filter to a new collection tube and add 150 μL of ABC buffer and 0.1–0.2 μg of trypsin to the filter.
11. Incubate the filter in a thermo mixer overnight at 37 °C.
12. Elute peptides from the filter by adding 200 μL of ABC buffer, spin and collect eluent to a fresh microtube. Repeat this step three times.

13. Dry the peptide solution using a SpeedVac.
14. Desalt peptides using the spinnable StageTips (*see* **Note 8**). After drying, the clean peptide samples are now ready for LC-MS/MS analysis (*see* Subheading 3.8).

3.7 In-Gel Digestion of Native-PAGE and SDS-PAGE Gel Slices

1. After proteins are resolved in a SDS-PAGE gel and visualized using Coomassie Blue staining (*see* Subheading 3.1), cut the region of interest into small cubes (~1 mm^3).
2. Destain gel slices until colors completely disappear.
3. Add 500 μL of 100% acetonitrile and gently shake for 20 min.
4. Repeat **step 3** 2–3 times until gel pieces shrink and become opaque.
5. Discard all liquid and then SpeedVac to completely dry.
6. Reduce the sample: add ~50 μL of 10 mM DTT (ensure full gel coverage) and incubate at 56 °C for 45 min.
7. Cool down sample tubes to room temperature.
8. Add 500 μL of acetonitrile to dehydrate gels, similar to the procedures in **steps 3** and **4**.
9. Alkylation of the sample: add ~50 μL of iodoacetamide solution to completely cover the gels and incubate at room temperature in the dark for ~20 min.
10. Repeat **steps 3** and **4** to dry the gels again.
11. Add 50–100 μL of trypsin solution to completely cover gel pieces and leave sample tubes in ice for 1–2 h. Afterward, check if the solution is absorbed. Add more if necessary (*see* **Note 9**).
12. Add 10–20 μL of ABC buffer to cover the gel slices.
13. Incubate the sample overnight at 37 °C with gentle shaking.
14. Extract peptides by adding 200 μL of extraction buffer and shake for 15–20 min.
15. Collect the supernatant into clean tubes and repeat the extraction steps (**steps 14** and **15**) three times.
16. Dry the peptide solution in a SpeedVac.
17. Desalt, using the spinnable StageTip protocol (*see* **Note 8**). After drying, the clean peptide samples are ready for LC-MS/MS analysis.

3.8 LC-MS/MS Analysis

1. Resuspend desalted peptides in 20 μL of HPLC solvent A.
2. Load 5–10 μL onto a nanoLC system connected directly to an electrospray source and a high-resolution mass spectrometer (for instance, Q Exactive, Thermo Scientific). The analytical C18 column (typical inner diameter 75 μm, length 15 cm) may be self-packed or purchased. Place a trap or guard column (for instance, C18 PepMap100, 300 μm × 5 mm, 5 μm, 100 Å)

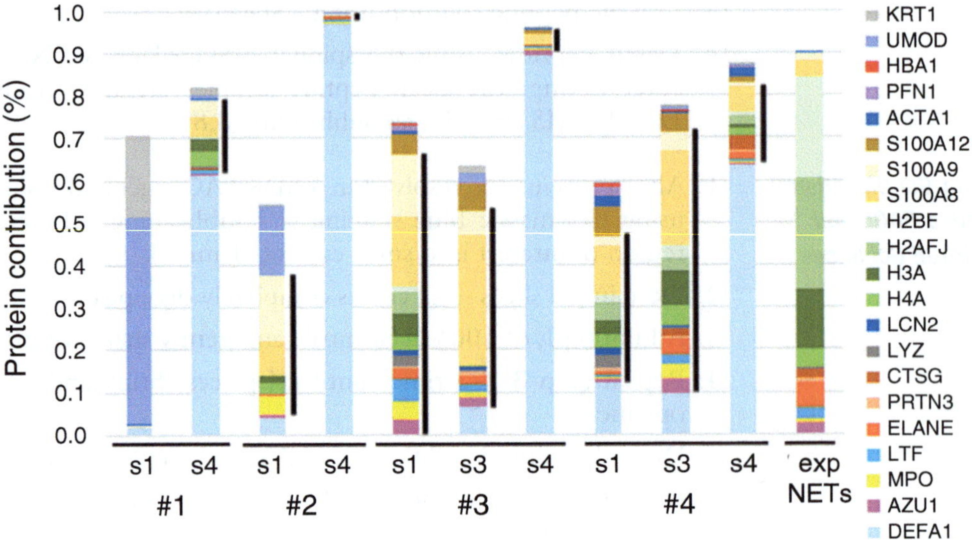

Fig. 5 LC-MS/MS-based protein abundance profiles of UP_{sol} fractions (designated here as s1, s2, s3, and s4) derived from stepwise extraction of four UP samples (#1–#4). The bar displayed on the right (**exp NETs**) represents the relative quantity of 13 proteins from in vitro generated NETs [5]. These proteins, in addition to several other proteins, are either likely relevant to NET formation (e.g., DEFA1) or are present in urine sediments based on epithelial secretion and exfoliation processes (e.g., uromodulin (UMOD) and keratin-1 (KRT1)). The 13 **exp NET** proteins are mostly antimicrobial effectors bound to the DNA fibers in NETs. In samples #2, #3, and #4, proteomic quantification of $UP_{sol}1$ to $UP_{sol}3$ fractions shows patterns with more similarity to those of **exp NETs**. The black line next to each multicolored bar indicates the contribution of those proteins (for each UP_{sol} fraction) that are dominant in **exp NETs**. Sample **#4** has the highest contribution of such proteins in UP_{sol3}, suggesting that NETs were more stable or abundant in this case compared to samples **#2** and **#3**. In contrast, sample **#1** shows no evidence of NETs. The bar heights (*y*-axis) represent the summed contribution of the 21 proteins depicted (listed in the panel on the right with the short name in the UniProt protein database) relative to total proteome. The quantities are based on the iBAQ quantification computed via use of the MaxQuant software tool. Each colored bar segment represents the quantity for an individual protein

ahead of the analytical column to protect pressure increase and clogging.

3. Depending on sample complexity, the length of the LC gradient may vary. A 150-min gradient (2–35% solvent B for 110 min) may apply with 200 nL/min for peptide separation.
4. Run Q Exactive mass spectrometer with the following settings: resolution of 70,000 in the Orbitrap; data-dependent acquisition (DDA) mode to automatically switch between MS and MS/MS acquisitions; top10 MS/MS at resolution 17,500 in the mass range 350–1800 *m/z*; normalize collision energy at 27. The settings for electrospray are as follows: spray voltage at 18 kV; capillary temperature 250 °C.

Figure 5 depicts outcomes of quantitative proteomic analyses associated with LC-MS/MS data for several sequential extraction steps from four urine samples.

4 Notes

1. To facilitate cell lysis, use a Misonex 3000 Sonicator (or equivalent instrument) with the water bath connection, not the probe connection. Set the program at amplitude 6 and proceed with on–off sonication time cycles as indicated in Subheading 3.1, **step 9** and Subheading 3.2, **step 14**.
2. An institutional review board (IRB) may need to review and approve collection of such samples. The IRB may grant an exemption if the specimens are considered medical waste (i.e., the specimens were collected for a different purpose and are deidentified). The specimens should be stored for a limited time at 4 °C. The time should ideally not exceed 15 h (the shorter the storage time, the more likely it is that NETs remain intact when experiments are started). Specimens can be stored in original collection vials.
3. The examples provided in Fig. 1 show the outcomes of urinary sediment resuspension in PBS. Three samples are shown where a viscous aggregate is not resuspended in PBS during the PBS incubation step. The aggregate maintains its structure and is more likely to contain NETs. Two samples are shown where the urinary sediment that was clumpy prior to resuspension in PBS homogenized to a considerable extent via incubation in PBS. They are less likely to contain NETs, but it cannot be ruled out in the latter case that NETs are beginning to degrade during the process of centrifugation and/or PBS resuspension.
4. The ES4 pellet is likely to contain additional proteins that are not very soluble, even in the presence of 1% SDS. Among them are neutrophil defensin 1 (DEFA1) and transmembrane domain proteins.
5. To detect the proteins that are present in NETs, we used anti-MPO heavy chain C-16 (sc-16128-R), anti-LTF (H-65, sc-25622), anti-ELANE (H-57, sc-25621), and anti-histone H4 (H-97, sc-10810).
6. Aliquot and freeze the lysate at −80 °C, if necessary.
7. To estimate protein concentration, load a 10-μL aliquot of the lysate onto SDS-PAGE along with 2 μg and 5 μg of BSA standards in the same gel. Stain the gel using Coomassie Brilliant Blue-G250. Estimate the total protein concentrations (amounts) based the summed band intensities per lane (for a given UP sample or UP_{sol} fraction) compared to band staining intensity for a known amount of BSA.
8. This protocol is the same as described before [15]. In our experience, the analytical column life time is longer when peptide extracts are desalted using the StageTip protocol as

well as on-line desalting with a C18 trap cartridge. Salt and other particulate matter in the LC system may damage switching valves and causes inconsistent backpressure. As troubleshooting such problems at nanoLC conditions is problematic, we have an off-line StageTip desalting step in addition to online desalting.

9. Depending on the size of gel pieces, first add ~50 μL of trypsin solution to completely cover the gels. Soak gels on ice for about 1 h. Fully soaked gels should look colorless (otherwise, white colors may be seen). Add 10–20 μL more trypsin solution if it completely disappears. Soak again for ~20 min. Extra trypsin solution may be pipetted out afterward. Add 30–50 μL of ABC solution on top of gels and then put sample tubes into a 37 °C incubator for overnight digestion.

Acknowledgments

This work was supported in part by the grant NIH-1R01GM103598 (National Institutes of Health, National Institute of General Medical Sciences).

References

1. Brinkmann V (2004) Neutrophil extracellular traps kill bacteria. Science 303:1532–1535
2. Papayannopoulos V, Metzler KD, Hakkim A, Zychlinsky A (2010) Neutrophil elastase and myeloperoxidase regulate the formation of neutrophil extracellular traps. J Cell Biol 191 (3):677–691. https://doi.org/10.1083/jcb.201006052
3. Fuchs TA, Abed U, Goosmann C, Hurwitz R, Schulze I, Wahn V, Weinrauch Y, Brinkmann V, Zychlinsky A (2007) Novel cell death program leads to neutrophil extracellular traps. J Cell Biol 176(2):231–241. https://doi.org/10.1083/jcb.200606027
4. Papayannopoulos V, Zychlinsky A (2009) NETs: a new strategy for using old weapons. Trends Immunol 30(11):513–521. https://doi.org/10.1016/j.it.2009.07.011
5. Urban CF, Ermert D, Schmid M, Abu-Abed U, Goosmann C, Nacken W, Brinkmann V, Jungblut PR, Zychlinsky A (2009) Neutrophil extracellular traps contain calprotectin, a cytosolic protein complex involved in host defense against *Candida albicans*. PLoS Pathog 5(10):e1000639. https://doi.org/10.1371/journal.ppat.1000639
6. Thammavongsa V, Missiakas DM, Schneewind O (2013) *Staphylococcus aureus* degrades neutrophil extracellular traps to promote immune cell death. Science 342 (6160):863–866. https://doi.org/10.1126/science.1242255
7. Metzler Kathleen D, Goosmann C, Lubojemska A, Zychlinsky A, Papayannopoulos V (2014) A myeloperoxidase-containing complex regulates neutrophil elastase release and actin dynamics during NETosis. Cell Rep 8(3):883–896. https://doi.org/10.1016/j.celrep.2014.06.044
8. Schauer C, Janko C, Munoz LE, Zhao Y, Kienhöfer D, Frey B, Lell M, Manger B, Rech J, Naschberger E, Holmdahl R, Krenn V, Harrer T, Jeremic I, Bilyy R, Schett G, Hoffmann M, Herrmann M (2014) Aggregated neutrophil extracellular traps limit inflammation by degrading cytokines and chemokines. Nat Med 20(5):511–517. https://doi.org/10.1038/nm.3547
9. Yu Y, Kwon K, Tsitrin T, Bekele S, Sikorski P, Nelson KE, Pieper R (2017) Characterization of early-phase neutrophil extracellular traps in urinary tract infections. PLoS Pathog 13(1): e1006151. https://doi.org/10.1371/journal.ppat.1006151
10. Schaffer JN, Norsworthy AN, Sun T-T, Pearson MM (2016) *Proteus mirabilis* fimbriae- and

urease-dependent clusters assemble in an extracellular niche to initiate bladder stone formation. Proc Natl Acad Sci U S A 113 (16):4494–4499. https://doi.org/10.1073/pnas.1601720113

11. de Buhr N, von Kockritz-Blickwede M (2016) How neutrophil extracellular traps become visible. J Immunol Res 2016:4604713. https://doi.org/10.1155/2016/4604713
12. Kenny EF, Herzig A, Krüger R, Muth A, Mondal S, Thompson PR, Brinkmann V, Bernuth HV, Zychlinsky A (2017) Diverse stimuli engage different neutrophil extracellular trap pathways. eLife 6:e24437. https://doi.org/10.7554/eLife.24437
13. Liu S, Su X, Pan P, Zhang L, Hu Y, Tan H, Wu D, Liu B, Li H, Li H, Li Y, Dai M, Li Y, Hu C, Tsung A (2016) Neutrophil extracellular traps are indirectly triggered by lipopolysaccharide and contribute to acute lung injury. Sci Rep 6:37252. https://doi.org/10.1038/srep37252
14. Brinkmann V, Abu Abed U, Goosmann C, Zychlinsky A (2016) Immunodetection of NETs in paraffin-embedded tissue. Front Immunol 7(513):513. https://doi.org/10.3389/fimmu.2016.00513
15. Yu Y, Smith M, Pieper R (2014) A spinnable and automatable StageTip for high throughput peptide desalting and proteomics. Protocol Exchange. https://doi.org/10.1038/protex.2014.1033

Chapter 22

Using Proteomics to Identify Inflammation During Urinary Tract Infection

Yanbao Yu and Rembert Pieper

Abstract

Urinary tract infections (UTIs) are one of the most common bacterial infections. Conventional approaches to diagnose these infections rely on microbial urine culture, urine sediment microscopy and basic molecular urinalysis tests, in combination with assessments of patient symptoms that are indicative of UTI. The last decade has seen a more widespread clinical use of standardized MALDI-TOF methods to identify UTI-causing microbial agents. Shotgun proteomics methods to determine the extent of inflammation and types of immune cell effectors in urine have not become part of routine clinical tests. However, such methods are useful to investigate UTI pathogenesis, identify difficult-to-culture pathogens and understand antimicrobial effector mechanisms. The present chapter describes these approaches in order to gain quantitative and qualitative insights into inflammation and immune responses in patients with UTI and simultaneously profile the causative agents. The methods are also applicable to examine catheter-associated UTIs and vaginal infections from urine samples. Protocols provided here pertain to direct analyses of clinical specimens including urine sediments and urethral catheter biofilms.

Key words Proteomics, Inflammation, Urinary tract infection, Urine, Mass spectrometry, Metaproteomics, FASP, LC-MS/MS, Pathogen identification

1 Introduction

Microbial organisms colonize the female/male periurethral area and the female vaginal epithelium and perivaginal area. Some of these microbes, mostly bacteria, ascend into the urethra and bladder and trigger immune responses in urothelial mucosal tissues and are pathogenic. The combination of rapidly growing pathogens in the urinary tract and proinflammatory defensive responses by the human host constitute a urinary tract infection (UTI). UTIs are a common bacterial infection, affecting approximately 150 million people worldwide each year [1]. Inability of the immune system to prevent ascension of pathogens into the ureters and kidneys leads to more severe infections, including pyelonephritis and urosepsis.

Melanie M. Pearson (ed.), *Proteus mirabilis: Methods and Protocols*, Methods in Molecular Biology, vol. 2021, https://doi.org/10.1007/978-1-4939-9601-8_22,

Certain medical conditions (e.g., spinal cord injury and anatomical abnormalities in the urinary tract) may require catheters, inserted medical devices, to allow patients to void their bladders. The risk of bacterial colonization and UTI is much higher in long term-catheterized patients and associated with bacterial persistence in the urinary tract. Such chronic infections lead to frequent use of antibiotic drugs, increased morbidity and higher risk of complications including death when other comorbidities exist. The public health burden associated with treatment of complicated UTIs is high (e.g., over $400 million annually) [2]. The main infectious agents are Enterobacteriaceae (e.g., *Escherichia coli*, *Klebsiella pneumoniae*, and *Proteus mirabilis*). Gram-positive pathogens (e.g., *Enterococcus faecalis* and *Staphylococcus saprophyticus*) and rare pathogens adapted to the lifestyle in biofilms have also been reported as causes of UTI and catheter-associated UTI (CAUTI) [3]. Most bacteria that are able to adhere to the urothelial surface and metabolize and grow using nutrient sources in urine will trigger innate immune system responses. If these responses are effective, colonization of the bladder by the pathogen is contained, and the infection may resolve quickly, with or without UTI symptoms. UTI treated with antibiotics is often resolved within a week of the emergence of symptoms.

Urine culture, urine microscopy and specific molecular tests to assess inflammation derived from leukocytes (e.g., leukocyte esterase test) and tissue injury (test for hematuria) are used to identify microbial pathogens or measure the extent of leukocyte infiltration and vascular injury semiquantitatively from clinical urine specimens [4]. In the last decade, MALDI-TOF and software tools standardizing pathogen identification from TOF spectral data have been introduced and clinically approved for diagnostic use in hospitals [5]. Shortcomings of standard diagnostic tests are a lack of accuracy in pathogen identification and an inability to quantify innate immune responses. Shortcomings of MALDI-TOF methods are the dependence on prior microbial culture which introduces bias toward uropathogens that grow fast on typically used lab culture media [6].

We have introduced liquid chromatography–tandem mass spectrometry (LC-MS/MS) methods and subsequent metaproteomic database search strategies directly applied to urine sediment samples to aid in the diagnosis and examine the pathogenesis of UTI and asymptomatic bacteriuria (ASB). This approach has been used to survey the immune responses and identify pathogens as compared to standard urinalysis [7], show that neutrophil infiltration is a common immune response in cases of ASB [8], quantify neutrophil protein effector release into urine in UTI cases in support of the notion of extensive neutrophil degranulation [9], and detect neutrophil extracellular traps (NETs) as a distinct defense mechanism during UTI [10]. We have also adapted proteomic sample preparation methods for urinary pellets to 96-well filter plates [11]. Others have reported the use of similar methods to the metaproteomic analysis of cell extracts from urethral catheters, thus demonstrating pathogen-innate immune system interactions [12]. In parallel, metagenomic profiling has increased the knowledge of the complexity of microorganisms present in the urinary tract during health and disease. Although controversial, the term urinary microbiome has been introduced as both bacteria and viruses have been identified in the absence of overt disease in the urinary tract [13].

Here, we present protocols describing the metaproteomic analysis of urine sediment samples related to UTI and CAUTI. This includes procedures to derive information on host immune responses to infectious agents. The wet lab protocols involve microbial cell lysis from urine sediments, filter aided sample preparation (FASP) and shotgun proteomics (LC-MS/MS). Computational methods for metaproteomic database searches and label-free proteome quantitation are also provided.

2 Materials

All solvents and buffers to prepare samples for LC-MS/MS experiments should be made of LCMS grade water. Tips and tubes should be those recommended for proteomics (e.g., low binding or maximum recovery tubes) (*see* **Note 1**).

2.1 Protein Extraction from Urine Sediments

1. 30 kDa cutoff filter-aided sample preparation (FASP) filter (e.g., Sartorius Vivacon 500 filter device, or Millipore Microcon YM30).
2. Benchtop centrifuge (e.g., Eppendorf 5415R or equivalent).
3. Misonix Sonicator 3000 Ultrasonic Cell Disruptor (*see* **Note 2**).
4. Precast SDS-PAGE gel (e.g., NUPAGE 4–12%).
5. Centrifugal vacuum concentrator (e.g., SpeedVac dryer, Thermo Scientific).
6. Thermal microtube mixer (e.g., Eppendorf ThermoMixer).
7. 10 kDa cutoff centrifugal filter, 4 mL volume (e.g., Amicon Ultra-4, Millipore).
8. Gel staining buffer: 1.0 g of Coomassie brilliant blue G250 in 1.0 L of 50% methanol, 10% acetic acid, and 40% H_2O.
9. BSA standard: 1 μg/μL of bovine serum albumin in H_2O.

10. Urea buffer: 8 M urea in 0.1 M Tris–HCl, pH 8.1. UA buffer should be prepared freshly each day.
11. SED buffer: 1% SDS, 5 mM Na-EDTA, 50 mM DTT.
12. TMN buffer: 40 mM Tris–HCl, pH 8.1, 5 mM $MgCl_2$, and 100 mM NaCl.
13. 2× SDS-PAGE sample buffer: 4% SDS, 200 mM DTT, 100 mM Tris–HCl (pH 6.8) and 0.2% bromophenol blue.
14. IAA solution: 0.05 M iodoacetamide in urea buffer.
15. ABC buffer: 50 mM ammonium bicarbonate in water.
16. Trypsin solution: 0.1 μg/μL of trypsin (sequencing grade, Promega) in ABC buffer (*see* **Note 3**).
17. CHO buffer: 100 mM sodium acetate (pH 5.5), 20 mM sodium metaperiodate, and 300 mM NaCl.

2.2 StageTip Desalting and LC-MS/MS

1. Empore C_{18} Extraction disk (3 M).
2. Pipette tip adaptors (The Nest Group, Inc., Southborough, MA).
3. Activation buffer: 100% methanol.
4. Wash and equilibration buffer: 0.5% acetic acid in water.
5. Elution buffer I: 0.5% acetic acid, 60% acetonitrile and 40% water.
6. Elution buffer II: 0.5% acetic acid, 80% acetonitrile, and 20% water.
7. LC solvent A: 0.1% formic acid in water.
8. LC solvent B: 0.1% formic acid in acetonitrile.
9. Trap column: Acclaim PepMap100, C_{18}, 5 μm, 100 Å, 100 μm × 2 cm.
10. Analytical column: PicoFrit, 75 μm × 10 cm, 5 μm BetaBasic C_{18}, 150 Å.
11. LC-MS/MS workstation (e.g., Ultimate 3000 nLC).
12. High-resolution mass spectrometer (e.g., Q-Exactive, Thermo Scientific).
13. MS data analysis software, such as Proteome Discoverer (version 1.4, Thermo Scientific) and MaxQuant/Perseus software suite (freely available at http://www.coxdocs.org/doku.php?id=maxquant:start).

3 Methods

3.1 Protein Extraction from Urine Sediments and Urethral Catheters

3.1.1 Extraction from Urine Sediments

1. Obtain urine specimen (*see* **Notes 4** and **5**).
2. Centrifuge at 3000 × *g* for 15 min to separate urine sediment (urinary pellet, UP) and urine supernatant (SU) fractions (*see* **Note 6**). Both fractions may be collected, but the main focus of this chapter is on the metaproteomic analysis of UP samples.
3. The UP sample may be stored at −80 °C or can be directly processed.
4. Add a sixfold volume of SED buffer (*see* **Note** 7) to the UP sample. If UP sample size is small (<20 μL), a minimum of 100 μL of buffer should be added.
5. Vortex vigorously to resuspend. Let the UP suspension sit for 10 min and vortex occasionally.
6. Sonicate the UP suspension in a Misonex ice–water bath at amplitude 6 in approximately ten 30 s on/off cycles (*see* **Notes 8** and **9**).
7. Let the UP lysate sit for 10 min and vortex occasionally.
8. Move the UP lysate to a heat block with boiling water and keep the lysate at a minimum of 95 °C for 3 min to continue cell lysis.
9. Cool down and let the UP lysate sit for 10 min; vortex occasionally.
10. Centrifuge the UP lysate at maximum speed (16,000 × *g*) for 15 min in a bench-top microtube centrifuge.
11. Transfer UP lysate supernatant to a clean microtube and discard the lysate pellet.
12. Freeze the UP lysate at −80 °C prior to use for protein analysis by SDS-PAGE (*see* Subheading 3.2) and on-filter digestion (*see* Subheading 3.3).

3.1.2 Extraction from Urethral Catheters

1. Collect a urinary catheter and cut into ~1 in. pieces with a sterile scalpel (*see* **Notes 4** and **10**).
2. Use tweezers to transfer pieces to a −80 °C freezer for storage.
3. Equilibrate a catheter piece to 20 °C.
4. Submerge the catheter piece in a polypropylene tube with 2–3 mL of CHO buffer. Use enough buffer to cover the segment.
5. Vortex for 1 min.
6. Incubate in an ultrasonic water bath for 10 min. The biofilm will begin to detach from the catheter surface. Scraping the catheter surface with a plastic spatula may increase recovery of biofilm materials.

7. Repeat **steps 5** and **6** once. Discard the catheter piece.
8. Centrifuge the tube at 3000 × *g* for 15 min to generate a catheter biofilm supernatant (CB_{sup}) and a catheter biofilm pellet (CB_{pel}). Retain both fractions.
9. Subject the CB_{pel} fraction to **steps 4–11** in Subheading 3.1.1.
10. Freeze the CB_{pel} lysate at −80 °C prior to use for protein analysis by SDS-PAGE (*see* Subheading 3.2) or on-filter digestion (*see* Subheading 3.3).
11. If the CB_{sup} fraction will be visualized using SDS-PAGE, proceed to Subheading 3.2.
12. Concentrate the CB_{sup} fraction to approximately 0.5 mL in a 4 mL volume, 10 kDa MWCO centrifugal filter unit via centrifugation at 5000 × *g* for 10 min.
13. Exchange the CB_{sup} fraction into 0.5 mL of PBS. This soluble protein fraction is prepared for proteomic analysis following an on-filter digestion protocol (*see* Subheading 3.3).

3.2 Visualization of Proteins in Lysates by SDS-PAGE

1. Thaw UP and CB_{pel} lysates and CB_{sup} supernatants (derived from **step 8** in Subheading 3.1.2). Add an equal aliquot of SDS-PAGE sample buffer and heat at 95 °C for 3 min.
2. Run SDS-PAGE using 5–10 μL aliquots of the UP lysate, the CB_{pel} lysate and/or the CB_{sup} soluble fraction. To estimate total protein concentration in these samples, load known amounts of BSA (1 μg, 2 μg and 5 μg) in adjacent SDS-PAGE lanes.
3. Stain the gel with Coomassie Brilliant Blue G250 and destain the gel following standard procedures [14].
4. From the overall intensity of gel bands, compared with the BSA standard band intensity, estimate the total protein amount for the set of UP and CB_{pel} lysates and CB_{sup} supernatants in the SDS-PAGE gel.

3.3 On-filter Digestion and StageTip Desalting of Lysates

1. Aliquot a volume equivalent to 20–50 μg of total protein into a FASP filter, and add 200 μL of urea buffer (*see* **Note 11**).
2. Spin at 10,000 × *g* for 15–20 min.
3. Repeat **step 2** one more time. The final volume should be reduced to less than 50 μL.
4. Add 100 μL of IAA solution into the filter and gently mix.
5. Incubate the sample in the dark for 20 min.
6. Centrifuge at 10,000 × *g* for 15–20 min to remove IAA solution.
7. Wash the filter once with 200 μL of urea buffer. Repeat **step 6** to remove extra IAA.

8. Wash the filter once with 200 µL of ABC buffer. Repeat **step 6**.
9. Transfer the filter to a new collection tube.
10. Add 150 µL of ABC buffer and 10 µL of trypsin solution (1 µg of trypsin) to the filter.
11. Incubate the filter in a thermal mixer overnight at 37 °C with gentle shaking (850 rpm).
12. To elute peptides from the filter, add 200 µL of ABC solution, spin, and collect flow through to a new tube. Repeat this step three times.
13. Dry the solution using a vacuum drier.
14. Desalt peptides using a C_{18}-based StageTip protocol [15] (*see* **Note 12**). Briefly, pack the C18 membrane into 200-µL pipette tips using a blunt-ended needle and a plunger.
15. Activate the tips in two steps, first with 200 µL of 100% methanol and then with 0.5% acetic acid in 80% acetonitrile and 20% water. The tips are now ready for peptide loading.
16. Resuspend dried peptide samples from **step 13** into 200 µL of wash and equilibration buffer.
17. Vortex for 10 min.
18. Spin in a microcentrifuge at maximal speed (16,000 × *g*) for 2 min.
19. Load samples onto StageTip, and spin at 1,200 × *g* for 2–3 min.
20. Collect the flow through and reload the StageTip three times.
21. Wash with 200 µL of wash and equilibration buffer.
22. Elute peptides with 200 µL each of Elution Buffer I and II, consecutively.
23. Dry peptides in a vacuum dryer.
24. Store samples at −80 °C for future use, or resuspend in 20 µL of LC solvent A for immediate LC-MS/MS analysis (*see* Subheading 3.4).

3.4 Protein Identification and Quantitation Using LC-MS/MS

1. Inject 5–10 µL of each sample into the LC-MS/MS system.
2. Perform LC-MS/MS analysis with an Ultimate 3000 nLC and Q Exactive mass spectrometer using a 150-min gradient (0–115 min, 2–35% B; 115–120 min, 35–80% B; 120–125 min, 80% B; 125–130 min, 80-2% B; 130–150 min, 2% B).
3. Acquire data on a Q Exactive mass spectrometer using a data-dependent mode. The full MS scan may be acquired at a resolution of 70,000 over a mass range of *m/z* 250 to 1800 with an automatic gain control (AGC) target of 10^6. Up to ten of the most abundant ions (Top10 method) will be subjected

to fragmentation by higher energy collisional dissociation (HCD) with a normalized collision energy of 27%.

4. Build a protein database by looking up common uropathogenic microbes in recent literature, and downloading their sequences from the UniProt Knowledgebase (http://www.uniprot.org/) (Table 1) (*see* **Notes 13** and **14**).
5. For protein database searches, use the Proteome Discoverer software (version 1.4) for general proteome identification, and MaxQuant-Andromeda software suite for proteome quantitation.
6. In MaxQuant software, apply default parameters to perform proteome quantitation.
7. Enable the LFQ (label-free quantitation) and iBAQ (intensity-based absolute quantitation) options embedded in the software. The LFQ data provide intensity of the integrated MS^1 peak area, and this is the relative protein quantification across all samples [16]. The iBAQ data are the sum of peak intensities of all the peptide for a given protein. It represents the approximate abundance of a protein within a given proteome.
8. In Proteome Discoverer software, define common search parameters such as enzyme specificity, missed cleavage site and modifications, and then perform a database search using the customized database as described in **steps 4** and **5**. The spectral counting method, according to the number of peptide-spectrum matches (PSMs) derived from Proteome Discover, provides a relative estimation of protein abundance [17] (Fig. 1).
9. Based on the iBAQ quantitation of individual proteins in **step 6**, systematically group proteins of interest in functional or cellular categories, such as innate immune response (Fig. 2) (*see* **Note 15**).

Figure 1 shows several metaproteomic profiles derived from the analyses of CB_{sup} and CB_{pel} samples (CAUTI) and UP samples (UTI). The proteomic profiles of catheter extracts can have a more diverse microbial species composition than UTI specimens. The latter typically contain a single, highly prevalent bacterial pathogen (Fig. 1, samples #3 and #4). The surveyed microbial proteome profiles for a given sample are indicative of the presence of one (or more) pathogens in the original urine sediment, and therefore serve as guidance to assess the extent of possible inflammation in the urinary tract in a distinct case (e.g., UTI or CAUTI).

However, Fig. 1 depicts only the sum of all human proteins and does not reveal the identity and function of human proteins present in a dataset. The quantitative contribution of common functional or cellular categories to the overall human proteome allows an

Table 1
Urinary metaproteome database

Species (strain)	# Annotated proteins	Taxon identifier
Acinetobacter baumannii BIDMC 57	3764	1439317
Actinobaculum massiliense ACS-171-V-Col2	1696	883066
Aerococcus urinae (strain ACS-120-V-Col10a)	1685	866775
Bacteroides fragilis (strain 638R)	4285	862962
Bifidobacterium scardovii	2500	158787
Bordetella hinzii	4574	103855
Candida albicans (strain WO-1)	5742	294748
Citrobacter koseri (strain ATCC BAA-895/CDC 4225-83 / SGSC4696)	5020	290338
Enterococcus faecium 10/96A	3261	1391465
Escherichia coli (strain UTI89/UPEC)	5199	364106
Gardnerella vaginalis (strain ATCC 14019/317)	1366	525284
Globicatella sp. HMSC072A10	2027	1739315
Klebsiella pneumoniae subsp. pneumoniae (strain ATCC 700721/ MGH 78578)	5127	272620
Lactobacillus gasseri (strain ATCC 33323/DSM 20243)	226	324831
Morganella morganii subsp. *morganii* KT	3510	1124991
Mycoplasma genitalium (strain ATCC 33530/G-37/NCTC 10195)	484	243273
Prevotella melaninogenica D18	2461	575612
Proteus mirabilis (strain HI4320)	3662	529507
Providencia stuartii (strain MRSN 2154)	4221	1157951
Pseudomonas aeruginosa (strain PA7)	6247	381754
Staphylococcus aureus (strain USA300/TCH1516)	420	451516
Staphylococcus saprophyticus subsp. *saprophyticus* (strain ATCC 15305/DSM 20229)	2408	342451
Veillonella parvula (strain ATCC 10790/DSM 2008/JCM 12972/ Te3)	1846	479436
Homo sapiens (reviewed)	20,411	9606

The database was built based on prior knowledge and our own findings (*see* **Note 13**). The number of protein sequences annotated for each proteome (as of October 2018) is listed. The taxonomy identifier can be used to download the latest protein sequences from UniProt Knowledgebase

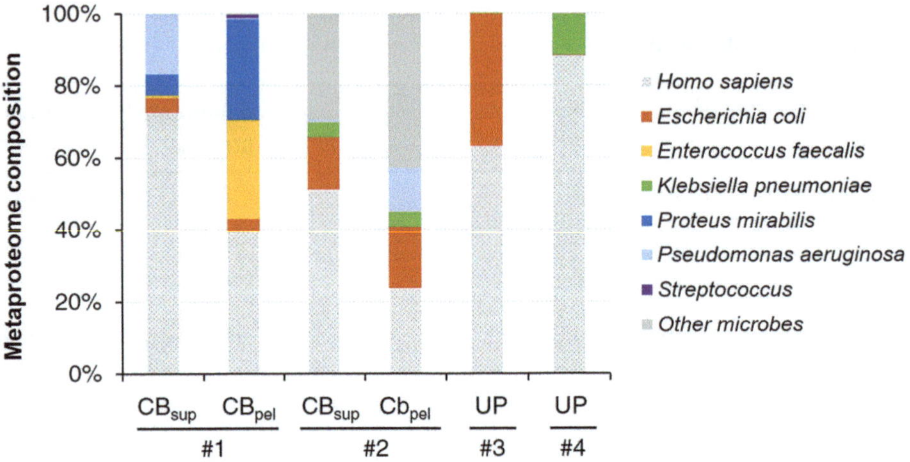

Fig. 1 Composition of metaproteomes derived from clinical urine specimens. Two fractions derived from catheter biofilms (CB_{sup} and CB_{pel}), each from two human subjects (#1, #2) diagnosed with CAUTI, and UP samples from two human subjects (#3, #4) diagnosed with UTI are illustrated. The contributions of microbial pathogens are based on peptide spectrum match (PSM) data using Proteome Discoverer software. In subject #2, microbes present in smaller relative quantities at the species level are combined into an "other microbe" segment

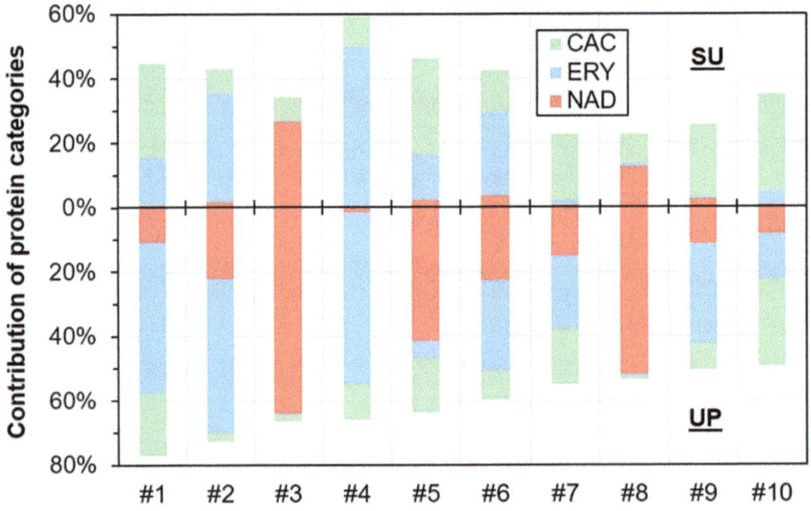

Fig. 2 Contribution of proteins representing three human cellular or functional categories to the human proteome profiled from urine samples with evidence of UTI. Both supernatant urine (SU) and urinary pellet (UP) fractions were analyzed. The iBAQ intensity of three categories of proteins, including *n*eutrophil *a*ctivation and *d*egranulation (NAD), *c*omplement system *a*ctivation and *c*oagulation (CAC) and *e*r*y*throcytes that may leak into urine due to vascular injury (ERY), were summed and divided by the total intensity of all human proteins to represent their mass contribution to the total proteome. Their individual contributions to the entire human proteome were plotted, with SU samples above the *x*-axis, and UP samples below

assessment of the extent and the components of an inflammatory response to the pathogen in a case of UTI and CAUTI. For example, highly abundant proteins with functions in inflammatory and innate immune responses yield the first stage of information on the extent of antimicrobial immune responses, such as calprotectin, immunoglobulins, and complement factors. A reasonable approach is to compare these contributions from samples with other evidence of infection versus those that do not. Figure 2 juxtaposes the contribution of three protein groups to UP samples and respective SU fractions derived from 10 cases of UTI.

To identify microbial species from any clinical specimen with a suspected infection, an unbiased database search approach using a pan-microbial protein database is attractive. However, the computing power usually is massive in order to search shotgun proteomics datasets using very large "pan-proteome" sequence databases. Therefore, using a conventional search algorithm such as Sequest HT in conjunction with a more limited database (23 microbes; 90,000 protein sequences) has been a compromise resulting in database search completion in a few hours for a typical size LC-MS/MS shotgun proteomics dataset. Recently, Liao and coworkers developed a web platform [18] that can perform protein searches against the entire human gut microbiome (9.8 million protein sequences). Clearly, more productive computational search strategies and algorithms are imminent and promise to advance the power of complex multiorganismal metaproteomic studies.

4 Notes

1. We recommend "Low Retention Pipet Tips" from Denville Scientific, Inc. and "Maxymum Recovery" tubes from Axygen, Inc.
2. To facilitate cell lysis, use a Misonex 3000 Sonicator (or equivalent instrument) with the water bath connection, not the probe connection.
3. Store aliquots of trypsin solution at −80 °C.
4. An institutional review board (IRB) may need to review and approve collection of such samples. The IRB may grant an exemption if the specimens are considered medical waste (i.e., the specimens were collected for a different purpose and are deidentified).
5. The urine sample collection should follow standard protocol (e.g., second morning, middle stream and clean catch). Minimize contamination during the collection of urine specimens, whether from a naturally voided sample or an aliquot from urine collected in a patient's catheter bag. Naturally voided

urine should be clean catch. Catheter bag urine should be from a sterile catheter port swiped clean with alcohol. The urine specimen may be stored for up to 6 h at 4 °C.

6. For proteomics experiments, we usually start with at least 5 mL of urine, and no maximum limit. If a very small volume (<2 mL) of urine was collected, first concentrate the sample using a 10-kDa MWCO centrifugal filter (e.g., Amicon Ultra-15), then proceed with direct on-filter digestion (large-scale FASP) [19].
7. In the case that Gram-positive bacteria with thick cell walls, such as *Streptococcus pneumoniae* and *Staphylococcus aureus*, are suspected to be present in urine samples, resuspend the urinary pellet samples in a volume of TMN buffer to have an approximate volume ratio of 1:10, usually less than 200 μL. Pipette the suspension up and down a few times in a 1.5 mL tube. Add lysostaphin and mutanolysin to a final concentration of 20 μg/mL. Mix gently to homogenize the enzymes and suspended cells. Incubate the sample in a 37 °C shaker-incubator. Take out the tube out every 30 min and check if a pellet collects at the bottom. If so, briefly vortex every 30 min. Complete digestion after a 3 h incubation.
8. Put sample tube into an ice–water bath. Set the program on amplitude 6 for six 20-s cycles with 20-s intermittent cooling steps.
9. For bacterial cells that are tough to lyse, such as Gram-positive microbes, additional lysis steps may be applied to the pellet to ensure that microbial organisms with robust cell walls are lysed. These customized lysis steps pertain to the digestion of microbial cell walls with enzymes specific for either fungal and Gram-positive bacterial organisms (e.g., lysozyme, mutanolysin, lysostaphin, and zymolase; *see also* **Note** 7). Use additional PBS wash steps to remove denaturing chemicals, choose a buffer optimal for enzyme activity, and proceed using residual cellular materials in **step 5** of Subheading 3.1.1.
10. For sample integrity, it is best to proceed with the processing of urine sediment and catheter samples as fast as possible. The frozen samples are suitable for proteomic sample preparation. These processing steps are suited to recover viable bacteria, but anaerobic bacteria and those sensitive to freeze–thaw cycles may have low viability when thawed.
11. Urea-containing buffers must be prepared freshly before use in the FASP protocol.
12. This is a modified and optimized protocol for our experiments. By using a pipette tip adaptor, we were able to process samples using a benchtop centrifuge instead of a manual syringe as described before [20]. This way, the processing can be

multiplexed and sped up. More details about the spinnable StateTip can be found here [15].

13. For instance, *Proteus mirabilis* strain HI4320 has a unique taxonomy identifier of 529,507, which can be found in the NCBI Taxonomy Browser (https://www.ncbi.nlm.nih.gov/taxonomy), and the number of protein sequences is 3662 (as of October 2018). Prior knowledge of the microbial composition of an unknown urine sample, such as 16S rRNA sequencing information, will help customize the meta-database. To facilitate future UTI-metaproteomics studies, we provide here a source table of usual as well as rare UTI pathogens, including their taxonomy IDs and the number of protein sequences (Table 1). Considering that a set of 20 to 25 pathogens includes causative agents of more than 95% of all UTI and CAUTI cases globally, it is unlikely that identification of a pathogen is missed in a metaproteomic analysis due to the absence of that organism in the search database.
14. Ideally, for a general metaproteomics analysis of urine specimens with unknown microbial composition, an unbiased search against a pan-microbial database or the entire UniProt database (>120 million sequences) is desired. Because many currently available search engines could not handle the overwhelming computational workload in this regard, the recently developed metaproteomics database search framework ProteoStorm can achieve a two-to-three orders-of-magnitude speedup over popular search engines [21].
15. For example, neutrophil elastase, azurocidin, and cathepsin G are highly enriched in neutrophils. Thus, proteins specifically enriched in a specific cell type (e.g., neutrophils) or emerging at the site of infection via secretory pathways (e.g., the complement system) or released into urine upon vascular injury (e.g., hemoglobin and erythrocyte-specific membrane proteins) can be separately categorized.

Acknowledgments

This work was supported in part by the grant NIH-1R01GM103598 (National Institute of General Medical Sciences).

References

1. Flores-Mireles AL, Walker JN, Caparon M, Hultgren SJ (2015) Urinary tract infections: epidemiology, mechanisms of infection and treatment options. Nat Rev Microbiol 13 (5):269–284. https://doi.org/10.1038/nrmicro3432
2. Jacobsen SM, Stickler DJ, Mobley HLT, Shirtliff ME (2008) Complicated catheter-associated urinary tract infections due to *Escherichia coli* and *Proteus mirabilis*. Clin Microbiol Rev 21(1):26–59. https://doi.org/10.1128/cmr.00019-07

3. Mohan B, Zaman K, Anand N, Taneja N (2017) *Aerococcus viridans*: a rare pathogen causing urinary tract infection. J Clin Diagn Res 11(1):DR01–DR03. https://doi.org/10.7860/JCDR/2017/23997.9229
4. Schmiemann G, Kniehl E, Gebhardt K, Matejczyk MM, Hummers-Pradier E (2010) The diagnosis of urinary tract infection: a systematic review. Dtsch Arztebl Int 107 (21):361–367. https://doi.org/10.3238/arztebl.2010.0361
5. Rahi P, Prakash O, Shouche YS (2016) Matrix-assisted laser desorption/ionization time-of-flight mass-spectrometry (MALDI-TOF MS) based microbial identifications: challenges and scopes for microbial ecologists. Front Microbiol 7:1359. https://doi.org/10.3389/fmicb.2016.01359
6. Carbonnelle E, Mesquita C, Bille E, Day N, Dauphin B, Beretti J-L, Ferroni A, Gutmann L, Nassif X (2011) MALDI-TOF mass spectrometry tools for bacterial identification in clinical microbiology laboratory. Clin Biochem 44(1):104–109. https://doi.org/10.1016/j.clinbiochem.2010.06.017
7. Yu Y, Sikorski P, Bowman-Gholston C, Cacciabeve N, Nelson K, Pieper R (2015) Diagnosing inflammation and infection in the urinary system via proteomics. J Transl Med 13 (1):111. https://doi.org/10.1186/s12967-015-0475-3
8. Yu Y, Zielinski MD, Rolfe MA, Kuntz MM, Nelson H, Nelson KE, Pieper R (2015) Similar neutrophil-driven inflammatory and antibacterial responses in elderly patients with symptomatic and asymptomatic bacteriuria. Infect Immun 83(10):4142–4153. https://doi.org/10.1128/iai.00745-15
9. Yu Y, Sikorski P, Smith M, Bowman-Gholston-C, Cacciabeve N, Nelson KE, Pieper R (2017) Comprehensive metaproteomic analyses of urine in the presence and absence of neutrophil-associated inflammation in the urinary tract. Theranostics 7(2):238–252. https://doi.org/10.7150/thno.16086
10. Yu Y, Kwon K, Tsitrin T, Bekele S, Sikorski P, Nelson KE, Pieper R (2017) Characterization of early-phase neutrophil extracellular traps in urinary tract infections. PLoS Pathog 13(1): e1006151. https://doi.org/10.1371/journal.ppat.1006151
11. Yu Y, Suh MJ, Sikorski P, Kwon K, Nelson KE, Pieper R (2014) Urine sample preparation in 96-well filter plates for quantitative clinical proteomics. Anal Chem 86(11):5470–5477. https://doi.org/10.1021/ac5008317
12. Lassek C, Burghartz M, Chaves-Moreno D, Otto A, Hentschker C, Fuchs S, Bernhardt J, Jauregui R, Neubauer R, Becher D, Pieper DH, Jahn M, Jahn D, Riedel K (2015) A metaproteomics approach to elucidate host and pathogen protein expression during catheter-associated urinary tract infections. Mol Cell Proteomics 14(4):989–1008. https://doi.org/10.1074/mcp.M114.043463
13. Magistro G, Stief CG (2018) The urinary tract microbiome: the answer to all our open questions? Eur Urol Focus 5(1):36–38. https://doi.org/10.1016/j.euf.2018.06.011; pii: S2405-4569(18)30159-7
14. Shevchenko A, Tomas H, Havlis J, Olsen JV, Mann M (2007) In-gel digestion for mass spectrometric characterization of proteins and proteomes. Nat Protoc 1(6):2856–2860
15. Yu Y, Smith M, Pieper R (2014) A spinnable and automatable StageTip for high throughput peptide desalting and proteomics. Protocol Exchange. https://doi.org/10.1038/protex.2014.1033
16. Tyanova S, Temu T, Cox J (2016) The MaxQuant computational platform for mass spectrometry-based shotgun proteomics. Nat Protoc 11(12):2301–2319. https://doi.org/10.1038/nprot.2016.136
17. Lundgren DH, Hwang SI, Wu L, Han DK (2010) Role of spectral counting in quantitative proteomics. Expert Rev Proteomics 7 (1):39–53. https://doi.org/10.1586/epr.09.69
18. Liao B, Ning Z, Cheng K, Zhang X, Li L, Mayne J, Figeys D (2018) iMetaLab 1.0: a web platform for metaproteomics data analysis. Bioinformatics 34(22):3954–3956. https://doi.org/10.1093/bioinformatics/bty466
19. Wiśniewski JR (2018) Filter-aided sample preparation for proteome analysis. In: Becher D (ed) Microbial proteomics: methods and protocols. Springer, New York, NY, pp 3–10. https://doi.org/10.1007/978-1-4939-8695-8_1
20. Rappsilber J, Mann M, Ishihama Y (2007) Protocol for micro-purification, enrichment, pre-fractionation and storage of peptides for proteomics using StageTips. Nat Protoc 2 (8):1896–1906
21. Beyter D, Lin MS, Yu Y, Pieper R, Bafna V (2018) ProteoStorm: an ultrafast metaproteomics database search framework. Cell Syst 7 (4):463–467. https://doi.org/10.1016/j.cels.2018.08.009

Chapter 23

Assessment of *Proteus mirabilis* Antigen Immunological Complexes by Atomic Force Microscopy

Wiesław Kaca, Joanna Gleńska-Olender, Iwona Konieczna, Józef Gawęda, and Sławomir Sęk

Abstract

Atomic force microscopy (AFM) is being increasingly used to directly measure protein interactions in nearly physiological environments. Here, protocols for atomic force microscopy (AFM) for visualization of antigen–antibody complexes are presented. The technique is used to demonstrate complexes formed by rheumatoid arthritis patient antibodies with lipopolysaccharide (LPS) isolated from *P. mirabilis* (O3) strain S1959 and a synthetic antigen (LPS epitope of 6 N-alpha-(D-galacturonoyl)-L-lysine residues).

Key words Atomic force microscopy, *Proteus mirabilis*, Lipopolysaccharide, Rheumatoid arthritis antibodies

1 Introduction

Analysis of biomolecular interactions is a key area of research in the fields of medicine, biology, and biotechnology. The most commonly used elements for molecular recognition are antibodies that play a key role in the immune system [1]. They recognize and specifically bind antigens, like the major component of Gram-negative bacterial lipopolysaccharides (LPS, endotoxin). A number of analytical methods have been developed for the detection of antibodies against LPS, such as immunoenzymatic approaches including ELISA, Western blot, and immunofluorescence [2]. However, immunoenzymatic methods indirectly detect antigen–antibody complexes and may end up with false positive/negative results [3, 4].

Atomic force microscopy (AFM) offers direct visualization of antigen–antibody complexes and enables a better understanding of the binding mechanisms at the molecular level [5], under conditions close to the physiological environment [6]. The application of AFM allows measuring the surface topography of

Melanie M. Pearson (ed.), *Proteus mirabilis: Methods and Protocols*, Methods in Molecular Biology, vol. 2021, https://doi.org/10.1007/978-1-4939-9601-8_23,

immunocomplexes and provides information on the size and shape of the structures occurring on the surface of the tested sample [7].

Here we present the application of AFM for assessment of immunocomplexes, using *P. mirabilis* (O3) S1959 LPS (antigen) with rheumatoid arthritis (RA) patient antibodies as an example. The O-polysaccharide part of O3 LPS is characterized by the presence of a noncarbohydrate constituent, L-lysine linked to the carboxyl group of D-galacturonic acid (Lys-GalA) [8, 9]. Lys-GalA is the serological epitope that binds both rabbit as well as RA patient antibodies [9, 10]. For precise confirmation of the role of the O-specific polysaccharide part of *P. mirabilis* O3 (S1959), AFM is used to examine an LPS synthetic epitope of 6 N-alpha-(D-galacturonoyl)-L-lysine residues copolymerized to polyacrylamide (PAA).

2 Materials

2.1 Reagents

1. Protein A IgG purification kit (e.g., Thermo Scientific).
2. Phosphate-buffered saline (PBS) buffer pH 7.4 (e.g., Sigma).
3. Piranha solution: concentrated H_2SO_4 and 30% H_2O_2 3:1 v/v.
4. Milli-Q (>18.2 MΩcm) ultrapure water.

2.2 Equipment

1. Atomic Force Microscope (AFM) 5500AFM instrument (Agilent Technologies Santa Clara, CA, USA).
2. Cantilevers (MAClevers Type VII, Keysight Technologies Santa Rosa, CA, USA)/PNP-DB silicon nitride cantilevers (NanoWorld AG) with a nominal spring constant of 0.06 N/m.
3. Mica-type plates (Ted Pella, Inc).
4. Automatic pipette.
5. Vortex.
6. Ultrasonic water bath.

2.3 Human Serum

A suitable antibody source is required for analysis of antigen–antibody complexes.

1. Studies using human serum must follow ethical guidelines and be approved by the local medical ethics committee.
2. Use ELISA to identify specific sera that have a strong reaction with the target antigen.

Here, selected RA patient serum is presented as an example, because it contains antibodies that have a strong reaction with *P. mirabilis* O3 (S1959) LPS, as detected by ELISA [10].

2.4 Antigens

Any type of purified LPS might be used as antigens for binding sera antibodies in immunochemical reactions. Here, smooth LPS from *P. mirabilis* (O3) strain S1959 was extracted by the Westphal phenol–water method and purified as described in [8, 11]. Briefly:

1. For LPS samples, cultivate bacterial strains in nutrient broth supplemented with 1% glucose.
2. Harvest bacterial cells at the end of the logarithmic phase of growth.
3. Kill bacteria using a 1% phenol solution in water, and centrifuge (3000 × *g*, 20 min 10 °C) three times with distilled water.
4. Lyophilize the bacterial pellet using a freeze dryer.
5. Mix 20 g of lyophilized bacterial cells with 600 mL of distilled water, and immerse in a glass container in a 65–67 °C water bath for 30 min. Mix until the suspension becomes homogenous.
6. Add 300 mL of 90% phenol (prepared from freshly prepared distilled crystalline phenol), and continue the extraction for 30 min at 65 °C. Cool the mixture in an ice-water bath to approximately 10 °C, then centrifuge for 30 min at 300 × *g*.
7. Separate the aqueous, milky upper phase containing crude LPS. Repeat the extraction on the sediment (phenol phase) two more times.
8. Pool the three water phases into a dialysis bag and, in a refrigerated room, dialyze against tap water for 2 days and in distilled water for 2 more days.
9. Concentrate dialyzed material by vacuum distillation. Purify a 1% solution of crude smooth and rough LPS in water from nucleic acid and proteins by ultracentrifugation (105,000 × *g*, 120 min). Repeat centrifugation two more times.
10. Remove nucleic acid contamination by treating 2 mg/g LPS with RNase and DNase in 10 mM Tris–HCl buffer (pH 7.6) for 18 h at room temperature. To prevent bacterial growth, add one drop of toluene.
11. Dialyze overnight against distilled water. Sediment LPS by ultracentrifugation (105,000 × *g*, 120 min) and lyophilize. This LPS preparation may contain up to 2% of proteins and a trace of nucleic acids (tested by absorption at 260 nm).

Alternatively, antigen may be synthesized. For example, a Lys-GalA-PAA antigen was synthesized by copolymerizing 2-acrylamidoethyl α-glycoside amides of D-GalA with lysine, representing a partial structure of *P. mirabilis* (O3) S1959 LPS O-polysaccharide, with polyacrylamide (PAA) [12].

2.5 Software

1. Pico Image modular AFM and SPM imaging and analysis post-processing software, for analysis of root-mean-square (RMS) surface roughness.

3 Methods

3.1 Purification of Serum to IgG Antibody Fraction

Purify the IgG antibody fraction from serum with a Protein A IgG purification kit, using the manufacturer's instructions (*see* **Note 1**). The protocol for this column-based purification is briefly described here.

1. Equilibrate the column by adding 3–5 mL of binding buffer.
2. Apply 5 mL of patient serum to the column with the bed and start collecting the flow-through into tube #1.
3. When the serum level has equilibrated with the upper surface of the bed, wash the column with 10 mL of binding buffer (in aliquots).
4. Collect the first 2 mL of the column flow-through into tube #2.
5. After washing the layers with column buffer, elute the immunoglobulin fraction by adding 5–10 mL of elution buffer.
6. Collect eluate, in 1 mL aliquots, into separate tubes.
7. Use a spectrophotometer to measure the absorbance at 280 nm to determine which fraction contains antibodies; use a mixture of 200 μL of elution buffer plus 20 μL of binding buffer as a blank. A reading of 2.48 at 280 nm indicates the tubes with IgG class antibodies.
8. Freeze IgG class fractions at −20 °C.

3.2 Sample Preparation

1. Prepare a solution of 0.5 mg/mL of the synthetic Lys-GalA-PAA or *P. mirabilis* (O3) S1959 LPS (*see* **Note 2**) antigens in PBS.
2. Prepare a solution of 0.5 mg/mL of purified RA patient serum in PBS.
3. These are the antigen and serum solutions used in Subheading 3.3, **steps 8** and **9**.

3.3 Atomic Force Microscopy (AFM)

1. Record all AFM images using Magnetic AC mode—MAC (*see* **Note 3**). Use silicon probes with a thin magnetic coating on the back side of the cantilever during imaging experiments.
2. Use the MAC levers VII with the nominal spring constant of 0.14 N/m and the resonance frequency in the range of 18–24 kHz. However, before each experiment calibrate the cantilevers using a thermal tune method, in order to obtain exact values of the cantilever spring constant.

3. Before experiments, clean the standard MAC mode fluid cell in piranha solution for 30 min and rinse with ultrapure water.
4. Before surface roughness analysis, optimize the AFM image quality by repetitive scanning of the sample.
5. Use a line by line leveling method to remove tilt from sample images.
6. Perform matrix filtering to erase noise and possible distortions.
7. Prepare samples by incubation of freshly cleaved mica discs with PBS aqueous solution at room temperature (*see* **Note 4**).
8. At the beginning of the study of surface topography using AFM, perform a scan of a clean mica disc immersed in an aqueous solution of PBS (600 μL).
9. Apply the test antigen to the mica disc (600–700 μL) and leave for about 40 min until the antigen adsorbs to the surface of mica (*see* **Note 5**).
10. Before AFM, rinse samples with PBS and equilibrate for 40 min in the AFM cell. Once the mica surface is covered with antigen, add about 5 μL of the purified serum IgG fraction, and scan the disc with the antigen–antibody complex.
11. Collect AFM images at room temperature in PBS buffer solution.
12. To begin imaging, follow these steps (*see* **Note 6**):
    (a) Insert the nose assembly into the scanner.
    (b) Insert a probe into the nose assembly.
    (c) Place the scanner in the microscope and connect its cables.
    (d) Align the laser on the cantilever.
    (e) Insert and align the detector.
    (f) Prepare the sample and mount the sample plate.
13. Adjust the video system to focus on the cantilever.
14. Use the manual screws for coarse approach.
15. Use the Close switch on the Head Electronic Box (HEB) for a final approach to bring the tip close to, but not touching, the sample.
16. In PicoView, choose Mode > ACAFM.
17. Choose Controls > AC Mode to open the ACAFM Controls window.
18. Set the Drive Mechanism to AAC.
19. Set the Drive% to 10%. This is the amplitude of the AC drive signal, stated as a percentage (0–100%) of the maximum available 10 V.

20. In the Servo window, set the Set point to 0 (the Set point must be zero in order to perform an Auto Tune with the HEB as the AC source).
21. Choose Controls > AC Mode Tune to open the AC Mode Tune and AC Tune windows.
22. The next step is to tune the oscillation signal to match the frequency of the particular cantilever. Use the controls in the AC Mode Tune window to sweep through a range of frequencies. The resultant plot should show one strong, sharp resonance peak. The cantilever's storage box should indicate the range in which the primary resonance frequency will be found.
23. In the upper Auto Tune area of the AC Mode Tune window, enter the Start and End frequencies (in kHz) for the tuning sweep. For a new or unknown cantilever, use the stated minimum and maximum frequencies given on the storage box. If the exact resonance frequency is known, you can use a smaller range to speed the tuning process.
24. Set the Peak Amplitude, the maximum desired amplitude of cantilever oscillation. Two volts is a typical starting value.
25. To ensure good engagement of the tip with the sample, set the oscillation frequency slightly below the actual resonance frequency of the cantilever. Enter an Off Peak value to offset the oscillation frequency from the cantilever's resonance frequency. A typical starting value is −0.200 kHz.
26. Click the Auto Tune button. The system will sweep several times through the range of frequencies, locating the peak oscillation amplitude within the range. The AC signal oscillation will be set to this value, taking into account the specified Offset.
27. Focus the cantilever in the video window.
28. Turn the video system focus knob toward you such that the tip goes just out of focus.
29. Press the Close switch to raise the sample until both the tip and sample are in focus.
30. In the Scan and Motor window, click the Motor tab.
31. Set the Stop At % to specify the percentage of total oscillation that represents "contact." For example, if the total oscillation amplitude is 2 V, and the Stop At value is set to 90%, the approach will stop when the oscillation is damped to 1.8 V.
32. Click the Approach button in PicoView's toolbar. The system will raise the sample until the amplitude is damped to the Stop At percentage. Because of air damping, oscillation typically decreases as the tip nears the sample. The software monitors

the rate of change of amplitude as well as the absolute value, so the final amplitude will not be exactly the Stop At percentage.

33. In the Servo window, set the I Gain and P Gain to 5%. These gains dictate how quickly the system will react to changes in amplitude.
34. In the Scan and Motor window, select the Scan tab.
35. Enter a scan speed of 1–2 ln/s and a resolution of 256.
36. Enter the size (in microns) and X Offset and/or Y Offset values to set the size and center of the scan.
37. In the Real Time Images window, make sure that Topography and Deflection are displayed.
38. In the Scan and Motor window, click the down blue arrow to initiate a scan starting at the top of the grid. Click the up blue arrow to initiate the scan from the bottom up. The image maps will begin to be rendered in the Real time images window.
39. To stop the scan cycle, click the red STOP circle that will replace the Up or Down arrow when you start a scan [13].
40. For each sample, collect $3 \times 3\ \mu m^2$ images at five different areas. Use these images to determine RMS surface roughness as a standard deviation of surface heights distribution (R_q) according to the equation:

$$R_q = \sqrt{\frac{1}{NM}\sum_{x=1}^{N}\sum_{y=1}^{M}(z(x,y) - \bar{z}(N,M))^2}$$

where N is number of points in a scan line, M is the number of scan lines, $z(x,y)$ is the difference in height at a given point (x, y), and $z(N,M)$ is arithmetic average height at given line point. Figure 1 shows the topography images of immobilized molecular films.

3.4 Assessing Immunocomplexes of *P. mirabilis* O3 (S1959) Antigens and RA Patient Antibodies Using AFM Topography

An example of interpretation of AFM topography using immunocomplexes of *P. mirabilis* O3 (S1959) antigens and RA patient antibodies is described here (Fig. 1a–d). A second set of images show synthetic Lys-GalA-PAA antigen binding to purified patient serum (Fig. 1e, f). Quantified AFM data are shown in Table 1.

1. Sera assembled layers (Fig. 1a, b, unpurified and purified serum controls) present a rather amorphous surface structure composed of aggregates. Small bright spherical objects are observed in both cases. For purified IgG antibodies from 005 RA serum, the diameter of aggregates varies in the range of 45–200 nm. Surface roughness (RMS) for purified IgG 005 RA serum was estimated to 0.80 nm (*see* **Note 7**). This observation would suggest that purified serum film shows a tendency to aggregate. The morphology of immobilized films

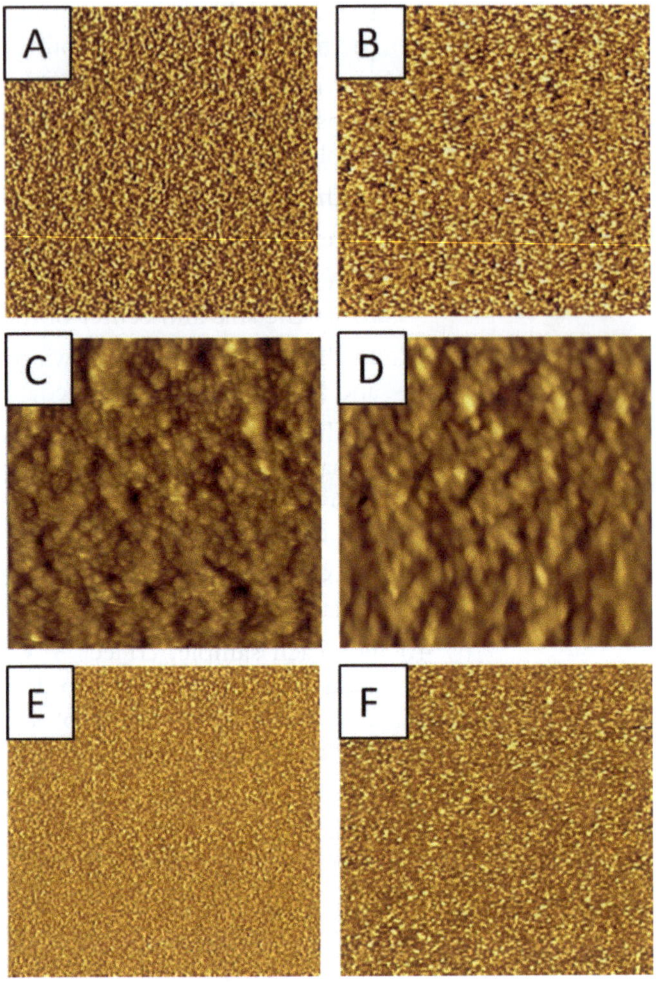

Fig. 1 3 × 3 μm^2 AFM images obtained for mica modified with (**a**) unpurified serum; (**b**) purified serum; (**c**) *P. mirabilis* LPS (O3) S1959; (**d**) purified serum + *P. mirabilis* LPS O3 (S1959); (**e**) Lys-GalA-PAA antigen; (**f**) purified serum + Lys-GalA-PAA antigen

cannot be related directly to the structure of underlying mica substrate, which is flat even in atomic scale.

2. The immobilization of LPS molecules results in the formation of amorphous films (observations of LPS alone, also a control; Fig. 1c). The features of the underlying substrate can also be distinguished as grooves forming triangular shapes. In this case, spherical features can be seen and the diameter of individual spots varies between 5 and 50 nm.
3. Figure 1d (purified serum + LPS) presents the results obtained for the *P. mirabilis* O3 LPS + purified sera pair. Figure 1d is a map of the topography of the sample. It shows randomly distributed species immobilized on a mica surface. The size distribution of the objects is quite broad, which indicates that

Table 1
Features of immunocomplexes of *P. mirabilis* O3 (S1959) antigens observed on bare mica[a] by AFM method

No.	Sample—Fig. 1	Surface roughness (RMS), nm	Aggregate diameter (nm)	Description of surfaces
1.	Unpurified serum (A)	0.58	30–75	Amorphous surface structure composed of aggregates
2.	Purified serum (B)	0.80	45–200	Amorphous surface structure composed of aggregates
3.	*P. mirabilis* LPS O3 (S1959) (C)	5.18	5–50	Molecules formed structure composed of aggregates
4.	Purified serum + *P. mirabilis* LPS O3 (S1959) (D)	2.69	10–100	Structure composed of aggregates
5.	Lys-GalA-PAA antigen (E)	0.17	35–70	Ordered structure small isolated aggregates can be recognized
6.	Purified serum + Lys-GalA-PAA antigen (F)	0.35	10–100	Tendency to form aggregates is reduced; only randomly distributed aggregates can be observed

[a]Reference value of RMS surface roughness for bare mica was 0.07 nm

some molecules formed aggregates. Most likely, this is related to the fact that some IgG molecules adopted an orientation which is inappropriate for binding with *P. mirabilis* O3 LPS. Moreover, a large fraction of molecules formed aggregates (*see* **Note 8**), which may also have hindered access to sensing molecules. Nevertheless, the AFM results confirm that specific binding between LPS O3 and IgG does occur.

4. RMS surface roughness (synthetic antigen by itself; another control; Fig. 1e) determined for bare mica (0.07 nm) or Lys-GalA-PAA antigen (0.17 nm) indicate that such substrates have a small contribution to the film's roughness.
5. The purified serum + synthetic antigen sample (Fig. 1f) suggests the presence of antigen–antibody complexes. AFM was applied to detect Lys-GalA-PAA binding by isolated IgG antibodies from RA serum (patient 005). In comparison to IgG alone (Fig. 1b), the tendency to form aggregates is reduced and more homogenous aggregates are observed, with sizes varying from 35 nm up to 70 nm. The aggregates are smaller and the surface of the sample is more homogeneous; therefore, it can be concluded that aggregation occurs to a lesser extent. That is reflected by RMS surface roughness of immunocomplexes of IgG and Lys-GalA-PAA antigen, which was estimated as 0.35 nm (*see* **Note 9**).

4 Notes

1. When purifying antibodies to the IgG class, remember that on all steps, the column flow will stop when the solution drains down to the disc at the top of the gel bed. This prevents the gel bed from drying out. However, do not leave a drained column uncapped for more than a few minutes.
2. Dissolve LPS in PBS in a 1.5 mL microcentrifuge tube. To properly dissolve the LPS antigen, vortex 5 times, then dissolve the LPS using an ultrasound bath for 15 min.
3. In Magnetic AC (MAC) Mode AFM, the back side of the cantilever is coated with magnetic material. A solenoid applies an AC magnetic field which is used to oscillate the cantilever. MAC Mode is typically cleaner and gentler than Acoustic AC Mode and is free from spurious background signals that are somewhat common when using AAC Mode. The benefits are particularly pronounced when imaging in liquid.
4. In order to prepare clean mica discs, press adhesive tape against the top mica surface and then peel off the tape to expose a fresh mica surface. Repeat this procedure as needed until the mica surface is smooth. The freshly cleaved mica should be used immediately.
5. Rinse the disc three times with aqueous PBS buffer and scan the mica surface with the antigen applied.
6. The procedure for imaging in MAC Mode is the same as for AC Mode, with these exceptions:
   (a) Connect the six-pin (MAC) end of the EC/MAC Cable to the six-pin connector on the sample plate. Connect the other end of the cable to the EC/MAC socket on the bottom of the microscope stand.
   (b) In the ACAFM Controls dialog box, choose MAC as the Drive Mechanism.
7. The surface roughness (RMS) of pure uncoated mica is 0.07 nm, which indicates a very small effect on the surface roughness of the tested sera.
8. RMS is the root mean square average of the profile height deviations from the mean line, recorded within the evaluation length. Its value can be used to quantify the surface roughness and hence evaluate extent of aggregation of the adsorbed species.
9. The topography map of the test sample with the accidental arrangement of particles on the surface of mica, and the quite wide distribution of sizes (in the range of 10–100 nm), can indicate/suggest that some molecules form aggregates.

Acknowledgments

The presented studies were supported by grant "Preludium 8, UMO-2014/15/N/NZ6/02505," National Research Center, Poland for JG-O, and BS 2018 UJK for WK.

References

1. Grieshaber D, MacKenzie R, Vörös J, Reimhult E (2008) Electrochemical biosensors - sensor principles and architectures. Sensors 8 (3):1400–1458
2. Volkov D, Strack G, Halámek J, Katz E, Sokolov I (2010) Atomic force microscopy study of immunosensor surface to scale down the size of ELISA-type sensors. Nanotechnology 21 (14):145503
3. Corry B, Uilk J, Crawley C (2003) Probing direct binding affinity in electrochemical antibody-based sensors. Anal Chim Acta 496:103–116
4. Waritani T, Chang J, McKinney B, Terato K (2017) An ELISA protocol to improve the accuracy and reliability of serological antibody assays. MethodsX 4:153–165
5. Dufrêne YF, Hinterdorfer P (2008) Recent progress in AFM molecular recognition studies. Pflügers Arch 456(1):237–245
6. Zhengjian L, Wang J, Chen G, Deng L (2010) Probing specific interaction forces between human IgG and rat anti-human IgG by self-assembled monolayer and atomic force microscopy. Nanoscale Res Lett 5(6):1032–1038
7. Gleńska-Olender J, Sęk S, Dworecki K, Kaca W (2015) A total internal reflection ellipsometry and atomic force microscopy study of the interaction between *Proteus mirabilis* lipopolysaccharides and antibodies. Eur Biophys J 44(5):301–307
8. Kaca W, Knirel YA, Vinogradov EV, Kotełko K (1987) Structure of the O-specific polysaccharide of *Proteus mirabilis* S1959. Arch Immunol Ther Exp 35(4):431–437
9. Gromska W, Mayer H (1976) The linkage of lysine in the O-specific chains of *Proteus mirabilis* 1959. Eur J Biochem 62(2):391–399
10. Gleńska-Olender J, Durlik K, Konieczna I, Kowalska P, Gawęda J, Kaca W (2017) Detection of human antibodies binding with smooth and rough LPSs from *Proteus mirabilis* O3 strains S1959, R110, R45. Antonie van Leeuwenhoek 110(11):1435–1443
11. Żarnowiec P, Czerwonka G, Kaca W (2017) Fourier transform infrared spectroscopy as a tool in analysis of *Proteus mirabilis* endotoxins. Methods Mol Biol 1600:113–124. https://doi.org/10.1007/978-1-4939-6958-6_11
12. Chernyak AY, Kononov LO, Kochetkov NK (1994) Glycopolymers from synthethic fragments (amides of α-D-galacturonic acid with amino acids) of *Proteus* O-antigens. J Carbohydr Chem 13(3):383–396
13. Agilent Technologies 5500 Scanning probe microscope user's guide. Agilent Technologies, Inc. 2008

Chapter 24

Considerations for Modeling *Proteus mirabilis* Swarming

Bruce P. Ayati

Abstract

In this chapter we provide some initial guidance to experimentalists on how they might go about creating mathematical representations of their systems under study. Because the interests and goals of different researchers can differ, we try to provide broad instruction on the creation and use of mathematical models. We provide a brief overview of some modeling that has been done with *Proteus mirabilis* colonies, and discuss the goals of modeling. We suggest ways that collaborative teams may communicate with one another more effectively, and how they can build more confidence in their model results.

Key words *Proteus mirabilis*, Swarm colony, Mathematical modeling, Collaborative science, Mathematical representations

1 Introduction

Interest in the mathematical modeling of *Proteus mirabilis* swarm colony development likely began with the publication of Rauprich et al. [20], which in turn led directly to the first known model of *P. mirabilis* swarming on laboratory media, by Esipov and Shapiro [12]. The focus of these experiments and models was on the bull's-eye concentric terrace patterns *P. mirabilis* creates on agar that is not too firm or too soft. In addition to the striking spatial regularity, it was found that there was also a temporal regularity—the time it took for each terrace to form in a given set of experimental conditions was nearly invariant.

The seminal modeling effort of Esipov and Shapiro used mathematical representations of three mechanisms to capture the spatial and temporal regularity: (1) quorum sensing of different cell types; (2) the lubrication of the swarmer "rafts" allowing cell motility on the surface of the agar (represented by a memory field or state variable); and (3) a differentiation/dedifferentiation cell cycle between surface-motile, filament cells called "swarmers," and non-surface-motile single cells called "swimmers." The resulting

Melanie M. Pearson (ed.), *Proteus mirabilis: Methods and Protocols*, Methods in Molecular Biology, vol. 2021,
https://doi.org/10.1007/978-1-4939-9601-8_24,

mathematical model had independent variables for time and space, and an age variable representing the swarmer cell cycle.

Following this work, other authors created simpler models that isolated and reduced the number of mechanisms. In [18], Medvedev, Kaper, and Kopell posited a somewhat sophisticated quorum sensing as the primary mechanism for generating the patterns, and represented this mechanism as the diffusivity in a system of reaction–diffusion and ordinary differential equations, removing any explicit representation of the differentiation–dedifferentiation cycle or the underlying fluid system. In [11], Czirók, Matsushita, and Vicsek used a more phenomenological representation of the colony radius. Frénod and Sire emphasized the fluid dynamics aspects of the swarm colony behavior [13].

Xue, Budrene, and Othmer modeled radial and spiral streaming patterns, rather than the bull's-eye patterns, that arise under different experimental conditions. They assumed some motility of swimmers in a surface colony, and used a hybrid discrete continuous model, rather than a fully continuous model [24]. Discrete rules-based activity on the part of the cell populations in the simulation essentially replaced the degenerate diffusion terms used in the partial differential equations in other approaches.

This author's work placed the primary emphasis on the differentiation and dedifferentiation cycle, and assumed less complexity in the quorum sensing underlying the choice of diffusion term [3, 5, 6]. As discussed in more detail in [6], these models were largely validated in their assumptions about the differentiation and dedifferentiation cycle by the measurements of Matsuyama et al. [17]. This author's models were fully continuous, and solved using what were state-of-the-art numerical methods and software for their time [2, 7, 8]. A somewhat more accessible exposition than the original numerical analysis papers was presented in [4]. Other, comparably effective approaches have since been developed (e.g., [10]).

The thing to take away from this cursory literature review is that different questions may call for different mathematical representations of *P. mirabilis*. This is especially the case in experimental systems where there is not necessarily a clinical or engineering goal in mind.

The remainder of this paper will focus on the goals of modeling, on different types of mathematical representations, on the importance of computational rigor and fidelity to the original equations being approximated (and how this may have affected the history of *P. mirabilis* modeling), on the types of data that are better suited for contemporary data-driven modeling and simulation, on some issues of uncertainty quantification (i.e., quality control), and on interpreting simulation results.

2 Goals of Mathematical Modeling

Mathematical modeling is a form of representation in the sense of representational art. When a mathematical model is created, a system is depicted in much the same way as someone painting a portrait. Some details are retained, some are left out, and among those that are retained some are emphasized more than others. *Ceci n'est pas une pipe*—a mathematical model of a pipe is not a pipe. Continuing with this analogy, a mathematical model may also, like a cubist painting, juxtapose elements in different ways. The perspective created by one representation (say that of how our eyes collapse three-dimensional space into two dimensions) may result in other representations seeming odd (e.g., Picasso's Guernica, Fig. 1).

Where mathematical representations begin to differ from representations in art is that we want to use them in some manner. Moving from high art to scientific illustration, it has been argued that wildlife illustrations, such as those by Audubon and Peterson, are more useful than photographs for use in field guides. From the preface of [19], "A drawing can do much more than a photograph to emphasize the field marks. A photograph is a record of a fleeting instant; a drawing is a composite of the artist's experience. The artist can edit out, show field marks to best advantage, and delete unnecessary clutter."

To paraphrase this for our purposes, a model should be a composite of the scientist's experience and interests. This suggests two questions. First, what do you think *P. mirabilis* is? An object of curiosity in its own right, a more accessible system for understanding something more complicated, such as morphogenesis in higher organisms, or an entity of immediate clinical or engineering interest? Is it a colony of individuals with emergent structure, or an organism in its own right?

Fig. 1 Pictorial representations of the Bombing of Guernica. Picasso's painting to the left and a photograph to the right. Guernica by Picasso is in the collection of Museo Reina Sofia, Madrid. Source page: http:/www.picassotradicionyvanguardia.com/08R.php (archive.org), fair use, https:/en.wikipedia.org/w/index.php?curid=1683114. The photograph is from the Bundesarchiv (Bild 183-H25224/Unknown/CC-BY-SA 3.0, CC BY-SA 3.0 de, https:/commons.wikimedia.org/w/index.php?curid=5434009)

Second, what do you want to learn from your model? Are you trying to link some phenomenon to an underlying mechanism more rigorously? Are you trying to predict an outcome of something in vivo? Do you want your model to guide your understanding, or be a reflection of your conceptualization?

The goals of a modeling effort manifest themselves in the choice of what is modeled and how those things are modeled.

3 Types of Representations

The types of useful representations available to a modeler are dictated by immutable aspects of the system of interest, such as size or number of individuals, and aspects tied to questions of interest concerning the system.

3.1 Partial Differential Equation Models

The types of representations reasonable for *P. mirabilis* can mostly be described, at the most basic level, as a vector-valued function $\mathbf{u}$ that takes a vector as an argument, $\mathbf{x}$.

For a system $\mathbf{u}(\mathbf{x})$ with m dependent variables $\mathbf{u} = (u_1, u_2, \ldots, u_m)$, and n independent variables $\mathbf{x} = (x_1, x_2, \ldots, x_n)$, increasing the number of independent variables will generally increase the computational (and analytical) effort put towards the model much more quickly than increasing the number of dependent variables.

In this section we will discuss fully continuous models. For a spatial system such as a *P. mirabilis* swarm colony, the starting point with respect to complexity would be a model that tracks both temporal and spatial changes in the swimmer and swarmer populations—tracking total cell numbers alone is not enough.

For example, if we assume radial symmetry in a fully continuous model, and ignore any explicit physiological structure in the swarmer-cell population, then we might choose to use a reaction--diffusion equation to model the swarmer cells. We will denote the swarmer cell population by u and the swimmer cell population by v. A radially symmetric reaction–diffusion equation for the swarmer cells would have the form

$$\underbrace{\frac{\partial u}{\partial t}(r,t)}_{\boxed{\text{I}}} = \underbrace{\frac{1}{r}\frac{\partial}{\partial r}\left(r\frac{\partial}{\partial r}D(u,v)u\right)}_{\boxed{\text{II}}} + \underbrace{f_u(u,v)}_{\boxed{\text{III}}}, \tag{1}$$

where we have used a diffusion that is more suitable for cell populations than Fickian diffusion [1]. Processes in differential equations models are additive: the rate of change of the swarmer cell population at a radius r and at a time t (term I) is equal to the net number of cells moving out of that spatial position (term II), and the creation or loss of any cells due to differentiation from swimmers or dedifferentiation to swimmers (term III).

If we neglect any motility on the surface by the swimmer cells, then their populations would be modeled by what are essentially ordinary differential equations parameterized by a spatial variable,

$$\frac{\partial u}{\partial t}(r,t) = f_v(u,v). \tag{2}$$

These equations would be coupled to initial conditions and a no-flux condition on the boundary of the domain of r.

If the question of interest is to understand how the differentiation–dedifferentiation cycle of individual swarmers may affect the overall spatial structure of the colony, we would need to add mathematical complexity to track this cycle. Overall complexity is an important consideration in choosing the types of representations in a model. We should avoid adding complexity when none is needed, but not shy away from it when it is.

Explicitly tracking the swarmer-cell size distribution in a fully continuous model requires an additional independent variable. Because the length of a swarmer cell is more or less a function of the amount of time since its differentiation, we can use age as the new independent variable. Our equations for the swarmer- and swimmer-cell populations, resp., become

$$\underbrace{\frac{\partial u}{\partial t}(r,a,t) + \frac{\partial u}{\partial a}(r,a,t)}_{\text{transport in the age-time domain}} = \frac{1}{r}\frac{\partial}{\partial r}\left(r\frac{\partial}{\partial r}D(u,v)u\right), \tag{3}$$

$$\frac{\partial u}{\partial t}(r,t) = f_v(u(r,\cdot,t),v), \tag{4}$$

where the "reaction" term f_u is now replaced by a boundary condition in the age variable, and the other reaction term, f_v, depends on the entire age distribution of u at a point in space-time (r, t). The main change is that the time derivative in Eq. 1 has been replaced by a total derivative in age and time in Eq. 3. This reflects the fact that age and time advance together: after 1 h, an individual swarmer cell is 1 h "older," i.e., has advanced an hour in the growth process. Alternatively, once can think of the time derivative as before, and take the age derivative $\partial u/\partial a$ to represent a sort of advection term with a dimensionless velocity of one.

In both the model with age structure and the model without, the specifics of the functions D, f_u, and f_v, and the boundary and initial conditions, need to be determined before any computations can take place.

The age-structured system in Eqs. 3–4 is significantly more complex than the system without age structure Eqs. 1–2—we have added an additional dimension to the problem. Only add this kind of complexity to a mathematical representation of a problem if it is critical to understanding the principal mechanisms underlying an observed phenomenon.

3.2 Discrete Models

Fully continuous models are not the only mathematical representations available to someone interested in modeling a phenomenon such as *P. mirabilis* swarming. Another popular choice is the family of hybrid discrete-continuous models, used, for example, in [24]. In these models, chemical concentrations are typically modeled using fully continuous, deterministic partial differential equations, and the behaviors of discrete individual cells are tracked explicitly according to a set of rules, usually stochastic. The choice and implementation of the discrete systems can vary quite a bit from modeler to modeler.

In simple terms, discrete versus continuous is the first choice a modeler needs to make among possible representations. Often the choice is determined by the scale of the problem. For *P. mirabilis* colonies on an agar surface, one can argue for either type of representation for the cell populations. Any chemical concentrations in a model would still need a continuous representation. Generally, fully continuous models are used when the system has a large number of cells. Discrete, stochastic representations are generally used when there are fewer numbers of cells. Hybrid discrete-continuous models are used in the case of lower cell numbers with explicit tracking of chemical concentrations. What constitutes "lower" depends heavily on computational resources. In [24], the simulations were of colonies with diameter less than 3 cm and maximum densities of 40,000 cells per square centimeter.

3.3 Mechanism vs. Phenomenon

Another, more nuanced choice is, for each interaction of interest in your system, whether to use a mechanistic or a phenomenological representation. These terms are inexact—one researcher's mechanism of interest is another's phenomenon. One approach is to use the parameters needed for a given representation as a guide. If those parameters are measurable, then a more mechanistic representation would be in order. If those parameters need to be estimated, a more phenomenological representation may be better.

What one can do with a model differs based on these choices. A virtual system that could be used to predict the outcomes of future experiments or conduct early stage trials would almost certainly need to be largely mechanistic, with parameters determined through direct measurement, as opposed to being estimated by some sort of matching of computed results with expected outcomes. These sorts of virtual systems are often the goal, but in practice very difficult to come by. Work towards building a virtual system could comprise mechanistic components with estimated parameters. It is important to remember that in these intermediate

Table 1
Parameter estimation and model type

Type\parameterization	Determined	Estimated
Mechanistic	Virtual system	Initial and intermediate efforts towards building a virtual system
Phenomenological	N/A	Lots in mainstream mathematical biology. Applications such as using clinical data in the clinic. Aid conceptual understanding

models not to take the estimated parameters to be the truth. In your model you may have left out important mechanisms, or added ones that don't really exist in the actual system, and thus what was estimated for a given may reflect more than just the influence of the mechanism it is tied to.

Many models in mathematical biology are mostly phenomenological—understanding that the duality between mechanism and phenomenon is really a continuum when it comes to mathematical representations, and that what constitutes one pole or the other is tied to the scale of interest. For example, the differentiation and dedifferentiation cycle of individual swarmer cells may be viewed as a mechanism underlying the concentric swarm colony phenomenon, but it is also a phenomenon that may be explained in part by mechanisms inside the cell.[1] The parameters for these models are generally estimated. Phenomenological models with estimated parameters are primarily useful for conceptualizing the system in more depth, and serving as a sanity check on verbal models that aim to describe what is happening in a given system. Phenomenological models fit to data may also be useful in using clinical data back in the clinic.

The parameters in a phenomenological model do not usually correspond to something directly measurable, and as such it doesn't make sense for them to be directly determined via experiments. This somewhat simplistic matrix description of the mechanism-phenomenon, determined-estimated dualities is illustrated in Table 1.

4 Communicating Across Disciplinary Boundaries

The biggest challenge in any collaboration between experimentalists in the life sciences and biomathematicians is the communication across the gap in shared knowledge. Although there are many areas

[1] A model that includes mathematical representations of what generates the dedifferentiation and dedifferentiation cycle and of how the swarmers generate the concentric rings would, in some circles, be referred to as "multiscale."

of the life sciences where a single person possesses sufficient expertise across disciplinary boundaries to avoid communication failures, for most collaborations communication is the biggest hurdle. With the exception of a few remarkable people, breadth of knowledge usually comes at the cost of depth—you can't be a sprinter and a sumo wrestler. In the author's experience, the most effective tool for bridging this gap has been schematics, referred to colloquially as wiring diagrams. In a recently completed collaboration on modeling inflammation in articular cartilage [14, 15], it can be argued that the most refined schematic was the true contribution of that effort, an easily understood statement about how the system works that can be used by others to create their own mathematical representations. The evolution of these schematics, first and last, is shown in Fig. 2.

Even with visual tools, communicating across a significant gap in core expertise may often involve frequent misunderstandings. We have found that iteration is the key to getting past this.

As mentioned above, perhaps the most important thing is being clear about what question you are trying to answer.

4.1 Your Data

The types of useful models you can create are tied closely to the kinds of data you have available. Once a team has a schematic with which they wish to move forward, the resulting model equations have parameters that need to be specified somehow. In an initial effort, many of these will be placeholder values, but the fewer placeholder values that are used, the better. For the case of *P. mirabilis*, it is likely that you are interested in quantities that change over time. The ability to provide your modelers with longitudinal data can dramatically improve the results of your collaboration, particularly if you can provide measurements between initiation of the experiment and its conclusion. It is understood that longitudinal data is often difficult to come by, given that extracting the data often involves ending the experiment, but it is the gold standard for this type of modeling and simulation.

Regardless, the old adage "garbage in, garbage out" applies. While some illustrative modeling is possible without good data, e.g., simple sanity checks on core assumptions, anything beyond that isn't.

5 Computational Methods and Rigor

Mathematical models written using differential equations have the properties of transparency and reproducibility. Pen-and-paper exact solutions (called "analytical solutions") to complex differential equations are generally not possible. Instead these solutions are approximated on a computer using what are called numerical

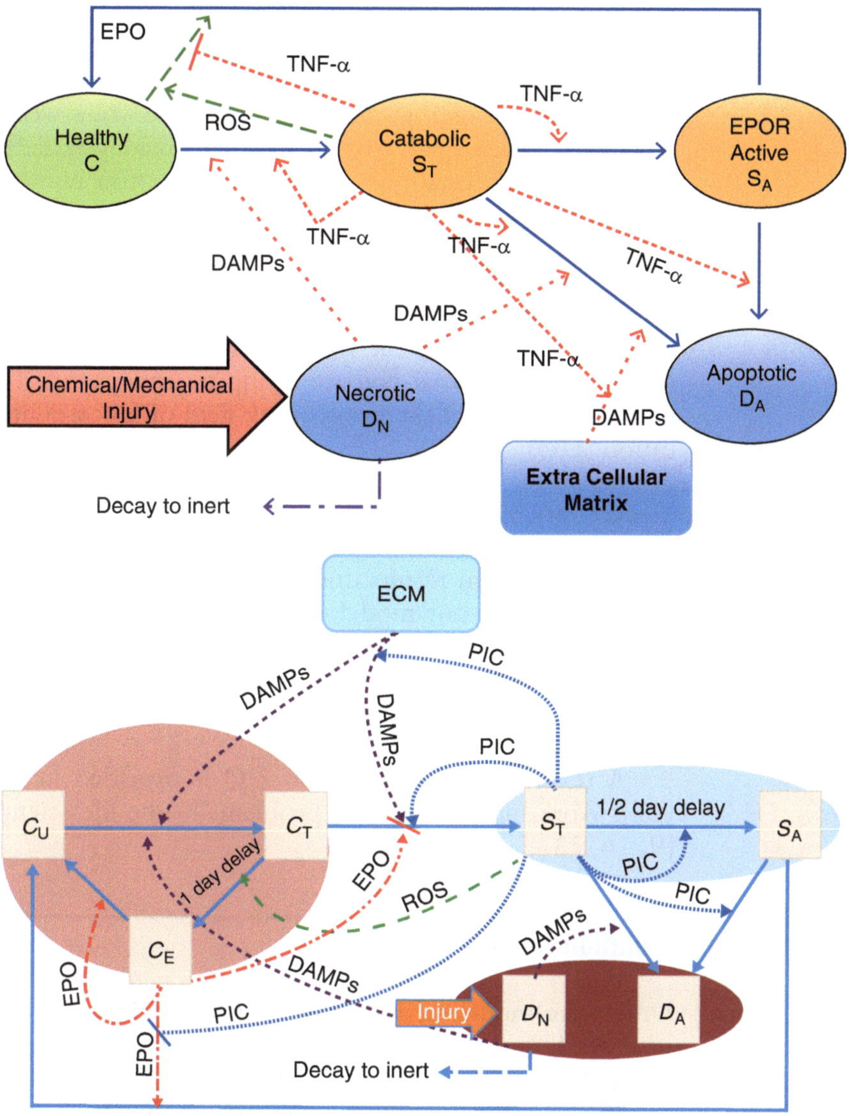

Fig. 2 Evolution of schematics for the modeling and simulation of articular cartilage lesion formation, showing first and last in the collaboration. The model results based on one schematic helped inform the following one. These two schematics are taken from Graham et al. [14] (top) and Kapitanov et al. [15] (bottom), resp. Permission granted by the copyright holder Bruce P. Ayati (this author) to republish these images

methods. So by reproducibility, we mean that if two parties approximate the solutions to the differential equations to high fidelity, even with different methods or software, they would get essentially the same answers. No aspects of the model are buried in software code.

Ideally, the computational results should be accurate mathematical solutions to the stated model equations. However, due to computational and other limitations, this seems not always to have been the case in the *P. mirabilis* modeling literature (*see* Appendix A in [3]). Fidelity to the original equations is necessary to have reproducibility in the absence of code sharing.

There is some benefit to using broadly available software, both to avoid computing garbage and to make your work easier to reproduce. Increasingly, software is made readily available for solving differential equations. The options are too numerous to list here, and depend heavily on your choice of programming language and model. For discrete models, one freely available and widely used package is CompuCell3D[2] [22].

Going beyond solving the equations accurately and as they were written, there is the issue of evaluating the relevance of the model results to what you are trying to understand. This is done under the rubric of the burgeoning field of Uncertainty Quantification (UQ).

5.1 Uncertainty Quantification

Uncertainty Quantification (UQ) is basically quality control for modeling and simulation. In its broadest sense, it aims to quantify the uncertainty in results due not just to how well you solve the mathematics in your model on a computer, but also how experimental errors and the like propagate through to the final results, which goes beyond standard statistical work on your data. Increasingly, some UQ is expected in contemporary work in modeling and simulation.

A textbook introduction to UQ is provided in Smith 2014 [21]. Work more specific to problems in the life sciences include [9, 16, 23].

6 Interpretation of Simulation Results

Unless the modeling and simulation effort had a straightforward goal, such as prediction of some quantity (e.g., biomass), some interpretation of the results is necessary. Was the aim to link a specific mechanism to an observation? If so, why does the model, with its attendant assumptions, back that up? Was the aim to obtain a systems-level understanding, perhaps illustrated by a schematic? In this case, which elements are essential and which are secondary? What is lost if the secondary mechanisms are removed?

How a collaborative team interprets the results is highly subjective. Nonetheless, it is an important part, perhaps the whole point, of doing modeling and simulation in the first place.

This author's models connected the spatial regularity to a regularity in the differentiation and dedifferentiation cycle of individual swarmer cells, and suggested that the collective cell motility and surface fluid dynamics were secondary considerations—retaining but de-emphasizing these mechanisms compared to what was

[2] Available at the time of writing at www.compucell3d.org.

done in [12]. Other authors (e.g., [18]) considered the specifics of the cycle to be less important, and left out the lengthening of the swarmers altogether.

These sorts of choices in models are not completely subjective. If the goal were to intervene in the system, the nature of the intervention would help determine the choices in the model, e.g., altering the differentiation and dedifferentiation cycle would call for a model that had that represented explicitly. If the intervention involved altering the fluid dynamics on the surface, then the model should explicitly represent the physics of the fluids. Similar sorts of considerations would be made for bioengineering to obtain some goal, such as maximizing biomass or surface coverage in a bioreactor.

References

1. Aronson DG (1985) The role of diffusion in mathematical population biology: skellam revisited. In: Mathematics in biology and medicine: proceedings of an international conference held in Bari, July 18–22, 1983, pp 2–6
2. Ayati BP (2000) A variable time step method for an age-dependent population model with nonlinear diffusion. SIAM J Numer Anal 37 (5):1571–1589
3. Ayati BP (2006) A structured-population model of *Proteus mirabilis* swarm-colony development. J Math Biol 52(1):93–114
4. Ayati BP (2007) Modeling and simulation of Age- and space-structured biological systems. In: Mahdavi K, Culshaw R, Boucher J (eds) Current developments in mathematical biology. World Scientific, Singapore, pp 107–130
5. Ayati BP (2007) Modeling the role of the cell cycle in regulating *Proteus mirabilis* swarm-colony development. Appl Math Lett 20 (8):913–918
6. Ayati BP (2009) A comparison of the dynamics of the structured cell population in virtual and experimental *Proteus mirabilis* swarm colonies. Appl Numer Math 59(3–4):487–494. https://doi.org/10.1016/j.apnum.2008.03.023
7. Ayati BP, Dupont TF (2002) Galerkin methods in age and space for a population model with nonlinear diffusion. SIAM J Numer Anal 40(3):1064–1076
8. Ayati BP Dupont TF (2005) Convergence of a step-doubling Galerkin method for parabolic problems. Math Comput 74(251):1053–1066
9. Collis J, Connor AJ, Paczkowski M, Kannan P, Pitt-Francis J, Byrne HM, Hubbard ME (2017) Bayesian calibration, validation and uncertainty quantification for predictive modelling of tumour growth: a tutorial. Bull Math Biol 79(4):939–974. https://doi.org/10.1007/s11538-017-0258-5
10. Coyle J, Nigam N (2016) High-order discontinuous Galerkin methods for a class of transport equations with structured populations. Comput Math Appl 72(3):1–17. https://doi.org/10.1016/j.camwa.2016.05.024
11. Czirók A, Matsushita M, Vicsek T (2001) Theory of periodic swarming of bacteria: application to *Proteus mirabilis*. Phys Rev E 63 (3):31911–31915
12. Esipov SE, Shapiro JA (1998) Kinetic model of *Proteus mirabilis* swarm colony development. J Math Biol 36:249–268
13. Frénod E, Sire O (2009) An explanatory model to validate the way water activity rules periodic terrace generation in *Proteus mirabilis* swarm. J Math Biol 59(4):439–466. https://doi.org/10.1007/s00285-008-0235-6
14. Graham JM, Ayati BP, Ding L, Ramakrishnan PS, Martin JA (2012) Reaction-diffusion-delay model for EPO/TNF-α interaction in articular cartilage lesion abatement. Biol Direct 7 (1):9–9. https://doi.org/10.1186/1745-6150-7-9
15. Kapitanov GI, Wang X, Ayati BP, Brouillette MJ, Martin JA (2016) Linking cellular and mechanical processes in articular cartilage lesion formation: a mathematical model. Front Bioeng Biotechnol 4(80):1–14. https://doi.org/10.1016/j.joca.2014.04.023
16. Laz PJ, Browne M (2010) A review of probabilistic analysis in orthopaedic biomechanics. Proc Inst Mech Eng H J Eng Med 224 (8):927–943. https://doi.org/10.1243/09544119JEIM739
17. Matsuyama T, Takagi Y, Nakagawa Y, Itoh H, Wakita J, Matsushita M (2000) Dynamic

aspects of the structured cell population in a swarming colony of *Proteus mirabilis*. J Bacteriol 182(2):385–393

18. Medvedev GS, Kaper TJ, Kopell N (2000) A reaction-diffusion equation with periodic front dynamics. SIAM J Appl Math 60 (5):1601–1638
19. Peterson RT (1989) A field guide to the birds: eastern and central North America. Houghton Mifflin Harcourt, Boston
20. Rauprich O, Matsushita M, Weijer CJ, Siegert F, Esipov SE, Shapiro JA (1996) Periodic phenomena in *Proteus mirabilis* swarm colony development. J Bacteriol 178(22):6525
21. Smith RC (2014) Uncertainty quantification: theory, implementation, and applications. SIAM, Philadelphia
22. Swat MH, Thomas GL, Belmonte JM, Shirinifard A, Hmeljak D, Glazier JA (2012) Multi-scale modeling of tissues using CompuCell3D. In: Methods in cell biology, vol 110. Elsevier, Amsterdam, pp 325–366
23. Van Schepdael A, Carlier A, Geris L (2016) Sensitivity analysis by design of experiments. In: Uncertainty in biology. Springer, Cham, pp 327–366
24. Xue C, Budrene EO, Othmer HG (2011) Radial and spiral stream formation in *Proteus mirabilis* colonies. PLoS Comput Biol 7(12): e1002332. https://doi.org/10.1371/journal.pcbi.1002332

Chapter 25

Transposon Insertion Site Sequencing in a Urinary Tract Model

Valerie S. Forsyth, Harry L. T. Mobley, and Chelsie E. Armbruster

Abstract

Transposon sequencing (Tn-seq) is a technique that combines quantitative next-generation sequencing and a saturating transposon mutant library for an organism of interest, and ultimately allows for quantitation of the relative abundance of all of the mutants under a given condition, such as during experimental infection. The massively parallel sequencing capabilities of this technique provide a significant advance over more traditional methods of screening transposon mutant pools or individually determining the fitness contribution of genes of interest. Here, we describe a method for generating a genome-saturating transposon mutant library in *Proteus mirabilis*, determining the appropriate number of mutants for inoculation in an experimental infection model, preparing transposon insertion junctions for Illumina sequencing, and downstream analysis of mapped DNA sequencing reads for estimation of the contribution of each gene in the genome to fitness during infection.

Key words Tn-seq, Southern blot, Urinary tract infection, Illumina sequencing, Transposon sequencing

1 Introduction

Transposon sequencing (Tn-seq) is a powerful tool for discovery of gene function in bacteria via loss-of-function screening. Tn-seq couples a library of random transposon mutants with next-generation sequencing (NGS) to evaluate the ability of mutants to grow and survive under a given condition on a genome-wide scale. First, a genome-saturating transposon library is constructed. Then, the mutant library is subjected to the condition of interest and bacterial genomes are harvested from both the input and output populations. The genomes are enriched for transposon insertion junctions, sequencing adapters are incorporated, and the resulting fragments are sequenced. After aligning the sequence reads to the bacterial genome, the number of reads pertaining to each transposon insertion event is compared between the input and output populations. This comparison thereby identifies genes for

Melanie M. Pearson (ed.), *Proteus mirabilis: Methods and Protocols*, Methods in Molecular Biology, vol. 2021,
https://doi.org/10.1007/978-1-4939-9601-8_25,

which disruption provides an advantage (increased number of output–input reads) or a defect (decreased number of output–input reads) in survival under a given culture condition. Due to the quantitative nature of NGS, the importance of each gene in relation to every other gene can be determined for the given condition. This technique can provide information on both protein-coding and noncoding genes, including sRNA. Tn-seq can also give insight to genomic regions essential for survival under normal culture conditions, including essential protein domains and regulatory regions of essential genes, depending on the level of saturation of the mutant library. Furthermore, generation of a transposon mutant library in an isogenic mutant strain allows for mapping of genetic interactions within a given condition.

Limitations to the technique include the requirement for a complete genome sequence for the bacteria of interest, as transposon sequences can be quite short and accurate alignment to the bacterial genome is a critical part of data analysis. The type of transposon used to generate the mutant library can influence the breadth of the library. The transposons commonly used to make mutant libraries are either Tn*5* or mariner transposon derivatives. Tn*5* has a weak preference for AT-rich regions [1, 2]. Mariner/Himar1 transposons have an absolute requirement for insertion at TA dinucleotides [3, 4]. This requirement leads to an underrepresentation of transposon mutants in regions where the genome has high GC-content. An additional limitation of Tn-seq is the increased sensitivity to stochastic loss of transposon mutants. Stochastic loss of a mutant will artificially reduce the number of reads recovered from that mutant in output pools. Steps should be taken to ensure the number of copies of each mutant in the pool is high enough to overcome the effects of stochastic loss. It is also important to choose a rich medium to generate the transposon library. It is conceivable that a gene may be essential for growth in the medium used to generate the library and therefore unable to be assayed in the test medium, even if it would have been nonessential in the test medium.

Several methodologies are available for both transposon delivery and enrichment of transposon insertion junctions. Transposons can be delivered via suicide plasmid by mating a donor with a recipient strain, provided the bacteria of interest is not able to produce functional proteins for initiation of the plasmid origin of DNA replication [5–7]. Transposons can also be delivered by transforming competent bacteria with transposomes (transposon coupled with purified transposase protein), delivered by phage [8], or in vitro using a purified transposase enzyme followed by natural transformation [9]. A critical step prior to sequencing the transposon library is enrichment of the genomic DNA samples for transposon insertion junctions to ensure that the majority of the sequencing reads obtained are informative. For all methods this involves PCR,

Table 1
Strategies for transposon insertion junction enrichment in different Tn-seq methods[a]

Amplification strategy	Specific transposon required?	Physical DNA shearing	Amplicon length	Other pros and cons	Reference
Type II restriction enzyme → adapter → PCR	Yes	No	Uniform, therefore low PCR bias	Increased mapping ambiguity due to short reads	Goodman et al. [5], Khatiwara et al. [21], van Opijnen et al. [22]
Shearing → adapter → PCR	No	Yes	Variable, therefore increased PCR bias	Large quantity of DNA required	Gawronski et al. [23], Langridge et al. [2]
Nested arbitrary PCR → nested PCR	No	No	Variable, therefore increased PCR bias	Potential target bias	Christen et al. [24]
Shearing → adapter → restriction enzyme → circularization → PCR	No	Yes	Variable, therefore increased PCR bias	Large quantity of DNA required Potential target bias Decreased nonspecific sequence reads	Gallagher et al. [25]
Shearing → C-tailing → PCR	No	Yes	Variable, therefore increased PCR bias	Large quantity of DNA required	Klein et al. [26]
Single primer extension → C-tailing → PCR	No	No	Variable, therefore increased PCR bias	Use of a host-adapted transposon to improve mutation frequency	Dawoud et al. [27]
Simple PCR of barcode regions with two universal primers	Yes	No	Uniform, therefore low PCR bias	Requires construction of transposons with random barcodes Simple PCR protocol High throughput	Wetmore et al. [28]

[a]Modified from [10], with permission

which can introduce bias toward enrichment of small fragments. For a full comparison of methods, *see* Table 1 [10]. The method presented below makes use of type II restriction digestion to generate uniformly sized fragments prior to amplification. It is important to note that this method requires a specific type of transposon and thus cannot be used in combination with other methodologies. Once a particular method has been chosen for enrichment of fragments it must be strictly adhered to, and the appropriate transposon is employed for generation of the mutant library.

In this chapter, we supply a method for performing Tn-seq in *P. mirabilis* type strain HI4320. The method for generating the transposon library has been modified from Goodman et al. [5] and is dependent on delivery of a suicide plasmid from a donor strain modified specifically for use in HI4320 [11]. Transposon mutagenesis of HI4320 is achieved by mating with a donor strain harboring the suicide plasmid. The resulting library is then verified for randomness and pools of mutants prepared for infection. Genomic DNA from input and output pools is harvested, enriched for transposon insertion junctions, and gel-purified. Illumina sequencing is performed on the resulting fragments and the relative contribution of each gene to fitness in the given condition is determined. Although this method outlines techniques that have been tailored for *P. mirabilis* HI4320, the information provided can be adapted for other bacterial strains or species.

2 Materials

Prepare all solutions in sterile, purified water. Filter-sterilize all supplements added to agar and growth media.

2.1 General Reagents

1. Low-salt Lysogeny broth (LB) (per liter): 10 g of tryptone, 5 g of yeast extract, and 0.5 g of NaCl. Autoclave to sterilize.
2. Low salt LB agar: LB with the addition of 15 g of agar/L. Autoclave to sterilize; cool to 55 °C; pour into Petri dishes, approximately 25 mL/plate.
3. Antibiotics: 15 μg/mL of tetracycline, 25 μg/mL of kanamycin, and 100 μg/mL of ampicillin. When specified, antibiotics should either be added to LB, or to molten LB agar that has been cooled to 55 °C.
4. 37 °C incubator and 37 °C shaker incubator.
5. Distilled water (dH_2O).
6. Spectrophotometer and 1.5 mL semimicrocuvettes.
7. 1.5 mL sterile microcentrifuge tubes and microcentrifuge capable of 16,000 × *g*.
8. −80 °C ultralow laboratory freezer, −20 °C freezer, and 4 °C refrigerator.
9. Crushed ice.
10. Tris–acetic acid–EDTA (TAE) buffer: 40 mM Tris, 1 mM EDTA, and 20 mM glacial acetic acid. Dissolve Tris and EDTA in water before adding glacial acetic acid.

11. Agarose gel electrophoresis supplies: electrophoresis chamber, DNA size ladder, and DNA detection method (e.g., ethidium bromide and ultraviolet light box).
12. 6× Gel loading dye: 30% (v/v) glycerol, 0.25% (w/v) bromophenol blue, 0.25% (w/v) xylene cyanol.
13. Phosphate-buffered saline (PBS), per L: 8 g of NaCl, 0.2 g of KCl, 1.44 g of Na_2HPO_4, 0.24 g of KH_2PO_4, pH 7.4. Autoclave to sterilize.
14. Low Tris–EDTA buffer (LoTE buffer): 3 mM Tris, 0.2 mM EDTA; pH 7.5 (*see* **Note 1**).
15. 70% ethanol.
16. Vortex.
17. Vacuum centrifuge.
18. Heat block.
19. Orbital shaker.
20. Thermocycler capable of cooling at 0.1 °C/s.
21. PCR tubes.
22. Magnetic particle concentrator (MPC) for 1.5 mL tubes and PCR tubes.

2.2 Generation of a Transposon Mutant Library

2.2.1 Mating

1. *Escherichia coli* S17-1λ*pir* + pSAM_AraC (Addgene plasmid #91569) (*see* **Note 2**).
2. *P. mirabilis* type strain HI4320.
3. Sterile cotton-tipped swabs.
4. Sterile 0.45 μm filter disks.
5. Sterile forceps.
6. 10 mM L-arabinose.
7. Plate spreader.

2.2.2 Preparation of Mutants for Verification of Random Insertions

1. Genomic DNA extraction method of choice.
2. NanoDrop microvolume spectrophotometer.
3. HindIII and compatible buffer.
4. 3 M sodium acetate, pH 5.2.
5. 100% ethanol prechilled to −20 °C.
6. 70% ethanol, room temperature.
7. PCR DIG probe synthesis kit (Roche).
8. Forward and Reverse primer to amplify antibiotic resistance cassette (*see* **Note 3**).
9. Plasmid preparation of pSAM_AraC.
10. 1% (w/v) agarose gel in TAE buffer.

11. 0.25 M HCl.
12. 0.4 M NaOH.
13. Paper towels.
14. Large glass dish.
15. 0.35 mm Whatman filter paper.
16. Positively charged nylon membrane, pore size 0.45 μm.
17. 15 mL culture tube or similarly sized round object.
18. 20× saline–sodium citrate (SSC): 3 M NaCl, 300 mM sodium citrate, pH to 7.0.
19. Heat-sealable hybridization bag.
20. Heat sealer.
21. Prehybridization solution: 11 mL of DIG EasyHyb (Roche), 11 μL of 10 μg/mL salmon sperm.
22. Hybridization oven.
23. 2× SSC with 0.1% sodium dodecyl sulfate (SDS).
24. 1× SSC with 0.1% SDS.
25. PBST: PBS with 0.1% Tween 20.
26. Dry milk.
27. Anti-DIG-AP (alkaline phosphatase conjugated) antibody.
28. Detection buffer: 0.1 M Tris–HCl, 0.1 M NaCl, pH 9.5.
29. CSPD substrate for alkaline phosphatase.
30. Autoradiography film.
31. Film developer.
32. *Proteus mirabilis* genomic DNA.

2.3 Infection Studies

2.3.1 Bottleneck Assessment to Determine Optimal Number of Transposon Mutants Per Pool

1. *P. mirabilis* HI4320.
2. *P. mirabilis* mutant without a fitness defect in the infection model of choice.
3. Female CBA/J mice, age 5–6 weeks acclimated for 7 days prior to experiment.
4. Anesthetic: 100 mg/mL of ketamine, 100 mg/mL of xylazine.
5. Sterile 1 mL disposable Luer lock tip syringes with clear barrel.
6. 26 G × ½ (0.45 mm × 13 mm) needles.
7. Harvard Apparatus Pump 11 Elite infuse-only single syringe pump.
8. Sterile 10 mL disposable Luer lock tip syringes with clear barrel.
9. 30 G × ½ (0.3 mm × 13 mm) needles.
10. Sterile polyethylene tubing: I.D. 0.28 mm (0.011″), O.D. 0.61 mm (0.024″), 30.5 m (100′).

11. Povidone iodine 10% solution.
12. Supplies for humane euthanasia of mice.
13. Supplies for organ homogenization.

2.3.2 Pooling of Mutants and Long-Term Storage

1. Tetracycline-resistant, kanamycin-resistant transposon mutants.
2. Sterile cotton-tipped swabs.
3. 50% glycerol.

2.3.3 Preparation of Pools for Inoculum

1. Pool of transposon mutants.
2. Female CBA/J mice, age 5–6 weeks acclimated for 7 days prior to experiment.
3. Anesthetic (per milliliter): 100 mg of ketamine, 100 mg of xylazine.
4. Sterile 1 mL disposable Luer-Lok™ tip syringes with clear barrel.
5. 26 G × ½ (0.45 mm × 13 mm) needles.
6. Harvard Apparatus Pump 11 Elite infuse-only single syringe pump.
7. Sterile 10 mL disposable Luer lock tip syringes with clear barrel.
8. 30 G × ½ (0.3 mm × 13 mm) needles.
9. Sterile polyethylene tubing: I.D. 0.28 mm (0.011″), O.D. 0.61 mm (0.024″), 30.5 m (100′).
10. Povidone iodine 10% solution.

2.3.4 Harvesting of Mutants Postinfection

1. Homogenized organs.
2. Sterile cotton-tipped swab.
3. 50 mL conical tubes.

2.4 Preparation of Transposon DNA for Illumina Sequencing

2.4.1 Preparation of Genomic DNA

1. Input and output pools of transposon mutants.
2. TE buffer: 10 mM Tris–HCl, 1 mM EDTA, pH 7.0.
3. 10% SDS.
4. 20 mg/mL of proteinase K.
5. 100 mg/mL of RNaseA.
6. 5 M NaCl.
7. CTAB–NaCl (per 100 mL): 4.1 g of NaCl, 10 g of cetrimonium bromide (CTAB). Dissolve 4.1 g of NaCl in 80 mL dH_2O. Add 10 g of CTAB incrementally while heating and stirring. Adjust final volume to 100 mL.
8. Heat block or water bath set to 65 °C.

9. 24:1 chloroform–isoamyl alcohol.
10. 25:24:1 phenol–chloroform–isoamyl alcohol.
11. Isopropanol.
12. 10 mM Tris–HCl pH 8.0.
13. NanoDrop or similar spectrophotometer.

2.4.2 Single-Stranded PCR

1. Platinum Pfx buffer.
2. 10 mM dNTPs.
3. 50 mM $MgCl_2$.
4. 1 pmol/μL Primer BioSamA, HPLC purified (*see* **Note 4**).
5. High-fidelity DNA polymerase (e.g., Platinum Pfx polymerase).
6. 200 μL PCR strip tubes.
7. Normalized gDNA from input and output pools of mutants.
8. gDNA from wild-type *P. mirabilis* HI4320.
9. Purified pSAM_AraC plasmid DNA.
10. Column-based PCR cleanup kit.

2.4.3 Hybridization to Streptavidin-Coated Beads

1. Dynabeads M-280 streptavidin.
2. Magnetic particle concentrator (MPC) for 1.5 mL tubes and PCR tubes.
3. Bind and wash buffer (2×): 2 M NaCl, 10 mM Tris, and 1 mM EDTA; pH 7.5.
4. Single-stranded PCR products.
5. Nonelectronic multichannel pipette P100.

2.4.4 Double-Stranded DNA Synthesis

1. 2.5 mg/mL hexanucleotide mix.
2. 10 mM dNTPs.
3. Klenow fragment (3′ → 5′ exo-).

2.4.5 Trimming of Fragments to Uniform Size

1. Buffer EB: 10 mM Tris–HCl, pH 8.5.
2. 100 μM M12_top and M12_bottom primers in buffer EB (*see* **Note 5**).
3. 1 M NaCl.
4. 32 mM *S*-adenosylmethionine (SAM).
5. MmeI and enzyme compatible buffer.

2.4.6 Preparation and Ligation of Double-Stranded Adapters

1. One LIB_AdaptT and one LIB_AdaptB, for each sample to be multiplexed in the same lane (*see* Table 2 for example sequences): 100 μM in buffer EB.
2. Buffer EB.

Table 2
DNA sequences of adapters with unique barcodes

Name	DNA sequence[a]
LIB_AdaptT_A	5′-TTCCCTACACGACGCTCTTCCGATCT**TTTT**NN-3′
LIB_AdaptT_B	5′-TTCCCTACACGACGCTCTTCCGATCT**ACCA**NN-3′
LIB_AdaptT_C	5′-TTCCCTACACGACGCTCTTCCGATCT**ACGT**NN-3′
LIB_AdaptT_D	5′-TTCCCTACACGACGCTCTTCCGATCT**AGGA**NN-3′
LIB_AdaptT_E	5′-TTCCCTACACGACGCTCTTCCGATCT**AGTC**NN-3′
LIB_AdaptT_F	5′-TTCCCTACACGACGCTCTTCCGATCT**ATCG**NN-3′
LIB_AdaptT_H	5′-TTCCCTACACGACGCTCTTCCGATCT**CATG**NN-3′
LIB_AdaptT_I	5′-TTCCCTACACGACGCTCTTCCGATCT**CCAA**NN-3′
LIB_AdaptT_J	5′-TTCCCTACACGACGCTCTTCCGATCT**CCCC**NN-3′
LIB_AdaptT_K	5′-TTCCCTACACGACGCTCTTCCGATCT**CGAT**NN-3′
LIB_AdaptT_O	5′-TTCCCTACACGACGCTCTTCCGATCT**GCTA**NN-3′
LIB_AdaptT_P	5′-TTCCCTACACGACGCTCTTCCGATCT**GGAA**NN-3′
LIB_AdaptT_Q	5′-TTCCCTACACGACGCTCTTCCGATCT**GGGG**NN-3′
LIB_AdaptT_R	5′-TTCCCTACACGACGCTCTTCCGATCT**GTAC**NN-3′
LIB_AdaptT_S	5′-TTCCCTACACGACGCTCTTCCGATCT**GTCA**NN-3′
LIB_AdaptT_U	5′-TTCCCTACACGACGCTCTTCCGATCT**TATA**NN-3′
LIB_AdaptT_V	5′-TTCCCTACACGACGCTCTTCCGATCT**TCAG**NN-3′
LIB_AdaptT_W	5′-TTCCCTACACGACGCTCTTCCGATC**TTCGA**NN-3′
LIB_AdaptT_Y	5′-TTCCCTACACGACGCTCTTCCGATCT**TTAA**NN-3′
LIB_AdaptT_Z	5′-TTCCCTACACGACGCTCTTCCGATCT**AAAA**NN-3′
LIB_AdaptT_AA	5′-TTCCCTACACGACGCTCTTCCGATCT**GAAG**NN-3′
LIB_AdaptT_CC	5′-TTCCCTACACGACGCTCTTCCGATCT**CCTT**NN-3′
LIB_AdaptT_DD	5′-TTCCCTACACGACGCTCTTCCGATCT**AACC**NN-3′
LIB_AdaptT_EE	5′-TTCCCTACACGACGCTCTTCCGATCT**TTGG**NN-3′
LIB_AdaptB_A	5′-**AAAA**AGATCGGAAGAGCGTCGTGTAGGGAA-3′
LIB_AdaptB_B	5′-**TGGT**AGATCGGAAGAGCGTCGTGTAGGGAA-3′
LIB_AdaptB_C	5′-**ACGT**AGATCGGAAGAGCGTCGTGTAGGGAA-3′
LIB_AdaptB_D	5′-**TCCT**AGATCGGAAGAGCGTCGTGTAGGGAA-3′
LIB_AdaptB_E	5′-**GACT**AGATCGGAAGAGCGTCGTGTAGGGAA-3′
LIB_AdaptB_F	5′-**CGAT**AGATCGGAAGAGCGTCGTGTAGGGAA-3′
LIB_AdaptB_H	5′-**CATG**AGATCGGAAGAGCGTCGTGTAGGGAA-3′

(continued)

Table 2
(continued)

Name	DNA sequence[a]
LIB_AdaptB_I	5′-**TTGG**AGATCGGAAGAGCGTCGTGTAGGGAA-3′
LIB_AdaptB_J	5′-**GGGG**AGATCGGAAGAGCGTCGTGTAGGGAA-3′
LIB_AdaptB_K	5′-**ATCG**AGATCGGAAGAGCGTCGTGTAGGGAA-3′
LIB_AdaptB_O	5′-**TAGC**AGATCGGAAGAGCGTCGTGTAGGGAA-3′
LIB_AdaptB_P	5′-**TTCC**AGATCGGAAGAGCGTCGTGTAGGGAA-3′
LIB_AdaptB_Q	5′-**CCCC**AGATCGGAAGAGCGTCGTGTAGGGAA-3′
LIB_AdaptB_R	5′-**GTAC**AGATCGGAAGAGCGTCGTGTAGGGAA-3′
LIB_AdaptB_S	5′-**TGAC**AGATCGGAAGAGCGTCGTGTAGGGAA-3′
LIB_AdaptB_U	5′-**TATA**AGATCGGAAGAGCGTCGTGTAGGGAA-3′
LIB_AdaptB_V	5′-**CTGA**AGATCGGAAGAGCGTCGTGTAGGGAA-3′
LIB_AdaptB_W	5′-**TCGA**AGATCGGAAGAGCGTCGTGTAGGGAA-3′
LIB_AdaptB_Y	5′-**TTAA**AGATCGGAAGAGCGTCGTGTAGGGAA-3′
LIB_AdaptB_Z	5′-**TTTT**AGATCGGAAGAGCGTCGTGTAGGGAA-3′
LIB_AdaptB_AA	5′-**CTTC**AGATCGGAAGAGCGTCGTGTAGGGAA-3′
LIB_AdaptB_CC	5′-**AAGG**AGATCGGAAGAGCGTCGTGTAGGGAA-3′
LIB_AdaptB_DD	5′-**GGTT**AGATCGGAAGAGCGTCGTGTAGGGAA-3′
LIB_AdaptB_EE	5′-**CCAA**AGATCGGAAGAGCGTCGTGTAGGGAA-3′

[a]Bold nucleotides indicate the 4 bp barcode. Each LIB_AdaptT should be paired with a corresponding LIB_AdaptB

3. 1 M NaCl.
4. 10× T4 DNA ligase buffer (typically supplied by manufacturer along with ligase).
5. 1× T4 DNA ligase buffer, ice-cold.
6. T4 DNA ligase.

2.4.7 Amplification of Final Fragments

1. DNA polymerase buffer (e.g., Platinum Pfx buffer).
2. 10 mM dNTPs.
3. 50 mM $MgCl_2$.
4. 5 μM LIB-PCR 5 (*see* **Note 6**).
5. 5 μM LIB-PCR 3 (*see* **Note 6**).
6. High-fidelity DNA polymerase (e.g., Platinum Pfx polymerase).

2.4.8 Size Selection of Final Fragments

1. 2% (w/v) agarose gel in TAE buffer.
2. High sensitivity, non-UV compatible nucleic acid dye.
3. 10× xylene cyanol DNA dye: 20% (wt/vol) Ficoll 400, 0.1 M disodium EDTA, 0.25% wt/vol xylene cyanol (*see* **Note** 7).
4. DNA ladder with markers at 100 and 200 bp.
5. Non-UV gel illuminator.
6. Clean razor blade.
7. Preweighed 1.5 mL microcentrifuge tube.
8. Agarose gel DNA cleanup kit.
9. QuBit or similar spectrophotometer.

2.5 Illumina Sequencing, Bioinformatic Analysis, and Statistics

1. Amplified final fragments.
2. TapeStation.
3. Illumina sequencer.
4. Reagents for cluster generation during sequencing.
5. Bacteriophage ΦX DNA.
6. *P. mirabilis* HI4320 chromosomal and plasmid sequence (NCBI accession numbers NC_010054 and NC_010555).
7. MapSAM, a Perl-based software available for download from supplemental data at https://www.cell.com/cell-host-microbe/fulltext/S1931-3128(09)00281-9.
8. Tnseq R package, available from the Comprehensive R Archive Network.

3 Methods

3.1 Generation of a Transposon Mutant Library

3.1.1 Mating

1. Inoculate 3 mL of LB medium containing kanamycin and ampicillin with *E. coli* S17 harboring pSAM_AraC and incubate at 37 °C with aeration (shaking at 225 rpm) until saturated, typically 16 h.
2. Dilute 1:40 in LB medium containing kanamycin and ampicillin, and incubate at 37 °C with aeration (shaking at 225 rpm) until cultures have reached mid-log phase (*see* **Note 8**).
3. Swab bacteria off agar plates spread with a fresh lawn of *P. mirabilis* HI4320 into 3 mL of LB medium (*see* **Note 9**).
4. Mix donor strain (*E. coli* S17 + pSAM_AraC) with recipient strain (*P. mirabilis* HI4320) in a 3:1 ratio of donor–recipient for a total volume of 1.5 mL in a microcentrifuge tube.
5. Concentrate the mixture by centrifugation at 16,000 × *g* for 1 min. Add 50 μL of fresh LB medium and incubate on benchtop for 5 min prior to gently resuspending (*see* **Note 10**).

6. Incubate for 60 min at 37 °C with aeration (shaking at 225 rpm).
7. Use sterile forceps to place a 0.45 μm filter disk onto an LB agar plate supplemented with 10 mM L-arabinose (to induce expression from the pBAD promoter) (*see* **Note 11**).
8. Spread the entire volume of the mating mixture using the side of a 200 μL pipette tip onto the 0.45 μm filter disk. Incubate at 30 °C upright for 2 h (*see* **Note 12**).
9. Transfer the filter to an LB agar plate with tetracycline and kanamycin and wash the bacteria off the filter using 100 μL of LB medium. Spread bacteria evenly over the surface of the plate and incubate at 37 °C overnight (*see* **Note 13**). Approximately 50–100 of the Tet^R Kan^R colonies should be screened for sensitivity to ampicillin (*see* **Note 14**).

3.1.2 Verification of the Transposon Library

It is critical to validate that the plasmid harboring the transposase is absent from the final library clones to prevent mobilization of the transposon during screening of the mutant library. Typical strategies include screening for susceptibility to appropriate antibiotics, phenotypic assays, and PCR. When using PCR to screen transposon mutants, design a primer pair homologous to the kanamycin cassette and within the transposon to test for transposon insertion. A second primer pair should amplify a region of the pSAM_AraC backbone to verify loss of the plasmid and lack of integration into the chromosome. *P. mirabilis* HI4320 transposon mutants should be tetracycline resistant, kanamycin resistant, and ampicillin sensitive. A previous study in HI4320 reported less than 0.001% ampicillin-resistant colonies [7]. HI4320 readily produces urease enzyme and expression or lack thereof can be assayed on urease segregation agar [12]. The library can be screened for mutants in urease production as an indicator that chromosomal transposon insertion events have occurred.

It is equally critical to verify the randomness of the transposon library and estimate the rate of occurrence of multiple transposon insertion events within a single chromosome. One method for achieving this is a Southern blot, which we provide a detailed protocol for below. Previous studies of *P. mirabilis* and *Klebsiella pneumoniae* Tn-seq reported the incidence rate of probe homology to multiple DNA fragments per sample, resulting from either multiple transposon insertion events or incomplete digestion, as ~13% and ~11% respectively [6, 7]. In each of these cases, the putative fitness factors identified were able to be reliably verified in a secondary screen.

1. Begin preparing the digoxigenin-labeled probe by designing a pair of 20 bp primers of similar melting temperature, homologous to the transposon such that a 500 bp product is generated (*see* **Note 3**).

2. Using a PCR DIG probe synthesis kit, set up the following PCR reactions:

Reagent	Labeled probe (μL)	Unlabeled probe (μL)
dH_2O	36.75	36.75
PCR buffer (tube 3)	5	5
PCR DIG synthesis mix (tube 2)	5	–
dNTP (tube 4)	–	5
Forward primer (100 μM)	1	1
Reverse primer (100 μM)	1	1
Polymerase (tube 1)	0.75	0.75
pSAM_AraC (30 pg/μL)	0.5	0.5

The numbers included correspond to the numbered reagent tubes provided in the PCR DIG probe synthesis kit.

3. Cycle the PCR reaction:
Initial denaturation 95 °C, 2 min
30 cycles:

Denaturation	95 °C, 30 s
Annealing	57 °C, 30 s (temperature dependent on primer design)
Extension	68 °C, 50 s

Final extension 68 °C, 7 min

4. Check 5 μL of the resulting product on a 1% agarose gel. If the probe is labeled well, the band will run noticeably slower than the unlabeled product. Store the probe at −20 °C.
5. To prepare transposon mutants for screening with Southern blot, inoculate LB medium supplemented with kanamycin with a single transposon mutant colony and incubate at 37 °C overnight with aeration (*see* **Note 15**).
6. Extract genomic DNA, taking steps to maximize product yield such that samples contain more than 5 μg in the minimum volume allowed by the extraction method of choice (*see* **Note 16**).
7. Determine the concentration of DNA in all samples and normalize to the lowest concentrated sample. Aim for greater than 40 ng/μL final concentration.
8. Digest entire volume of normalized genomic DNA and control DNA with HindIII in enzyme compatible buffer, adding half

the calculated amount of enzyme, and aliquot 50 μL into multiple PCR tubes. Enzyme efficiency is enhanced by setting up multiple aliquots of the same digestion reaction (*see* **Notes 15** and **17**).

Example Reaction Setup:

10× Enzyme buffer	20 μL
gDNA	125 μL
HindIII	2 μL
dH_2O	51 μL
Total	198 μL

9. Digest for 3 h at 37 °C, aliquot 49.5 μL into separate PCR tubes, add an additional 0.5 μL of HindIII enzyme to each tube, and continue incubation overnight (*see* **Note 18**).
10. Concentrate the digested DNA to a volume that is easily loaded into an agarose gel by precipitating with ethanol: first, add 1/10 volume of 3 M sodium acetate, pH 5.2, to the gDNA digest. Mix gently.
11. Add 2.5 volumes (calculated after sodium acetate addition) of cold 100% ethanol. Mix gently and incubate at −80 °C for at least 30 min to precipitate DNA.
12. Pellet precipitate at 16,000 × *g* for 5 min and remove the supernatant gently by drawing off the liquid with a pipette from the side of the tube opposite the pellet, starting near the top and moving downward. A white pellet should be visible at this step for all transposon samples.
13. Add 1 mL of room temperature 70% ethanol, and invert the tube several times to mix.
14. Repeat **step 12**.
15. Remove all traces of ethanol from the samples by evaporation on the benchtop or in a vacuum centrifuge.
16. Gently resuspend pellet in 18 μL of dH_2O and heat for 5 min at 65 °C to aid in resuspension. Avoid vortexing to resuspend at this step to prevent DNA shearing. Store digested genomic DNA at 4 °C.
17. The first step of alkaline transfer during Southern blot is to perform gel electrophoresis of the entire volume of digested, concentrated genomic DNA from transposon mutants, wild type and pSAM_AraC on a 1% agarose gel.
18. Stain the gel with ethidium bromide and image with UV fluorescence (*see* **Note 19**). Bands on the gel should be crisp and evenly distributed throughout the length of the lane; if

not, genomic DNA digestion and gel electrophoresis steps should be repeated.

19. Rinse the gel in dH_2O, then incubate in 250 mL of 0.25 M HCl at room temperature for 20 min with gentle agitation on an orbital shaker (*see* **Note 20**).
20. Rinse the gel with dH_2O, then incubate in 250 mL of 0.4 M NaOH at room temperature for 20 min with agitation to denature the double stranded DNA.
21. Refold or cut paper towels to form a neat stack at least 3 cm thick in a glass dish making sure that the width of the towels is greater than the agarose gel (Fig. 1).
22. Place four pieces of dry 0.35 mm-thick Whatman filter paper cut to the size of the agarose gel on top of the stack of paper towels. Wet a fifth piece of Whatman filter paper in 0.4 M NaOH and layer on top of the stack (Fig. 1).
23. Wet positively charged nylon membrane cut to the size of the agarose gel in dH_2O and add to the stack (Fig. 1). Make sure to remove any air bubbles by rolling a 15 mL culture tube dampened in dH_2O over the surface of the stack. It is critical that air bubbles be removed from all layers in the transfer stack (*see* **Note 21**).
24. Add the agarose gel to the stack with wells facing up (to minimize air bubbles) and smooth over the membrane to ensure even contact with the membrane using the same technique as in **step 23** (Fig. 1).
25. Add three pieces of Whatman filter paper, cut to the size of the agarose gel and dampened in 0.4 M NaOH, on top of the agarose gel (Fig. 1). Again remove any air bubbles that may have formed between layers.
26. Add two pieces of Whatman filter paper cut to the width of the agarose gel and length large enough to span the distance from the gel stack to a fluid reservoir and wet with 0.4 M NaOH. Fill this fluid reservoir with 500 mL of 0.4 M NaOH (Fig. 1).

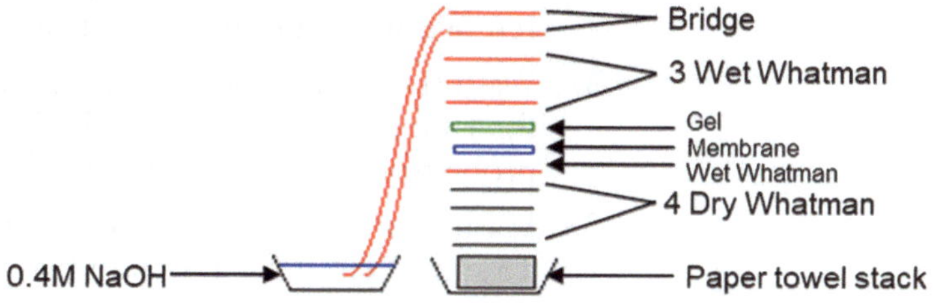

Fig. 1 Assembly of transfer stack for Southern blot. The transfer stack should be assembled as pictured to facilitate migration of the digested DNA from the agarose gel into the nylon membrane via capillary action of buffer from the reservoir into the dry Whatman and dry paper towel

27. Allow transfer to proceed at room temperature for at least 2 h (*see* **Note 22**).
28. Disassemble the transfer stack and rinse the membrane in room temperature 2× SSC (*see* **Note 23**).
29. Dry the membrane completely. Drying is more efficient if the membrane can be hung to facilitate air circulation.
30. Hybridize the probe to the membrane by first, rehydrating the membrane in 2× SSC warmed to 65 °C.
31. Transfer the membrane to a heat-sealable hybridization bag that is 2–3 cm larger than the membrane on all sides and add 74 μL of prehybridization solution per cm^2 of membrane, warmed to 65 °C. For a 10 cm × 15 cm membrane, use 11 mL of prehybridization solution. The salmon sperm DNA in the prehybridization solution serves as a blocking agent and prevents nonspecific binding of the probe to the membrane (*see* **Note 24**).
32. Remove as much air as possible from the hybridization bag and seal the bag. Insert the hybridization bag into the hybridization tube and use laboratory tape to adhere the corners of the bag to the tube to facilitate even distribution of the prehybridization solution as the hybridization tube turns in the oven (*see* **Note 25**).
33. Incubate 1 h at 65 °C with rotation in a hybridization oven.
34. Denature the DIG-labeled probe from **step 3**, by incubating at 95 °C for 10 min.
35. Add 2 μL of denatured probe per mL of prehybridization solution into the hybridization bag, reseal and continue incubation of the membrane in the hybridization oven at 65 °C overnight. The probe will bind to any complementary sequences on the membrane.
36. The next steps serve to reduce nonspecific binding of the probe to membrane bound DNA. First, remove the membrane from the hybridization bag and wash in 250 mL of room temperature 2× SSC for 5 min. Repeat for a total of two washes to remove any unbound probe from the membrane.
37. Wash the membrane in 250 mL of 60 °C 2× SSC + 0.1% SDS for 15 min. Repeat for a total of two washes to disassociate the probe from sequences with low homology.
38. Wash membrane in 250 mL of 60 °C 1× SSC + 0.1% SDS for 15 min. Repeat for a total of two washes (*see* **Note 26**).
39. Rinse membrane in PBS.
40. Begin detection of the bound probe by incubating the membrane in PBST +5% milk at room temperature with shaking for 1 h to block nonspecific binding of antibody to the membrane (*see* **Note 27**).

41. Incubate membrane in a 1:5000 dilution of anti-DIG-AP antibody in PBST +5% milk at room temperature with shaking for 1 h.
42. Wash membrane in PBST for 5 min to remove unbound antibody. Repeat for a total of 4 washes.
43. Equilibrate membrane in detection buffer for 5 min at room temperature.
44. Mix 20 μL of CSPD and 1.98 mL of detection buffer and add solution to a sealable bag slightly larger than the membrane on all sides. Transfer the membrane to the bag.
45. Ensure even contact between CSPD and the membrane by removing any air bubbles from the bag. Seal bag and incubate membrane at room temperature for 5 min.
46. Squeeze any extra buffer out of the bag and reseal. Incubate at 37 °C for 10 min.
47. Expose film to the membrane until desired intensity of bands is achieved and develop film (*see* **Note 28**). The size of the fragments detected by the probe can be determined by measuring the distance of the band from the top of the membrane and comparing that to the DNA ladder from the image of the agarose gel with UV ruler.

Detection of the probe should reveal a single distinct band per lane. If more than one band is detected, it can mean one of three things: (1) enzyme digestion was incomplete, (2) the genome of the transposon mutant tested has more than one transposon insertion, or (3) the genomic DNA was harvested from more than one colony (Fig. 2). The bands produced should represent differently sized fragments in each lane. This is an indicator that the transposon is inserted in a different region of the chromosome in each mutant strain; thus, when the genomic DNA is randomly digested, there is a high probability that any resulting fragments containing transposon sequence are different sizes. If the bands are all similarly sized, the transposon has inserted nonrandomly into the genome. If the fragment size is similar to the band generated from the pSAM_AraC control, then the plasmid harboring the transposon has not been lost from the recipient strain.

3.2 Infection Studies

Prior to conducting Tn-seq, the optimal number of transposon mutants must be determined for the organism of interest in the infection model of interest. This typically includes a bottleneck assessment to determine if there are constraints in the infection model that may sharply limit population size. For instance, if the anatomy of the ureters is such that only a small proportion of bacteria colonizing the bladder can ascend the ureters to reach the kidneys, this would result in a founding population initially

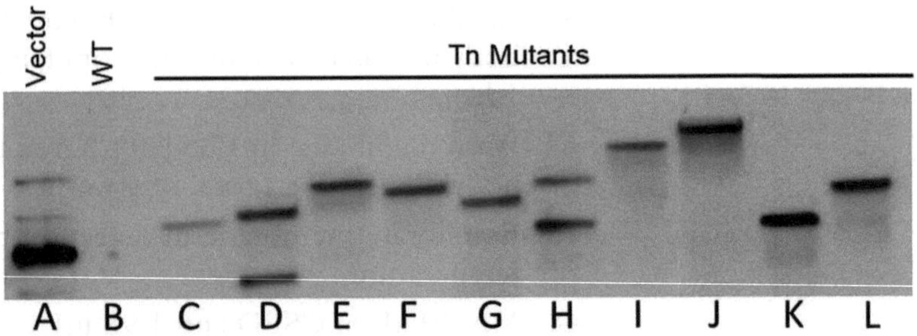

Fig. 2 Southern blot of transposon mutants indicates random mutagenesis. A representative Southern blot of isolated transposon delivery vector (lane A), genomic DNA isolated from the wild-type strain to be mutagenized (lane B), and transconjugants with appropriate antibiotic resistance profiles (lanes C–L). DNA was digested with HindIII and Southern blotting was performed using a digoxigenin-labeled probe against the resistance cassette harbored within the transposon. Variation in fragment size is indicative of transposon insertion into random chromosomal locations, and the presence of a single band within a given lane indicates that a single transposon insertion event occurred. Double bands (lanes D and H) are indicative of either incomplete digest or multiple insertion events. None of the fragments in lanes C–L were similarly sized to the fragment in lane A indicating that the transposon delivery vector is suicidal in the strain of interest

colonizing and expanding in the kidneys. For Tn-seq analysis, this would lead to the false interpretation that other mutants are defective for persistence within the kidney, when in fact they never had a chance to reach the kidneys to establish infection. Bottleneck assessments typically involve testing various ratios of a tagged bacterial strain and wild-type strain to determine if the proportion of tagged:wt is similar in the organs of interest postinfection to the proportion of tagged:wt in the input inoculum. For Tn-seq, the goal is to achieve the greatest dilution of tagged strain to wt strain, while maintaining a similar ratio as was present in the input (indicating lack of a significant bottleneck).

3.2.1 Bottleneck Assessment to Determine Optimal Number of Transposon Mutants Per Pool

1. Inoculate 5 mL of LB broth with WT *P. mirabilis* HI4320, and 5 mL of LB + antibiotic with a tagged strain (such as a mutant that contains a selectable marker and is not expected to have a fitness defect in the infection model of choice).
2. Incubate cultures for 10 h at 37 °C with aeration (shaking at 225 rpm) (*see* **Note 29**).
3. Adjust cultures to a cell density appropriate for the desired infection type and combine in a 1:1 ratio as well as a series of ratios for determination of minimum colonization density (lowest bacterial burden observed in any mouse) and maximum number of transposon mutants per pool (example ranges are 1:100 mutant to WT, 1:1000, 1:10,000, and 1:100,000).

4. Inoculate mice, and plate an aliquot of each inoculum on LB agar with and without antibiotic selection to quantify the starting ratio of mutant to WT (*see* **Note 30**).
5. After the desired period of time (e.g., 96 h for *P. mirabilis* HI4320 in the CAUTI model), humanely euthanize the mice, remove organs of interest, homogenize, and plate appropriate dilutions on LB agar with and without antibiotic selection to quantify the minimum colonization density of each organ and the competitive index of mutant to WT (*see* **Note 31**).
6. To determine the maximum number of mutants that can be pooled while minimizing stochastic loss from bottlenecks, divide the observed minimum colonization density for the organ of interest (this is the lowest CFU recovered from the organ) by the mutant to wild-type ratio which produced a competitive index within +/− 100 in bottleneck experiments. For example, if the suitable mutant to wild type ratio is 1:100, there must be 100 copies of each mutant within the pool to avoid a bottleneck. If the minimal colonization density in the organ of interest is 100,000 CFU, the maximum number of mutants that can be included in the pool without being subject to a bottleneck is 100,000/100, or 1000 mutants.

3.2.2 Pooling of Mutants and Long-Term Storage

1. Pools of mutants should be generated in an appropriate size to accommodate any potential bottlenecks that may occur in the model system of choice. For a murine model of catheter associated urinary tract infection with *P. mirabilis* strain HI4320, pools of 10,000 mutants were generated (*see* **Note 32**).
2. Enumerate the number of tetracycline-resistant, kanamycin-resistant colonies post-mating (**step 9** in Subheading 3.1.1) and select an appropriate number of plates for the size of the pool to be generated. Depending on the total number of mutants to be assayed and the pool size allowed by the model system of choice, several individual mutant pools may need to be generated.
3. Pool colonies by swabbing as many bacterial cells as possible off each plate using a cotton swab moistened in PBS, resuspend bacteria in a small volume of PBS (typically 5 mL), and mix well to ensure the mixture of mutants is homogenous (*see* **Note 33**).
4. Adjust the optical density of the pool to $OD_{600} = 4.0$ (~4×10^9 CFU/mL) with LB and mix 1:1 with 50% glycerol, again making sure the mixture is homogenous. Store each mutant pool in several 1 mL aliquots at −80 °C.

3.2.3 Preparation of Mutant Pools for Inoculum

1. Thaw an appropriate number of 1 mL aliquots of transposon mutant pools on ice. For instance, we conducted infection studies with 2 pools of 10,000 mutants for *P. mirabilis* HI4320 in the CAUTI model.

2. Dilute thawed pools 1:10 in LB with kanamycin.
3. Incubate at 37 °C with aeration (shaking at 225 rpm) for 10 h (*see* **Note 34**).
4. Adjust cultures to a cell density appropriate for the desired infection type. Inoculate mice, and dilution plate an aliquot of each inoculum on LB agar with kanamycin to quantify the starting input (*see* **Note 30**). Reserve five 1 mL aliquots of each input mutant pool, pellet at 8000 × *g* for 3 min and store pellets at −20 °C or proceed directly to preparation of transposon DNA for Illumina sequencing.

3.2.4 Harvesting of Mutants Post-infection

1. After the desired period of time (e.g., 96 h for *P. mirabilis* HI4320 in the CAUTI model), humanely euthanize the mice, remove organs of interest, and homogenize in 2 mL of PBS.
2. Remove 100 μL to dilute and plate for determination of colonization density.
3. Plate the entirety of the remaining homogenate 125 μL at a time onto LB agar plates with kanamycin. Usually 16 plates per organ are needed (*see* **Note 35**).
4. Incubate at 37 °C overnight.
5. Using a sterile cotton swab, remove all bacterial growth from an individual plate. Dislodge the bacteria from the swab by swirling into a 50 mL conical tube prefilled with sterile PBS (*see* **Note 36**).
6. Repeat for each additional plate from a single organ into the same tube. If intending to use barcoding to distinguish individual organs from individual mice, be sure to swab the stacks of plates from each individual homogenate into separate conical tubes.
7. Vortex well to maintain a homogenous mixture of bacteria in each conical, and aliquot 1 mL of the suspended mix into each of five 1.5 mL microcentrifuge tubes. The five replicate aliquots are to ensure sufficient material for replicate DNA extractions if necessary (*see* **Note 37**).
8. Centrifuge at 16,000 × *g* for 3 min to pellet; remove PBS.
9. Store pellet at −20 °C, or proceed directly to extraction of genomic DNA.

3.3 Preparation of Transposon DNA for Illumina Sequencing

The following protocol is a slight modification to that of Goodman et al. [11]. Transposon insertion junctions of gDNA are enriched from the input and output pools via amplification with single stranded PCR utilizing a biotinylated primer. The single stranded products are then bound to streptavidin coated magnetic beads, allowing for their purification from contaminating genomic DNA. A second strand is then synthesized with DNA polymerase I Klenow fragment and random hexanucleotide primers. The resulting double-stranded fragments are trimmed to uniform size by MmeI digest, which is intended to reduce the introduction of size bias during PCR amplification. Using this digestion strategy, the fragments capture one end of the transposon and 16 bp of chromosomal DNA at the insertion junction (*see* **Note 38**). MmeI digestion results in a 2 bp overhang, which is exploited for ligation of double stranded adapters containing both sample specific barcodes that facilitate demultiplexing, and sequences required for cluster generation on the Illumina flowcell. Ligated fragments are then amplified by PCR with primers homologous to the transposon (containing Illumina specific sequence) and the sample specific barcoded adapter, resulting in products that are unbound to the streptavidin-coated beads. The final PCR products are then purified on an agarose gel followed by gel extraction.

3.3.1 Preparation of Genomic DNA

1. Thaw pelleted input and output pools of bacteria (from **step 4**, Subheading 3.2.3, and **step 9**, Subheading 3.2.4, respectively) and resuspend pellets in TE buffer, Q.S. to 567 μL (*see* **Note 39**).
2. Add 30 μL of 10% SDS, 3 μL of 20 mg/mL proteinase K, and 7 μL of 100 mg/mL RNaseA. Mix well and incubate for 1.5–2 h at 37 °C until lysis has occurred. Appearance of the solution will transition from turbid culture to relatively clear lysate.
3. Add 100 μL of 5 M NaCl and mix well.
4. Add 80 μL of CTAB–NaCl and mix well. Incubate at 65 °C for 10 min (*see* **Note 40**).
5. Add 0.8 mL of 24:1 chloroform–isoamyl alcohol and invert to mix. Centrifuge at 16,000 × *g* for 5 min (*see* **Note 41**).
6. Carefully remove the top layer to a new 1.5 mL microcentrifuge tube with a P200 pipette, making sure not to contact any white precipitate that may have formed between the aqueous and organic layers (*see* **Note 42**).
7. Repeat **steps 5** and **6**. Any samples that may have been separated into multiple tubes can be combined as long as the total volume is less than 750 μL.
8. Add 25:24:1 phenol–chloroform–isoamyl alcohol to fill the microcentrifuge tube and invert to mix. Centrifuge at 16,000 × *g* for 5 min (*see* **Note 43**).
9. Very carefully remove the top layer to a new 1.5 mL microcentrifuge tube, making sure not to contact any white precipitate that may have formed between the aqueous and organic layers. Any contact with the organic layer at this point will transfer contaminant phenol into the final product and should be avoided. The aqueous solution should be clear. If it is not,

repeat **steps 8** and **9**, combining any samples in multiple tubes as long as the total volume is less than 750 μL.

10. Add at minimum 0.6 volume of isopropanol and invert to mix. A stringy precipitate should form. Incubate at −20 °C for at least 20 min or overnight.
11. Centrifuge at 16,000 × *g* for 5–10 min.
12. Remove isopropanol from pellet and resuspend in 70% ethanol.
13. Centrifuge at 16,000 × *g* for 5 min. Remove supernatant. Remove residual ethanol with a vacuum centrifuge. Alternatively, incubate tubes open on benchtop overnight.
14. Resuspend in 50 μL of 10 mM Tris–HCl pH 8.0 (*see* **Note 44**).
15. Determine the concentration of extracted genomic DNA (a spectrophotometer capable of reading microliter volumes, e.g., NanoDrop, is preferred) and normalize with dH_2O to 2 μg in 80 μL. Store at 4 °C or proceed to next section (*see* **Note 45**).

3.3.2 Single-Stranded PCR

1. Make a master mix of the following reagents on ice for the number of reactions to be performed plus one additional reaction (*see* **Note 46**).

Reagent	Volume per reaction (μL)
Pfx buffer	10
10 mM dNTPs	2
50 mM $MgCl_2$	2
Primer BioSamA (1 pmol/μL)	5
Pfx polymerase	1

2. Add 20 μL of master mix to a 200 μL PCR tube, then add 80 μL of normalized extracted genomic DNA from **step 15** in Subheading 3.3.1.
3. Split the reaction into two tubes with 50 μL in each PCR tube.
4. Perform the following cycles in a thermocycler:
   (a) 1 cycle: 94 °C, 15 s
   (b) 50 cycles: 94 °C for 15 s and 68 °C for 1 min.
5. Combine tubes containing similar samples and perform a PCR cleanup using a column based kit according to the manufacturer's instructions. Elute in 50 μL of elution buffer from kit of choice.

3.3.3 Hybridization to Streptavidin-Coated Beads

1. Resuspend streptavidin-coated beads and add 32 μL multiplied by the number of samples to a single 1.5 mL microcentrifuge tube. If the volume is greater than 1 mL, use multiple tubes (*see* **Note 47**).
2. Place the tube on the magnetic particle concentrator (MPC) for 1.5 mL tubes for 2 min.
3. Once the supernatant is clear, carefully remove the supernatant by pipetting from the tube wall opposite the magnet while keeping the tube on the MPC.
4. Remove the tube from the MPC and add 1 mL of 1× bind and wash buffer. Gently resuspend by pipetting (*see* **Note 48**).
5. Repeat **steps 2–4** twice for a total of three washes with 1× bind and wash buffer.
6. Remove the supernatant and resuspend in 52 μL × number of samples of 2× bind and wash buffer.
7. Aliquot 50 μL of suspended beads into PCR strip tubes with one tube per sample. Beads settle quickly, so agitate the suspension prior to each aliquot.
8. Add the entire volume of cleaned PCR product from **step 5**, Subheading 3.3.2, to the appropriate tube.
9. Incubate at room temperature with gentle mixing for 30 min.
10. Place the tube on the MPC for PCR tubes for 2 min.
11. Use a hand operated (nonelectronic) multichannel pipette to carefully remove the supernatant by using one hand to steady the tubes on the MPC and the other to pipette from the top downward along the tube wall opposite magnetic strip. Any beads lost in the following steps will reduce yield and disturbing the beads while on the MPC should therefore be avoided.
12. Remove the tube from the MPC and add 100 μL of 1× bind and wash buffer; pipette gently to resuspend (*see* **Note 49**).
13. Repeat **steps 10–12** twice, except resuspend beads in 100 μL of LoTE buffer instead of 1× bind and wash buffer.
14. Store fragment-bound beads at 4 °C overnight or proceed to the next section.

3.3.4 Double-stranded DNA Synthesis

1. Incubate fragment-bound beads at 95 °C for 2 min and chill quickly to 4 °C to denature PCR products. A thermocycler is preferred because exact temperatures and incubation times can be determined by the user.
2. During incubation, prepare the following as a master mix on ice for the number of reactions to be performed plus one additional reaction.

Reagent	Volume per sample (μL)
dH_2O	16
2.5 mg/mL hexanucleotide mix	2
10 mM dNTPs	1
Klenow (exo-)	1

3. Place the denatured products from **step 1** on the MPC for 2 min, carefully discard the supernatant, remove the tube from the MPC, and gently resuspend in 20 μL of the master mix from **step 2** per sample.
4. Incubate at 37 °C for 30 min. Mix by gently flicking the tubes every 10–15 min. Take care not to get any volume lodged in the lid of the tube (*see* **Note 50**).
5. Add 100 μL of LoTE buffer, concentrate beads on the MPC for 2 min, and carefully discard the supernatant.
6. Repeat **step 5** for a second wash.
7. Resuspend beads in 100 μL of LoTE buffer.
8. Store fragment-bound beads at 4 °C overnight or proceed to the next section.

3.3.5 Trimming of Fragments to Uniform Size

1. Mix the following in a PCR tube (*see* **Note 51**):

Reagent	Volume (μL)
M12_top (100 μM in EB)	15
M12_bottom (100 μM in EB)	15
1 M NaCl	1.5

2. In a thermocycler, anneal the nucleotides by heating to 95 °C for 5 min, followed by slow cooling to 4 °C at a rate of 0.1 °C/s.
3. Aliquot annealed products into six PCR tubes, 5 μL each, and store at −20 °C (*see* **Note 52**).
4. During incubation, prepare the following reagents in order and on ice as a master mix for the number of reactions to be performed plus one additional reaction (*see* **Note 53**):

Reagent	Volume per sample (μL)
dH_2O	16.8
10× enzyme buffer	2
32 mM SAM	0.08
M12 dsDNA	0.2
MmeI	1

5. Concentrate beads from **step 8**, Subheading 3.3.4, on the MPC for 2 min, carefully discard the supernatant, remove the tube from the MPC, and gently resuspend each sample in 20 μL of the master mix prepared in **step 4**.
6. Incubate at 37 °C for 1 h. Mix by gently flicking the tubes every 15 min. Take care not to get any volume lodged in the lid of the tube.
7. Add 100 μL of LoTE buffer to samples, collect the beads with the MPC for 2 min, then carefully remove and discard the supernatant.
8. Repeat **step 7**.
9. Resuspend the MmeI digested beads in 100 μL of LoTE buffer.
10. Store fragment bound beads at 4 °C overnight or proceed to the next section.

3.3.6 Preparation and Ligation of Double-Stranded Adapters

A unique double stranded adapter must be prepared for all samples that will be ultimately be multiplexed in a single lane for Illumina sequencing. Each adapter contains a 4 bp barcode that allows reads to be assigned to individual samples. It is important that the barcodes for each of the adapters chosen for samples that will be multiplexed together are nucleotide balanced with each nucleotide represented in equal number at each base pair as the sequencer reads (*see* **Note 54**). This prevents cluster registration failure and improves accuracy during data analysis.

1. Mix the following in a PCR tube with a different barcode per sample up to the number of samples intended to be multiplexed (*see* **Note 55**).

Reagent	Volume (μL)
LIB_Adapt**T**_(barcode) 100 μM in EB	15
LIB_Adapt**B**_(barcode) 100 μM in EB	15
1 M NaCl	1.5

2. Anneal oligonucleotides by incubating at 95 °C for 5 min followed by slow cooling to 4 °C at a rate of 0.1 °C/s.
3. Aliquot annealed products into six PCR tubes, 5 μL each, and store at −20 °C (*see* **Note 56**).
4. Dilute dsDNA adapters to 5 μM by adding 45 μL of ice cold 1× T4 DNA ligase buffer. A unique adapter should be used for each sample that will be sequenced in the same flow cell lane.
5. Prepare the following reagents on ice as a master mix for the number of reactions to be performed plus one additional reaction:

Reagent	Volume (μL)
dH_2O	16.4
10× T4 DNA ligase buffer	2
T4 DNA ligase	1

6. Concentrate the beads from **step 10**, Subheading 3.3.5 on the MPC for 2 min, carefully discard the supernatant, remove the tube from the MPC, and gently resuspend each sample in 19.4 μL of the master mix prepared in **step 5**.
7. Add 0.6 μL of 5 μM dsDNA sequencing adapter, making sure to add unique adapters where appropriate. Record the individual barcode associated with each sample.
8. Incubate at 16 °C for 1 h, gently flicking the tubes every 15 min. Take care not to get any volume lodged in the lid of the tube. A thermocycler is preferred because exact temperatures and incubation times can be determined by the user.
9. Add 100 μL of LoTE buffer to each sample, then concentrate the beads on the MPC for 2 min and carefully discard the supernatant.
10. Repeat **step 9** for a total of 3 washes.
11. Store fragment bound beads in 100 μL LoTE at 4 °C overnight or proceed to the next section.

3.3.7 Amplification of Final Fragments

1. Prepare the following reagents in order and on ice as a master mix for the number of reactions to be performed plus one additional reaction (*see* **Note 57**):

Reagent	Volume (μL)
dH_2O	31.5
10× Pfx buffer	10 (final concentration is 2×)
10 mM dNTPs	2
50 mM $MgCl_2$	2
5 μM LIB—PCR 5	2
5 μM LIB—PCR 3	2
Pfx polymerase	0.5

2. Concentrate the beads from **step 11**, Subheading 3.3.6 on the MPC for 2 min, carefully discard the supernatant, remove the tube from the MPC, and gently resuspend each sample in 50 μL of PCR master mix on ice.
3. Perform the following cycles in a thermocycler:

(a) 1 cycle: 94 °C, 2 min

(b) 18 cycles: 94 °C, 15 s; 60 °C, 1 min; 68 °C, 2 min

(c) 1 cycle: 68 °C, 4 min.

4. Store fragment-bound beads at 4 °C overnight or proceed to the next step.
5. Concentrate the beads on the MPC and transfer the supernatant to a new PCR tube (*see* **Note 58**).
6. Add 100 μL of LoTE buffer to the remaining beads and store at 4 °C (*see* **Note 59**).

3.3.8 Size Selection of Final Fragments

1. As the PCR is running, prepare a 2% (wt/vol) agarose gel with a non-UV nucleic acid dye, making sure to allow two lanes per sample and two lanes for the DNA ladder (*see* **Note 60**).
2. Concentrate the PCR products from **step 6** in Subheading 3.3.7 in a vacuum centrifuge until the volume is low enough to easily load into a single lane in an agarose gel.
3. Add xylene cyanol DNA dye to each sample to 1× final concentration (*see* **Note 7**).
4. Prepare a DNA ladder with bands that migrate near 100 and 200 bp as per manufacturer instructions.
5. Load the DNA ladder onto the first and last lanes of the gel.
6. Load each sample into its own lane in the gel, leaving one lane between each sample blank to facilitate excision of the resolved bands.
7. Run the gel at an appropriate voltage for the size of the gel and buffer composition. Electrophoresis should continue until the bromophenol band in the DNA ladder has migrated to the bottom quarter of the gel.
8. Visualize the gel on a non-UV gel illuminator to prevent damage to the DNA fragments (*see* **Note 60**). The positive control sample containing pSAM_AraC should have a single bright band at ~125 bp. The negative control containing *P. mirabilis* genomic DNA will likely result in a smear of larger sized bands. Sample lanes should contain a band of identical size to that in the positive control lane. Sample lanes may also contain some smearing or an additional band either slightly larger or slightly smaller than that in the positive control lane (*see* **Note 61**).
9. Excise a single band from each sample lane that is identical size to that found in the positive control lane into a preweighed 1.5 mL microcentrifuge tube, taking care to minimize the amount of agarose in the gel fragment.
10. Extract the DNA from the gel fragments using a gel clean up kit, making sure that any ethanol-containing buffer is

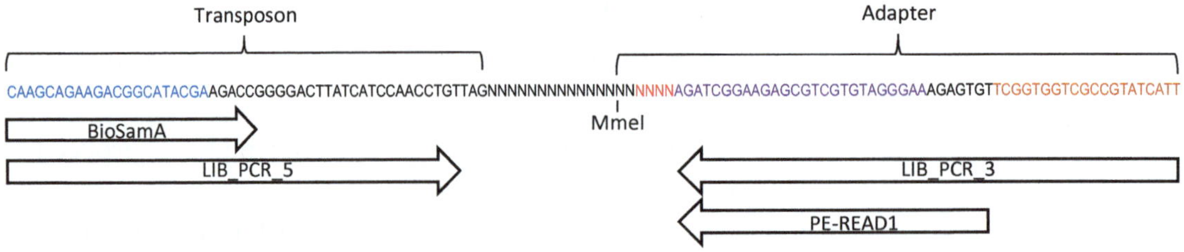

Fig. 3 Schematic of amplified transposon insertion junctions. Diagramed here is the top strand of the final PCR products after trimming transposon insertion junctions with MmeI, ligating to sample-specific barcoded adapters (red, barcode; purple, sequence added by LIB_AdaptT/B) and amplifying the final fragments with a primer including P5 (orange). Final fragments will generate DNA sequence beginning with the 4 bp barcode (red) followed by 15 bp of chromosomal DNA (N's) that can be quantified and mapped to the wild-type chromosome. The P7 (blue) Illumina flowcell binding sequence, required for cluster generation, is included within the transposon sequence. Arrows represent the direction of amplification of the various primers used to generate final fragments and during Illumina sequencing. The MmeI restriction site is also depicted

completely removed from the sample. Final fragments should be eluted in 32 μL of the kit-provided elution buffer or water (*see* **Note 62**).

11. Quantify the sample on a QuBit or similar spectrophotometer. Any samples less than 10 ng/μL should be combined with similar samples from a second round of final fragment preparation and gel extraction.

Final products will contain P5 and P7 sites for annealing to the Illumina flow cell, transposon inverted repeat sequence, chromosomal sequence, a unique 4 bp barcode and sequence homologous to the PE-READ1 primer used to generate reads during a sequencing run (Fig. 3).

3.4 Illumina Sequencing, Bioinformatic Analysis, and Statistics

The specifics of the Illumina sequencing and analysis process will vary depending on available sequencing capabilities and experimental preferences. This section provides general points to consider.

3.4.1 Sequencing

1. In preparation for sequencing, the DNA should undergo quality control on a TapeStation or similar instrument to determine the size and quantity of all fragments within each sample (*see* **Note 63**).
2. Using the TapeStation data for the fragment size of interest, samples should be normalized to the same concentration. The optimal loading concentration and number of samples that can be multiplexed per lane will vary by sequencing platform and coverage of the library.
3. An equal volume of each sample to be run in the same lane should be multiplexed together and 15% bacteriophage ΦX

DNA added. The addition of bacteriophage DNA aids in the accuracy of the base calling software in the presence of fragments with low nucleotide diversity.

4. This Tn-seq protocol generates short, uniformly sized fragments that can be sequenced as short, single end reads with an index read and has been successfully sequenced using V4 Illumina reagents using a HiSeq-2500 sequencer with High-Output and single end 50 settings to produce 50 bp of DNA sequence downstream of the Illumina PE-READ1 sequencing primer. Illumina sequencing chemistry has been updated recently and it is suggested that final fragment composition be verified with your DNA sequencing facility prior to sample submission.

3.4.2 Bioinformatic Analysis

1. *P. mirabilis* HI4320 contains a large plasmid. Therefore, to align insertion sites to the genome, the chromosome and plasmid sequences (NCIB accession numbers NC_010054 and NC_010555) can be combined into a single sequence, with 1000 N's between them, and the gene coordinates for NC_010555 adjusted accordingly.
2. There are several pipelines available for trimming and filtering reads, debarcoding, and aligning to the genome sequence, and calling insertion sites. The MapSAM analysis pipeline [11] can be applied on the raw sequencing reads to perform read filtration, transposon nucleotide removal, debarcoding, alignment to the combined *P. mirabilis* genome sequence described in **step 1**, and insertion calling.
3. An additional script should be written for mapping to the *P. mirabilis* genome to exclude insertions in the 3′ 20% of each gene (*see* **Note 64**). Thus, only 80% of each gene toward the transcriptional start site is used as the effective gene region.
4. The output of this pipeline provides the positions of all unique insertion sites in the effective gene region, the number of insertions at each site in the effective gene region, the dinucleotide bases of each insertion site, and the total number of TA dinucleotides in the effective gene region (whether or not they have insertions).
5. The standard MapSAM pipeline includes a cap to only use insertion sites with >3 reads. This cap can be removed to allow for statistical modeling of insertion sites with any number of reads.

3.4.3 Statistical Analysis

1. The statistical methods required for analysis depend on the experimental setup. The protocol described here is an overview of the TnseqDiff function in the R package Tnseq [13], which can be installed from the Comprehensive R Archive Network (CRAN).

2. A general first step is to correct the input samples for potential read count bias caused by the DNA replication process. To do this, a mean count for each genomic region can be estimated from the LOESS function to obtain a bias factor for each region, and the observed insertion counts at any particular location can then be divided by the bias factor for their genomic region.
3. A Bayesian mixture model can be used to estimate the rate for insertion counts in each gene, assuming the counts follow a Poisson distribution.
4. Absolute essential genes that were required for growth under the library prep conditions can be estimated by having a low rate of insertion counts in the input samples, while nonessential genes will have a higher rate of insertion counts. As Mariner transposons have a strict requirement for inserting at TA dinucleotides, the rate of insertion counts at non-TA sites (which occurs at a very low level and represents background error) can be used to determine the background level for each gene. If the insertion rate at the TA sites within a gene is similar to the background insertion rate at non-TA sites, the gene is estimated to be essential, and JAGS [14] can be used to obtain a posterior probability for the likelihood of each gene being essential.
5. To identify fitness factors, cutoffs can be set based on the number of reads for each insertion site in the input pool and the number of unique insertion sites within genes (*see* **Note 65**). The sum of the reads across all insertion sites within a gene can also be used as an additional cutoff.
6. A confidence distribution (CD) function [15] can be constructed to collect fitness evidence for each insertion site by comparing read counts in the input samples to those in the output samples.
7. To account for potential overdispersion of the count data that can occur from the sequencing technique, a precision weight can be estimated for each observation from the mean–variance relationship of the log-counts and incorporated into linear modeling [16].
8. The CD function can be constructed using the slope (log fold-change) and estimates (mean, standard error, and degrees of freedom) from the linear model.
9. Insertion-level CD functions can then be combined to provide a representative fitness assessment for a given gene, and a *P*-value can be derived from the combined CD function to infer the fitness contribution of that gene [15]. These *P*-values should then be further adjusted for multiple hypothesis correction [17].
10. For a gene to be considered a fitness factor, it must have a *P* value less than 0.05, and have a fold change for input/output of <0.05 or >2.0.

3.5 Validation of Fitness Genes

Validation of genes indicated as important for fitness should be performed to rule out the occurrence of potential bias at any of the infection or fragment preparation steps. Typically, validation is performed by making clean mutants in the gene of interest and coinfecting them with the isogenic wild-type strain in a 1:1 mix using the same parameters for the infection as those used during the infection with the transposon mutant pool (*see* Subheading 3.2). Most commonly, mutants in *P. mirabilis* are generated using targetron mutagenesis (described in detail in Chapter 7), which inserts an intron into the gene of interest. If an ordered library of transposon mutants was curated when the transposon library was generated, coinfections can be performed with mutants harboring transposon insertions at different locations within the gene of interest, potentially allowing for identification of specific domains important for fitness. When performing a cochallenge experiment, the competitive index can be calculated for the mutant compared to the wild-type strain by differential plating and back calculation of the CFU for each strain. The formula for competitive index (CI) is as follows:

$$CI = \frac{\text{Strain A output/Strain B output}}{\text{Strain A input/Strain B input}}$$

If the $\text{Log}_{10}\text{CI} = 0$, neither strain has an advantage over the other, and therefore the mutated gene has no effect on fitness. If the $\text{Log}_{10}\text{CI} > 0$, strain A has an advantage over B. If the $\text{Log}_{10}\text{CI} < 0$, strain B has an advantage over A. The results of the cochallenge may not be reflective of the Tn-seq results because the mutant strain is present in much greater quantity as compared to that in the pool of transposon mutants. Additionally, the mutant phenotype may be complemented by an excreted product from the wild-type strain. These possibilities can be ruled out by conducting independent challenge experiments, that is, infecting one set of mice with the wild-type strain and a second set of mice with only the mutant strain to compare their ability to establish an infection (*see* Chapter 15 for independent challenges in an ascending UTI model and Chapter 17 for independent challenges using a CAUTI model).

4 Notes

1. The decreased molarity of Tris and EDTA in this buffer compared to standard TE buffer is used to prevent interference with downstream enzymatic reactions.
2. pSAM_AraC is a modified version of pSAM_EC used to generate a transposon library in *E. coli* strain F11 [18]. The plasmid contains the arabinose promoter and AraC from pBAD to drive transposase expression, and kanamycin resistance within the transposon to select for mutagenized *P. mirabilis* HI4320. Information on how pSAM_AraC was generated can be found

in Armbruster et al. [7]. pSAM_AraC is available for purchase through Addgene. The pSAM backbone was designed such that it can be easily modified for use in various organisms [5]. The important components of a transposon delivery vector are origin of replication, transposase induction system, and the antibiotic markers within the transposon and on the plasmid backbone. It is critical that the origin of replication be unable to replicate in the recipient strain of interest to prevent production of multiple copies of the transposon within conjugated cells, which may result in multiple transposon insertions or further mobilization of transponsons once inserted into the genome. For the R6K origin of replication this means that the strain of interest must not contain a *pir* gene. It is also important to have tight regulation of transposase expression to prevent mobilization of the transposon into the genome of the donor strain and to maintain transposon stability once inserted into the recipient strain. Both the donor and recipient strain should be tested for sensitivity to the antibiotic markers within the transposon and the plasmid backbone to allow for plasmid maintenance within the donor strain, selection of transposon mutants, and verification that the suicide plasmid restricts growth in the recipient strain.

3. For generation of a probe homologous to the kanamycin resistance cassette within the Mariner/Himar1 transposon, we used Forward primer 5′ ACA AGA TGG ATT GCA CGC AG 3′ and Reverse primer 5′ CTG ATG CTC TTC GTC CAG AT 3′.
4. The sequence of BioSamA is: 5′-Bio-TEG-*CAA GCA GAA GAC GGC ATA CGA AGA CC*-3′. The primer includes a biotin tag and a tetraethylene glycol spacer (TEG) to reduce steric interactions with the streptavidin coated beads during enzymatic steps. The italicized sequence corresponds to 26 bp of the transposon inverted repeat containing the P7 sequences homologous to the Illumina flowcell. The primer binds to the inverted repeats of the transposon; thus, chromosomal sequence can be obtained from either end of the transposon insertion. BioSamA must be HPLC-purified and diluted to 1 pmol/μL working stock. BioSamA should be blasted against the *P. mirabilis* (or other species) genome of interest to verify a lack of homology.
5. Sequences of M12_top: 5′-CTG TCC GTT CCG ACT ACC CTC CCG AC-3′, and M12_bottom: 5′-GTC GGG AGG GTA GTC GGA ACG GAC AG-3′.
6. Sequences of LIB-PCR 5: 5′-CAA GCA GAA GAC GGC ATA CGA AGA CCG GGG ACT TAT CAT CCA ACC TGT-3′, and LIB-PCR 3: 5′-AAT GAT ACG GCG ACC ACC GAA CAC TCT TTC CCT ACA CGA CGC TCT TCC GAT CT-3′.

7. Bromophenol blue dye is excluded from the loading dye because it migrates at a similar size to that of the final fragments and may partially obscure the band in the gel.
8. Mid-log is defined at OD_{600} 0.4–0.6. *P. mirabilis* generally takes about 2 h to reach mid-log phase when an overnight culture is diluted 1:40 and incubated at 37 °C with aeration (shaking at 225 rpm).
9. To make a bacterial lawn, spread 50 μL of saturated liquid culture of *P. mirabilis* HI4320 evenly onto LB agar and incubate at 37 °C overnight.
10. In **steps 4** and **5**, include an appropriate number of tubes based on the size of the library you intend to create. A single 50 μL concentrated mating mixture generates around 400 mutants. The ratio of donor to recipient strain should be optimized if using other species.
11. Multiple filters can be placed on the same agar plate to reduce the total number of agar plates required.
12. Conjugation on filters has been successful with up to 24 h incubation at 30 °C. Optimize incubation time for the strain of choice.
13. Carefully optimize the CFU per plate to prevent the formation of bacterial lawns; single colonies should be visible. If an ordered library is desired, colonies should be dispersed enough to allow recognition of individual colonies by colony picking software, typically around 500 colonies per plate. Protocols for creating ordered libraries of transposon mutants can be found in Goodman et al. and Vandewalle et al. [11, 19]. Ordered libraries decrease the time required to validate potential fitness genes and allow for validation with the identical insertion event that resulted in a significant decrease in fitness in the condition of interest.
14. In this step, tetracycline is used to select for the recipient strain and eliminate the donor strain from the population. Most *P. mirabilis*, including HI4320, are natively tetracycline resistant; however, this feature should be confirmed for the target strain. Kanamycin is used to select for bacteria that have had a transposon insertion event. The kanamycin-resistant population may also include bacteria that have not lost pSAM_AraC independent of a transposon insertion event. Ampicillin is used to select for bacteria that have not lost pSAM_AraC. If pSAM_AraC is not suicidal in the strain of interest, an alternative method should be used to generate the transpson library (e.g., electroporation with transposomes).
15. Select 20–50 transposon mutants to compare at the same time, and include WT genomic DNA as a negative control and

miniprepped pSAM_AraC as a positive control. It is important to inoculate with a single distinct colony to determine the transposon insertion location of individual mutants.

16. Consult the manufacturer's protocol for strategies to maximize DNA yield during genomic DNA extraction. If a column-based kit is used, some strategies include: optimization of input CFU to prevent blockage of the column, increasing incubation time during lysis, and heating elution buffer to 55 °C prior to elution. Ideally, there is a single copy of transposon sequence per chromosome available for the DIG labeled probe to hybridize with. Therefore, it is important to maximize the yield of chromosomal DNA to generate a detectable signal from hybridized probe.
17. It is important to ensure complete digestion of genomic DNA into randomly sized fragments; thus, enzymes with a high frequency of restriction sites in your genome of interest should be selected. The enzymes of choice should not cut within the transposon to prevent the appearance of multiple bands after probe detection. The selected enzyme should linearize the plasmid backbone to facilitate differentiation of digested fragments indicative of transposon mutants that have not lost the mating plasmid from those that may have incorporated the plasmid into the chromosome. HindIII is appropriate for use in *P. mirabilis* because the recognition site is 33% GC; thus, digestion of genomic DNA results in generally useful fragments that can be distinguished by Southern blot. For the plasmid control, a separate set of reactions with different volumes should be prepared; typically 5 μL of a plasmid miniprep is sufficient. Much less DNA is needed due to the fact that all fragments contain the transposon, whereas in the genomic DNA samples there is ideally only one fragment with transposon per genome; therefore, the amount of nonreactive DNA present in each sample is much higher. The negative control WT genomic DNA should be treated the same as transposon mutant samples in regards to DNA concentration and restriction digest.
18. This additional step is important to ensure complete digestion of genomic DNA. DNA that is not completely digested will result in multiple bands per lane after final exposure and cannot be differentiated from transposon mutants with multiple transposon insertions per genome.
19. If gel is imaged with a UV ruler and a DNA ladder is included, the final size of probe-hybridized fragments can be estimated after development of the Southern blot.

20. If DNA loading dye with bromophenol blue and xylene cyanol is used, the dye color will change from blue and purple to yellow and green, respectively.
21. Nylon membrane (e.g., Hybond-N+) should be handled with gloves at all times to prevent contamination with human DNA. When assembling the transfer stack, clip or mark one corner of the membrane to facilitate correct orientation of the membrane postexposure.
22. Transfer can also be done overnight. The paper towels under the stack should dampen over time if the transfer stack has been assembled correctly to allow for capillary action of the buffer into the paper towel.
23. The agarose gel should be compressed evenly after transfer. Any areas that are unevenly compressed indicate where air bubbles have formed, and DNA has not been transferred efficiently in these areas.
24. The key consideration in this step is to ensure that the amount of prehybridization solution used covers the membrane completely, taking into account the intended final concentration of the labeled probe suggested in the manufacturer's protocol.
25. The hybridization bag is used to minimize the total volume of prehybridization solution required to wet the membrane, thus reducing the amount of labeled probe needed. Alternatively, the membrane can be incubated directly in the hybridization tube in a larger volume of prehybridization solution as allowed by the amount of labeled probe generated.
26. The concentration and temperature of SSC can be varied to increase or decrease the threshold of homology for probe binding. Increased temperature and decreased salt concentration is more stringent, such that only perfectly matching DNA on the filter is able to bind to labeled probe.
27. It is critical to completely dissolve the dry milk in PBST before incubating with the membrane. If the milk is not dissolved it will result in punctate spots on the membrane.
28. Typical exposure times are 15 min, 1 h, 2 h, and overnight. In general, the signal from the probe is not very robust and ideal exposures are achieved with long incubations of membrane and film. Alternatively, the signal from bound probe can be detected with any chemiluminescence method that utilizes an alkaline phosphatase cleavable substrate.
29. The purpose of a 10 h incubation for bottleneck assessment is to mimic the conditions that will be used for infection with the transposon library pools. This timepoint was selected for

P. mirabilis HI4320 Tn-seq to decrease potential selection against mutants during stationary phase.

30. For further details on mouse UTI models that could be used with Tn-seq, *see* Chapter 16 for cochallenge in an ascending UTI model, and Chapter 17 for a CAUTI co-challenge model.
31. Minimum colonization density is the lowest CFU/gram tissue, observed postharvest, in any organ for mice that were successfully inoculated with bacteria. The competitive index (CI) can be calculated from the CFU/gram tissue of each strain determined with differential plating using the following formula:

$$\mathrm{CI} = \frac{\text{Strain A output/Strain B output}}{\text{Strain A input/Strain B input}}$$

32. Mutant libraries should contain an appropriate number of copies of each mutant within the infecting dose to overcome any effects of stochastic loss of mutants that may occur within the model of choice.
33. It is extremely important that the mixture be homogeneous to ensure similar frequencies of each mutant in comparable aliquots.
34. Incubation of the inoculum mutant pools for 10 h prior to infection prevents selection against any mutants that have defects during stationary phase growth, but not during growth on agar plates during the initial library-building step or during exponential growth in rich medium.
35. The entirety of the organ must be plated and all bacteria harvested in order to maintain the ratios of each mutant in the output pool.
36. The volume of PBS used is dependent on the CFU/mL for each organ. Highly colonized organs will generate near lawns of bacteria and should be resuspended in about 30 mL of PBS. Less colonized organs with individual colonies on each plate may only require 5 or 10 mL of PBS. The goal is to estimate the amount of PBS needed such that 1 mL of the resuspension generates a similarly sized pellet for all organs at **step 8**. If any organs are colonized below the threshold determined during bottleneck assessments, they should be terminated at this step. Any mutants able to colonize in these organs will have artificially increased output/input ratios due to complete elimination of the majority of mutants within the pool.
37. It is extremely important that the mixture be homogeneous to ensure similar frequencies of each mutant in comparable aliquots.
38. Mapping of Illumina reads must be unambiguous. 16 bp of chromosomal DNA may not be suitable for mapping of all

genomes; however, this method has been used successfully in *P. mirabilis* [7], *Providencia stuartii* (unpublished data), and *K. pneumoniae* [6].

39. If the pellet is particularly large, add 189 μL of resuspended pellet into 2 additional tubes for a total of 3 tubes. Add 378 μL of additional TE buffer to each tube such that total volume in each tube is 567 μL.
40. CTAB–NaCl must be gently heated to at least 42 °C prior to addition to reduce viscosity and aid in pipetting.
41. Chloroform is a hazardous chemical and should only be added to samples within a fume hood; waste should be disposed of appropriately. After addition of chloroform, the top aqueous solution contains nucleic acids as well as contaminating sugars and salts. The bottom organic solution contains proteins and lipids.
42. If too many bacteria are used at the start of the procedure, the aqueous (top) layer will be too viscous to pipette. It is important to utilize the alternate steps in **Note 39** to reduce the size of the bacterial pellet.
43. Phenol–chloroform solutions are hazardous, and should only be added to samples within a fume hood; waste should be disposed of appropriately. To prevent oxidation, phenol–chloroform–isoamyl alcohol is packaged with an aqueous solution, which forms a distinct layer on top of the phenol solution. When adding phenol–chloroform–isoamyl alcohol it is important to take volume from the bottom of the container to avoid transferring the aqueous layer.
44. If pellets get too dry or there is a large quantity of DNA it can be difficult to resuspend in Tris–HCl. Gentle heat with intermittent mixing can aid in resuspension.
45. Some genomic DNA samples can be quite concentrated; however, no less than 1 μL of genomic DNA should be used to normalize samples. Therefore, some samples may have greater than 2 μg of genomic DNA.
46. Include a negative control reaction with wild-type HI4320 DNA, and a positive control reaction with pSAM_AraC to allow nonspecific amplification products to be distinguished from bona fide transposon insertion junctions during the gel purification step. It is advised to use PCR strip tubes as this facilitates high throughput sample preparation. The preparation of transposon insertion junction typically takes 2–3 days so it is recommended to prepare the maximum number of samples possible at a given time.
47. The use of magnetic beads allows for purification of attached DNA from different reaction mixes including PCR, restriction

digest and ligation without the need for specialty kits and minimizes loss of product by avoiding the use of a column. Magnetic beads also facilitate increased throughput not achievable with traditional purification methods.

48. Bind and wash buffer stock is 2× and thus must be diluted for this step.

49. Some beads will form hydrostatic interactions with the walls of pipette tips, so some loss of beads during mixing steps is unavoidable.

50. If any volume gets lodged in the lid of the tube, perform a quick spin in a minicentrifuge and resume resuspension. If left uncorrected, buffer or enzyme incompatible with upcoming steps may carry over in the tube.

51. The DNA fragment M12 sequence is a recognition site for the type II restriction enzyme MmeI. Type II restriction enzymes commonly require more than one recognition site for endonuclease activity. MmeI cleavage is 70% as efficient when a single recognition site is provided as compared to when two sites are provided [20]. Thus, M12 provides an in trans second recognition site to maximize cleavage efficiency. The M12 fragment can be made ahead of time to decrease total time for this section.

52. When not using freshly made M12 fragments, it is important to thaw frozen aliquots on ice to prevent dissociation. Once thawed, fragments should not be refrozen.

53. Enzyme buffer (NEBuffer 4) and SAM are included with MmeI when ordered from New England BioLabs.

54. As an example, the following four barcodes are balanced at each base and could be successfully multiplexed together: ATCG, TAGC, CGAT, and GCTA.

55. The adapters can be made ahead of time to decrease total time for this section.

56. It is critical to maintain the adapter on ice after annealing. When not using freshly made adapters, it is important to thaw frozen aliquots slowly on ice to prevent dissociation. Once thawed, fragments should not be refrozen.

57. To save time, the PCR master mix can be prepared during the ligation of the double stranded DNA sequencing adapters in **step 8** of Subheading 3.3.6.

58. The supernatant contains the intended product because in **step 3** the adapter ligated and barcoded fragments are amplified using primers that are homologous to the entire adapter sequence up to the barcode; therefore, a single forward primer and reverse primer are required. The resulting full length

fragments are no longer biotinylated and therefore do not bind to the streptavidin coated beads.

59. A second round of PCR using the same beads can be performed if the final concentration of gel purified fragments is lower than 10 ng/μL.

60. Gel purification of the final products not only allows visualization of successful reactions, but prevents primer dimers and incorrect products from contributing to the final concentration of appropriate final fragments. A non-UV dye (e.g., GelRed or SYBR Safe), along with a non-UV method of visualization, is used to prevent DNA thymine dimer formation in the final fragments, which would interfere with sequencing.

61. If there are no bands in any lane, repeat the final PCR step using the beads saved during amplification of final fragments to verify if any reaction components were unintentionally omitted. If the second PCR fails, preparation of transposon insertion junctions should be repeated starting with a fresh preparation of genomic DNA. If there are bands in the positive control lane, but absent or weak bands in sample lanes, a contaminant from the phenol–chloroform extraction of genomic DNA could have interfered with an enzymatic step. In this case, prepare fresh genomic DNA, taking care not to transfer any material from the organic phase along with the aqueous phase during extraction. Alternatively, ligated adapters may not have been properly annealed. Sample preparation can be repeated using technical replicates of the same sample with different barcoded adapters to determine if the adapter or the input DNA is the cause. It is possible to prepare transposon insertion junctions using double the amount of genomic DNA and/or double the final concentration of BioSamA primer during single strand PCR to increase yield.

62. An aliquot of the final fragments can be Sanger sequenced to ensure that the P5 and P7 sites required to anneal fragments to the Illumina flowcell are present, barcode sequences are free of errors and that the region containing chromosomal DNA is random (all nucleotides will be present in similar amounts, no single nucleotide should dominate and base calling software should fail). The Monarch DNA gel extraction kit (New England BioLabs) typically provides higher yield than other purification methods we have tried.

63. TapeStation results typically estimate fragment size at 50–75 bp larger than the expected fragment size of ~125 bp.

64. Insertions in the 3′ 20% of a gene have a lower likelihood of disrupting the gene in a way that would result in a dysfunctional protein when translated.

65. Transposons have a region of identical sequence at both the 3′ and 5′ ends called an inverted repeat. As a result, primers that have homology to the inverted repeat can amplify from either end of the transposon, resulting in final fragments containing chromosomal sequence on either side of the transposon insertion site. This fact can be applied as a quality control measure when aligning Illumina sequence to the chromosome because an equal number of sequences with homology upstream and downstream of the insertion site should be generated during Illumina sequencing.

References

1. Barquist L, Langridge GC, Turner DJ, Phan M-D, Turner AK, Bateman A, Parkhill J, Wain J, Gardner PP (2013) A comparison of dense transposon insertion libraries in the *Salmonella* serovars Typhi and Typhimurium. Nucleic Acids Res 41(8):4549–4564. https://doi.org/10.1093/nar/gkt148
2. Langridge GC, Phan M-D, Turner DJ, Perkins TT, Parts L, Haase J, Charles I, Maskell DJ, Peters SE, Dougan G, Wain J, Parkhill J, Turner AK (2009) Simultaneous assay of every *Salmonella* Typhi gene using one million transposon mutants. Genome Res 19 (12):2308–2316. https://doi.org/10.1101/gr.097097.109
3. Lampe DJ, Grant TE, Robertson HM (1998) Factors affecting transposition of the *Himar1 mariner* transposon *in vitro*. Genetics 149 (1):179–187
4. Rubin EJ, Akerley BJ, Novik VN, Lampe DJ, Husson RN, Mekalanos JJ (1999) *In vivo* transposition of *mariner*-based elements in enteric bacteria and mycobacteria. Proc Natl Acad Sci U S A 96(4):1645–1650
5. Goodman AL, McNulty NP, Zhao Y, Leip D, Mitra RD, Lozupone CA, Knight R, Gordon JI (2009) Identifying genetic determinants needed to establish a human gut symbiont in its habitat. Cell Host Microbe 6(3):279–289. https://doi.org/10.1016/j.chom.2009.08.003
6. Bachman MA, Breen P, Deornellas V, Mu Q, Zhao L, Wu W, Cavalcoli JD, Mobley HLT (2015) Genome-wide identification of *Klebsiella pneumoniae* fitness genes during lung infection. mBio 6(3):e00775. https://doi.org/10.1128/mBio.00775-15
7. Armbruster CE, Forsyth-DeOrnellas V, Johnson AO, Smith SN, Zhao L, Wu W, Mobley HLT (2017) Genome-wide transposon mutagenesis of *Proteus mirabilis*: Essential genes, fitness factors for catheter-associated urinary tract infection, and the impact of polymicrobial infection on fitness requirements. PLoS Pathog 13(6):e1006434. https://doi.org/10.1371/journal.ppat.1006434
8. Santiago M, Matano LM, Moussa SH, Gilmore MS, Walker S, Meredith TC (2015) A new platform for ultra-high density *Staphylococcus aureus* transposon libraries. BMC Genomics 16(1):252. https://doi.org/10.1186/s12864-015-1361-3
9. Hendrixon DR, Akerley BJ, DiRita VJ (2001) Transposon mutagenesis of *Campylobacter jejuni* identifies a bipartite energy taxis system required for motility. Mol Microbiol 40 (1):214–224. https://doi.org/10.1046/j.1365-2958.2001.02376.x
10. Kwon YM, Ricke SC, Mandal RK (2016) Transposon sequencing: methods and expanding applications. Appl Microbiol Biotechnol 100(1):31–43. https://doi.org/10.1007/s00253-015-7037-8
11. Goodman AL, Wu M, Gordon JI (2011) Identifying microbial fitness determinants by insertion sequencing using genome-wide transposon mutant libraries. Nat Protoc 6 (12):1969–1980. https://doi.org/10.1038/nprot.2011.417
12. Farmer JJ, Hickman FW, Brenner DJ, Schreiber M, Rickenbach DG (1977) Unusual *Enterobacteriaceae*. "*Proteus rettgeri*" that "change" into *Providencia stuartii*. J Clin Microbiol 6(4):373–378
13. Zhao L, Anderson MT, Wu W, Mobley HLT, Bachman MA (2017) TnseqDiff: identification of conditionally essential genes in transposon sequencing studies. BMC Bioinformatics 18 (1):326. https://doi.org/10.1186/s12859-017-1745-2
14. Plummer M (2003) {JAGS}: a program for analysis of {Bayesian} graphical models using {Gibbs} sampling. In: Proceedings of the 3rd International Workshop on Distributed

Statistical Computing, Vienna, Austria. citeulike-article-id:10263507

15. Singh K, Xie M, Strawderman WE (2005) Combining information from independent sources through confidence distributions. Ann Stat 33(1):159–183. https://doi.org/10.1214/009053604000001084
16. Smyth GK (2004) Linear models and empirical bayes methods for assessing differential expression in microarray experiments. Stat Appl Genet Mol Biol 3:Article3. https://doi.org/10.2202/1544-6115.1027
17. Storey JD, Tibshirani R (2003) Statistical significance for genomewide studies. Proc Natl Acad Sci U S A 100(16):9440–9445. https://doi.org/10.1073/pnas.1530509100
18. Wiles TJ, Norton JP, Russell CW, Dalley BK, Fischer KF, Mulvey MA (2013) Combining quantitative genetic footprinting and trait enrichment analysis to identify fitness determinants of a bacterial pathogen. PLoS Genet 9(8):e1003716. https://doi.org/10.1371/journal.pgen.1003716
19. Vandewalle K, Festjens N, Plets E, Vuylsteke M, Saeys Y, Callewaert N (2015) Characterization of genome-wide ordered sequence-tagged *Mycobacterium* mutant libraries by Cartesian pooling-coordinate sequencing. Nat Commun 6:7106. https://doi.org/10.1038/ncomms8106
20. Morgan RD, Bhatia TK, Lovasco L, Davis TB (2008) MmeI: a minimal Type II restriction-modification system that only modifies one DNA strand for host protection. Nucleic Acids Res 36(20):6558–6570. https://doi.org/10.1093/nar/gkn711
21. Khatiwara A, Jiang T, Sung S-S, Dawoud T, Kim JN, Bhattacharya D, Kim H-B, Ricke SC, Kwon YM (2012) Genome scanning for conditionally essential genes in *Salmonella enterica* serotype Typhimurium. Appl Environ Microbiol 78(9):3098–3107. https://doi.org/10.1128/aem.06865-11
22. van Opijnen T, Bodi KL, Camilli A (2009) Tn-seq: high-throughput parallel sequencing for fitness and genetic interaction studies in microorganisms. Nat Methods 6:767. https://doi.org/10.1038/nmeth.1377
23. Gawronski JD, Wong SMS, Giannoukos G, Ward DV, Akerley BJ (2009) Tracking insertion mutants within libraries by deep sequencing and a genome-wide screen for *Haemophilus* genes required in the lung. Proc Natl Acad Sci U S A 106(38):16422–16427. https://doi.org/10.1073/pnas.0906627106
24. Christen B, Abeliuk E, Collier JM, Kalogeraki VS, Passarelli B, Coller JA, Fero MJ, McAdams HH, Shapiro L (2011) The essential genome of a bacterium. Mol Syst Biol 7(1):528. https://doi.org/10.1038/msb.2011.58
25. Gallagher LA, Shendure J, Manoil C (2011) Genome-scale identification of resistance functions in *Pseudomonas aeruginosa* using Tn-seq. mBio 2(1):e00315–e00310. https://doi.org/10.1128/mBio.00315-10
26. Klein BA, Tenorio EL, Lazinski DW, Camilli A, Duncan MJ, Hu LT (2012) Identification of essential genes of the periodontal pathogen *Porphyromonas gingivalis.* BMC Genomics 13(1):578. https://doi.org/10.1186/1471-2164-13-578
27. Dawoud TM, Jiang T, Mandal RK, Ricke SC, Kwon YM (2014) Improving the efficiency of transposon mutagenesis in *Salmonella* enteritidis by overcoming host-restriction barriers. Mol Biotechnol 56(11):1004–1010. https://doi.org/10.1007/s12033-014-9779-4
28. Wetmore KM, Price MN, Waters RJ, Lamson JS, He J, Hoover CA, Blow MJ, Bristow J, Butland G, Arkin AP, Deutschbauer A (2015) Rapid quantification of mutant fitness in diverse bacteria by sequencing randomly bar-coded transposons. mBio 6(3):e00306–e00315. https://doi.org/10.1128/mBio.00306-15

Index

Melanie M. Pearson (ed.), *Proteus mirabilis: Methods and Protocols*, Methods in Molecular Biology, vol. 2021, https://doi.org/10.1007/978-1-4939-9601-8, © Springer Science+Business Media, LLC, part of Springer Nature 2019

MIX
Papier aus verantwortungsvollen Quellen
Paper from responsible sources
FSC® C105338

If you have any concerns about our products,
you can contact us on
ProductSafety@springernature.com

In case Publisher is established outside the EU,
the EU authorized representative is:
Springer Nature Customer Service Center GmbH
Europaplatz 3, 69115 Heidelberg, Germany

Printed by Libri Plureos GmbH
in Hamburg, Germany